AUTHORS

Blümlein J., DESY-Zeuthen, Platanenallee 6, 15735 Zeuthen, Germany

Boorstein J., Enrico Fermi Institute, University of Chicago, 5640 S. Ellis Ave., Chicago, IL 60637, U.S.A

Braun V.M., NORDITA, Blegdamsvej 17, 2100 Copenhagen Ø, Denmark

Brodsky S.J., Stanford Linear Accelerator Center, Stanford University, Stanford, California 94309, U.S.A.

Burkardt M., Department of Physics, New Mexico State University, Las Cruces, New Mexico 88003-0001, U.S.A.

Dalley S., Department of Applied Mathemetics and Theoretical Physics, Silver Street, Cambridge CB3 9EW, England

Dalley S., Theory Division, CERN, CH-1211 Geneva 23, Switzerland

El-Khozondar H., Department of Physics, New Mexico State University, Las Cruces, New Mexico 88003-0001, U.S.A.

Glazek D., Institute of Theoretical Physics, Warsaw University, ul. Hoza 69, 00-681 Warsaw, Poland

Heinzl T., Institut für Theoretische Physik, Universität Regensburg, 93040 Regensburg, Germany

Hüfner J., Institut für Theoretische Physik, Philosophenweg 19, 69120 Heidelberg, Germany

Klauder J.R., Departments of Physics and Mathematics University of Florida, Gainesville 32611, U.S.A.

Klevansky S.P., Institut für Theoretische Physik, Philosophenweg 19, 69120 Heidelberg, Germany

Kutasov V., Department of Physics of Elementary Particles, Weizmann Institute of Science, Rehovot, Israel

Lenz F., Institute for Theoretical Physics III, University of Erlangen-Nürnberg, Staudtstr. 7, 91058 Erlangen, Germany

Lusanna L., Sezione INFN di Firenze L.go E.Fermi 2 (Arcetri), 50125 Firenze, Italy

Mankiewicz L., Institute for Theoretical Physics TU-München, 85747 Garching, Germany

Marchesini G., Dipartimento di Fisica, Università di Milano INFN, Sezione di Milano, Italy

Marnellius R., Institute of Theoretical Physics, Chalmers University of Technology, Göteborg University, 412 96 Göteborg, Sweden

McCartor G., Department of Physics, Southern Methodist University, Dallas, Texas 75275, U.S.A.

Miller G.A., Department of Physics, Box 351560, University of Washington, Seattle, WA 98195-1560, U.S.A, and Stanford Linear Accelerator Center, Stanford University, Stanford, California 94309, U.S.A, and National Institute for Nuclear Theory, Box 35150, University of Washington, Seattle, WA 98195-1560 U.S.A

Moshe M., Department of Physics Technion - Israel Institute of Technology, Haifa 32000 Israel

Neveu A., Laboratoire de Physique Mathématique, Université Montpellier II, CNRS, 34095 Montpellier, France

Niemi A.J., Department of Theoretical Physics, Uppsala University P.O.Box 803, 75108 Uppsala, Sweden

Ogura A., Institut für Theoretische Physik, Philosophenweg 19, 69120 Heidelberg, Germany

Pang Y., Department of Physics, Colombia University, New York 10027, U.S.A, and Department of Physics, Brookhaven National Laboratory, Upton 11973, U.S.A

Parisi G., Dipartimento di Fisica, Università La Sapienza and INFN Sezione di Roma Piazzale Aldo Moro, Roma 00187, Italy

Pauli H.C., Max-Planck-Institut für Kernphysik, Postfach 103980, 69029 Heidelberg, Germany

Pinsky S., Department of Physics, The Ohio State University, Columbus, OH 43210, U.S.A.

Rehberg P., Institut für Theoretische Physik, Philosophenweg 19, 69120 Heidelberg, Germany

Ren H., Department of Physics, Rockfeller University, New York 10021, U.S.A

Robertson D.G., Department of Physics, The Ohio State University, Columbus, OH 43210, U.S.A.

Royon C., CEA, DAPNIA, Service de Physique des Particules, Centre d'Études de Saclay, France

Sonnenchein J., School of Physics and Astronomy, Beverly and Raymond-Sackler Faculty of Exact Sciences, Tel-Aviv University, Ramat-Aviv, Israel

Stirling W.J., Departments of Mathematical Sciences and Physics, University of Durham, Durham DH1 3LE, England

NEW NON-PERTURBATIVE METHODS AND QUANTIZATION ON THE LIGHT CONE

Les Houches School, February 24 - March 7, 1997

Editors

P. GRANGÉ, A. NEVEU
H.C. PAULI, S. PINSKY, E. WERNER

Springer-Verlag Berlin Heidelberg GmbH

Centre de Physique des Houches

Books already published in this series

1 Porous Silicon Science and
Technology
Jean-Claude VIAL and
Jacques DERRIEN, Eds. 1995

2 Nonlinear Excitations in
Biomolecules
Michel PEYRARD, Ed. 1995

3 Beyond Quasicrystals
Françoise AXEL and
Denis GRATIAS, Eds. 1995

4 Quantum Mechanical
Simulation Methods for
Studying Biological Systems
Dominique BICOUT and
Martin FIELD, Eds. 1996

5 New Tools in Turbulence
Modelling
Olivier MÉTAIS and
Joel FERZIGER, Eds. 1997

6 Catalysis by Metals
Albert Jean RENOUPREZ and
Hervé JOBIC, Eds. 1997

7 Scale Invariance and
Beyond
B. DUBRULLE, F. GRANER
and D. SORNETTE, Eds. 1997

Book series coordinated by Michèle LEDUC

**Editors of "New Non Perturbative Methods
and Quantization on the Light Cone" (N° 8)**

P. Grangé, A. Neveu (LPM, Montpellier, France), H.C. Pauli (MPI, Heidelberg, Germany),
S. Pinsky (OSU, Colombus, USA) and E. Werner (Univ. Regensburg, Germany)

ISBN 978-3-540-64520-7   ISBN 978-3-662-08973-6 (eBook)
DOI 10.1007/978-3-662-08973-6

© Springer-Verlag, Berlin Heidelberg 1998

Originally published by EDP Sciences, Les Ulis; Springer-Verlag, Berlin, Heidelberg in 1998.

Thies M., Institute for Theoretical Physics III, University of Erlangen-Nürnberg, Staudtstr. 7, 91058 Erlangen, Germany

Trittman U., Max-Planck-Institut für Kernphysik, Postfach 10 39 80, 69029 Heidelberg, Germany

Van Baal P., Isaac Newton Institute for Mathematical Sciences, 20 Clarkson Road, Cambridge CB3 0EH, England, and Institut-Lorentz for Theoretical Physics, University of Leiden, P.O.Box 9506, 23000 RA Leiden, The Netherlands

Van de Sande V., Institut für Theoretische Physik III, Staudstrasse 7, 91058 Erlangen, Germany

Verbaarschot J., Department of Physics, SUNY at Stony Brook, Stony Brook, NY 11794, U.S.A

Vogt A., Institut für Theoretische Physik, Universität Würzburg, Am Hubland, 97074 Würzburg, Germany

Wegner F., Institut für Theoretische Physik, Ruprecht-Karls-Universität, Philosophenweg 19, 69120 Heidelberg, Germany

Yamawaki K., Department of Physics, Nagoya University, Nagoya 464-01 Japan

Zwanziger D., Physics Department, New York University, New York 10003, U.S.A

PREFACE

The aim of this volume is to present the major contributions given at the session **"New non-perturbative methods and quantization on the light cone"**, held in **Les Houches (France)** from February 24 to March 7, 1997.

The genesis of light-cone QCD was the 1949 work by P.A.M. Dirac in which he showed that there were several distinct ways to formulate and quantize a Hamiltonian system; these were later extended to quantum field theory. Among them is what we now call light-cone or light-front field theory (LFFT). The first real application of these ideas was in the mid 1960s when Fubini and Furlan used the infinite momentum frame, which is very closely related to LFFT, in the context of current algebra.

The first application of the method to Gauge theories appeared in the early 1970s but it was not until the mid 1980s that this method emerged as an approach for solving QCD and became a sub-specialty in its own right. The unique property of LFFT that is a corner stone of this approach is that the ground state used in perturbation theory is also the ground state of the full interacting theory. This is a unique property of LFFT and provides a key advance over other approaches, where one has to struggle with a very complicated ground state. Since the mid 1980s there have been a growing number of people who have used LFFT methods to attack the problem of QCD or other strongly coupled theories. The sub-specialty of LFFT has become as diverse as theoretical physics itself. Research in this area now ranges from formal discussions of uses of the light-cone gauge to lattice calculations but the bulk of the current research is centred on non-perturbative solutions of QCD and other strongly coupled gauge theories, and it covers theoretical nuclear physics and theoretical particle physics with about equal footing in both fields.

In the theory of the strong interaction the determination of fundamental quantities, such as hadron masses, requires large-scale numerical, non-perturbative methods. The partonic composition of hadrons as reflected in structure functions and form factors are extremely difficult to obtain. Phenomena such as confinement and chiral symmetry breaking are understood only in a qualitative sense which is not sufficient to allow for meaningful calculations that can be tested by experiment.

Recently, new renormalization techniques and numerical methods have been developed for Hamiltonian formulation, which opens up new ways of investigating quantum chromodynamics, the fundamental theory of hadrons. The use of light-cone coordinates introduces essential simplifications and allows for a non-perturbative determination of an effective Hamiltonian for hadrons.

An essential step was recently made (1990-1992) in recognizing the role of zero-mode field operators as a signature of non-perturbative physics. In scalar field theory these zero modes lead to a clear understanding of the phase transition mechanism despite the triviality of the light cone vacuum. However, there are a number of fundamental unsolved problems besetting the general solution of gauge field theories. The most debated ones concern issues of non-trivial topologies, zero-mass field theories, chiral symmetry breaking and the interplay of zero modes, renormalization and Gauge fixing. Other topics that are covered range from fundamental questions of quantization of constrained systems to the effective Hamiltonians and the use of the renormalization group to demonstrate confinement. Many of these ideas are further supported by detail phenomenological applications of light-cone dynamics. These subjects form the core of this volume and will be the central issues in the future developments of the field.

The revival of Dirac's approach has boosted many research activities in strong interaction physics in the USA, Europe, and Japan. In the recent past topical meetings have gathered physicists working in the field. Despite being in a row of predecessors, the meeting of Les Houches was original for it facilitated the presentation of new non-perturbative methods specific to light-cone quantization and their confrontation with other more established treatments of hadronic physics and field theories to their mutual benefit. By the presentation and discussion of the achievements and challenges, the meeting aimed at strengthening international collaboration. The variety in the nationalities of the participants reflects a world wide and expanding interest in the field.

The history of these meetings and workshops is:
1991 Max Plank Institute Heidelberg; Aspen Center of Physics
1992 Southern Methodist University Dallas; Telluride Summer Research Institute
1993 PSI Zurich; Gran Sasso Italy
1994 Institute for Nuclear Theory, Seattle; Warsaw Poland
1995 Regensburg Germany; Telluride Summer Research institute
1996 UNESCO Institute at Iowa State University
1997 Les Houches, France

The contributions are ordered according to the way the sessions were held. Subjects sometimes overlapped and the heading of sessions provided a convenient presentation for the organizers. They would like to thank the convenors for their involvement in the preparation of the different sessions and the collecting of contributed papers.

As a chairman of the workshop I thank all the people who have helped in organizing this meeting. I thank Drs A. Neveu and J. Zinn-Justin for their commitment and interest. This session would not have taken place without their support. Mrs Josette Cellier and the staff of Les Houches are gratefully acknowledged for their careful and cheerful administrative collaboration.

P. GRANGÉ
Chairman

with the Organizing Committee:
H.C. Pauli (Co-Chair), A. Neveu, S. Pinsky, E. Werner

International Advisory Committee

A. Bassetto, *INFN Padova*

S.J. Brodsky, *SLAC*

Y. Frishman, *Inst. Weizman*

St. Glazek, *Warsaw U.*

J. Hiller, *Duluth*

G. McCartor, *SMU Dallas*

G.A. Miller, *Seattle*

H.C. Pauli, *MPI Heidelberg*

R.J. Perry, *Columbus*

S. Pinsky, *Columbus*

J.P. Vary, *Ames*

E. Werner, *Regensburg*

K.G. Wilson, *Columbus*

D. Wyler, *Zürich*

Organizing Committee

P. Grangé, *Montpellier (Chair)*

A. Neveu, *Montpellier*

H.C. Pauli, *MPI Heildelberg (Co-Chair)*

S. Pinsky, *Columbus*

E. Werner, *Regensburg*

CONTENTS

INTRODUCTION

CHAPTER I: Effective Hamiltonian and Renormalization Group (*Convenor: R. Perry*)

LECTURE 1

Renormalization of Hamiltonians

by Stanislaw D. Glazek

LECTURE 2

Spin Glasses and the Renormalization Group

by G. Parisi

LECTURE 3

Hamiltonian Flow in Condensed Matter Physics

by F. Wegner

CHAPTER II: Quantization of Constrained Systems
(*Convenor: M. Marinov*)

CHAPTER III: Results in 3 + 1 Dimensions
(*Convenor: H.C. Pauli*)

CHAPTER V: Light Cone Quantization Under Scrutiny
(Convenor: J. Zinn-Justin)

CHAPTER VI: Gauge Theories and Topological Issues
(Convenor: Y. Frishman)

LECTURE 18

More on Screening and Confinements in 2D QCD

by J. Sonnenchein

LECTURE 19

Intermediate Volumes and the Role of Instantons

by Pierre van Baal

LECTURE 20

Renormalization in the Coulomb Gauge

by D. Zwanziger

LECTURE 21

Stable Knotlike Solitons

by Antti J. Niemi

CHAPTER VII: Structure Functions. Theory and Experiments *(Convenor: P. Chiappetta)*

CHAPTER VIII: Phenomenological Applications
(*Convenor: G. McCartor*)

CHAPTER IX: Condensates
and Chiral Symmetry Breaking (*Convenor: S. Pinsky*)

LECTURE 32

Chiral Symmetry and Light-Cone Wave Functions

by T. Heinzl

LECTURE 33

QCD at Finite Extension.

by F. Lenz and M. Thies

LECTURE 34

Technicalities of the Zero Modes in the Light-Cone Representation

by G. McCartor

LECTURE 35

Zero Mode and Symmetry Breaking on the Light Front

by Koichi Yamawaki

INTRODUCTION

1. Historical background

In his original paper Dirac [1] showed that the uniqueness of the non-relativistic hamiltonian description is lost in the relativistic case. For massive particles several initial surfaces are possible which cut the world lines only once. Among them the light front one has the largest stability group [2] : seven generators leave invariant the hypersurface $\tau = t + z = 0$:

- P_x, P_y, generators of transverse translation,
- $P_z + P_t$, combined generator of translation in the (z, t) direction,
- R_z, generator of the rotations around the z axis in the light reference frame (LRF),
- Λ, generator of boosts in the τ direction,
- $Q_z(Q_y)$, sum of generators of boosts in the $x(y)$ directions and of rotation around the $y(x)$ axes.

In this LRF where all momenta tend to infinity Weinberg [3] was first to show that in scalar field theories all problematic vacuum fluctuations did not contribute. This early indication of the possible triviality of the vacuum in this quantization framework was clarified by Bardacki and Halpen [4], Chang and Ma [5], and subsequent works. However an important difficulty quickly appeared : a light-front frame in which all particles travel with the speed of light is not of the Lorentz type since there is no finite Lorentz transformation which transform this light-like frame in a rest-frame. Domokos [6] was the first to notice that a system in the light cone frame is a constrained system: the equation of motion for fields independent of $x^0 + x^3$ does not contain anymore the temporal derivative and becomes a constraint. Hence one has to rely on quantization procedures specific to constrained systems, first elaborated by Dirac [7,8].

2. Poincaré algebra on the light cone

Light cone coordinates (LCC) $X^\mu = (x^+, x^1, x^2, x^-)$ are defined by the transformation of a Minkowski space vector $\tilde{X}^\mu = (x^0, x^1, x^2, x^3)$ to a vector X^μ.

$$X = C\tilde{X},$$

with

$$C = \begin{pmatrix} 1 & 0 & 0 & 1 \\ 0 & 1 & 0 & 0 \\ 0 & 0 & 1 & 0 \\ 1 & 0 & 0 & -1 \end{pmatrix}, \tag{2.1}$$

i.e.

$$\begin{cases} x^\pm &= x^0 \pm x^3 \\ x^\perp &= (x^1, x^2) \end{cases}. \tag{2.2}$$

The x^+ coordinate is chosen as the light cone time and determines the time evolution of the system.

The metric tensor is then :

$$[g_{\mu\nu}] = \begin{pmatrix} 0 & 0 & 0 & 1/2 \\ 0 & -1 & 0 & 0 \\ 0 & 0 & -1 & 0 \\ 1/2 & 0 & 0 & 0 \end{pmatrix}. \tag{2.3}$$

Hence

$$x.y = x_\mu y^\mu = \frac{1}{2}x^+ y^- + \frac{1}{2}x^- y^+ - x^i y^i \quad (i = 1, 2), \tag{2.4}$$

and

$$\begin{cases} \partial_- &= \dfrac{\partial}{\partial x^-} = \frac{1}{2}\partial^+ \\ \partial_+ &= \dfrac{\partial}{\partial x^+} = \frac{1}{2}\partial^- \end{cases} \tag{2.5}$$

The stability group is made of the Poincaré generators F which leave invariant the surface $x^+ = 0$, i.e. they are such that

$$x^{+*} = x^+ + [x^+, F] = x^+ = 0. \tag{2.6}$$

Hence

$$[x^+, F] = 0.$$

Let $P^\mu = p^\mu$, $M^{\mu\nu} = x^\mu p^\nu - x^\nu p^\mu$. On the light cone one finds

$$\begin{aligned}
&[x^+, P^i] = [x^+, P^+] = [x^+, M^{+i}] = [x^+, M^{ij}] = 0 \quad (i, j = 1, 2), \\
&[x^+, M^{+-}] = -2ix^+ = 0, \\
&[x^+, P^-] = -2i, \\
&[x^+, M^{-i}] = -2ix^i.
\end{aligned} \tag{2.7}$$

The seven kinematic generators which define the stability group are then the three generators of the spatial translation, P^1, P^2, P^+, the M^{12} generator and the three Lorentz boosts M^{+1}, M^{+2}, M^{+-}. This is the essential difference with the three Lorentz boosts M^{01}, M^{02}, M^{03} which are dynamical generators. Here the dynamical ones are the generator of time translation, P^-, and the two transverse rotations M^{-1}, and M^{-2}.

Following Dirac [1] imposing the condition $x^+ = 0$ gives an indefined p^-. Hence all dynamical generators must be independent of p^-. This is achieved by writting

$$P^\mu = p^\mu + \lambda^\mu (p^2 - m^2),$$

$$M^{\mu\nu} = x^\mu p^\nu + x^\nu p^\mu + \beta^{\mu\nu}(p^2 - m^2), \tag{2.8}$$

and fixing λ^μ and $\beta^{\mu\nu}$ in such a way as to eliminate p^-. One finds :

$$P^i = p^i \; ; \; P^+ = p^+ \; ; \; M^{+i} = -x^i p^+ \; ; \; M^{+-} = -x^- p^+ \; ; \; M^{ij} = x^i p^j - x^j p^i,$$

and

$$P^- = \frac{p^j p^j + m^2}{p^+} \; ; \; M^{-i} = x^- p^i + x^i \frac{p^j p^j + m^2}{p^+}, \tag{2.9}$$

which are the three Hamiltonians. The infrared singularity at $p^+ = 0$ appears explicitly as a major problem of the theory.

3. Vacuum structure on the light cone

The spectrum of the energy-momentum operator P^μ is contained in the half positive light cone : one has $P^2 \geq 0 \quad P^0 \geq 0$. In the conventional frame P^0 is the energy but in the light cone frame the positivity conditions implies that the energy P^- and the longitudinal momentum P^+ are positive :

$$P^2 \geq 0 \Rightarrow (P^0)^2 - (P^3)^2 \geq (P^\perp)^2 \geq 0,$$

i.e.

$$P^0 \geq |P^3|, \tag{3.1}$$

hence

$$P^+ = P^0 + P^3 \geq |P^3| + P^3 \geq 0,$$

and

$$P^- = P^0 + P^3 \geq |P^3| - P^3 \geq 0. \tag{3.2}$$

The particular state with $P = 0$ i.e. $P^\mu = 0, \forall \mu = 0, 1, 2, 3$, is invariant under all Poincaré transformations. It corresponds to the trivial vacuum state and is unique. In the conventional framework this vacuum is modified by the interaction and the true vacuum of a continuum field theory does not overlap with the trivial vacuum.

The on-shell relation for a free particule tells us that

$$p^+ p^- - (p^\perp)^2 = m^2,$$

that is

$$p^+ = \frac{m^2 + (p^\perp)^2}{p^-}. \tag{3.3}$$

The physical state of a free particle with $m \neq 0$ and $p^+ = 0$ is excluded as it would correspond to a particle with an infinite energy. Hence any physical state of a free particle with a non zero mass in always characterized by $p^+ \neq 0$. Moreover the spectral condition tells us that $p^+ \geq 0$, hence p^+ is always > 0. The quantum numbers of the trivial vacuum are such that $p = 0$ $(p^- = p^+ = p^\perp = 0)$. A physical vacuum with a non-zero energy p^- and a longitudinal momentum $p^+ = 0$ cannot exist. On the light cone the vacuum is always the trivial vacuum : it is not polarized and modified by the interaction.

This reasoning is valid for scalar fields. For fermionic theories the situation is much more complicated and, it is fair to say, not yet fully understood. Indications on how to proceed in the fermionic case are furnished by the study of the Schwinger model which can be solved on the light-cone [9,10]. There one obtains a nontrivial vacuum for the fermions which is however much simpler to construct than in the conventional approach.

4. Signature of nonperturbative effects (in LCFT.)

If, as it was stated above, the vacuum is trivial in scalar field theories, the immediate question arises : where does one finds the nonperturbative effects embedded in the nontrivial vacuum of the conventional treatment ? This was a puzzle for quite a long time and led to the opinion that LCQFT was only valid in the perturbative regime. The situation got clarified in the early nineties with the discovery of the role of zero modes of field operators as the carriers of nonperturbative physics [11-14].

In order to simplify the presentation we treat as an example the case of scalar theory in 1+1 dimension with a ϕ^4-interaction. The theory is defined by the Lagrangian

$$\mathcal{L} = \frac{1}{2}\partial^+\phi\partial^-\phi - \frac{m}{2}\phi^2 - \frac{\lambda}{4!}\phi^4 \tag{4.1}$$

which yields as equation of motion

$$\partial^+\partial^-\phi - m^2\phi - \frac{\lambda}{3!}\phi^3 = 0 \tag{4.2}$$

In order to separate the vacuum sector (quantum number $p^+ = 0$) from the particle sector (quantum number $p^+ > 0$) one introduces projection operators P and Q on these sectors (acting in the space of field operators) :

$$\Omega = P * \phi \quad ; \quad \varphi = Q * \phi \quad ; \quad \phi = \Omega + \varphi \quad ; \quad P + Q = 1. \tag{4.3}$$

Applying them to the eq. of motion (4.2) one obtains the two coupled equations

$$\partial^+\partial^-\varphi - m^2\varphi - \frac{\lambda}{3!}Q * (\Omega + \varphi)^3 = 0. \tag{4.4}$$

$$m^2\Omega + \frac{\lambda}{3!}P * (\Omega + \varphi)^3 = 0 \tag{4.5}$$

The second one which contains no time derivative is apparently not an equation of motion but a constraint, which we call θ for future use.

In order to define the theory completely we impose periodic boundary conditions in a spatial "volume" of length 2L :

$$\phi(-L) = \phi(+L).$$

In Fourier space this leads to a discretization of momentum quantum numbers :
$k_n^+ = \frac{2\pi}{L}.n$, $n = 0, 1, 2, ...$ and it allows for a simple, straightforward definition
of the projectors P and Q :

$$\begin{cases} P * \phi(k_n^+) = \phi(k_n^+) & ; \quad n = 0 \\ P * \phi(k_n^+) = 0 & ; \quad n = 1, 2, 3, ..., \end{cases}$$

$$\begin{cases} Q * \phi(k_n^+) = 0 & ; \quad n = 0 \\ Q * \phi(k_n^+) = \phi(k_n^+) & ; \quad n = 1, 2, 3, ..., \end{cases}$$

The explicit form of P is defined by

$$P * f = \frac{1}{2L} \int_{-L}^{+L} f(x)dx, \tag{4.6}$$

which changes (4.4, 4.5) into :

$$\partial^+ \partial^- \varphi - m^2 \varphi - \frac{\lambda}{3!}[\varphi^3 + Q * (\varphi^2 \Omega + \varphi \Omega \varphi + \Omega \varphi^2) + Q * (\varphi \Omega^2 + \Omega \varphi \Omega + \Omega^2 \varphi)] = 0, \tag{4.7}$$

$$\theta = m^2 \Omega + \frac{\lambda}{3!} \frac{1}{2L} \int_{-L}^{+L} [\varphi^3(x) + \varphi^2(x)\Omega + \varphi(x)\Omega\varphi(x) + \Omega\varphi^2(x)]dx = 0. \tag{4.8}$$

The general strategy for the solution of these coupled equations is to develop φ and Ω as a Haag series in terms of sums of normal ordered products of field operators $\varphi_0(x)$, where $\varphi_0(x)$ is the free field solution defined by
$(\partial^+ \partial^- - m^2)\varphi_0(x) = 0$. This is best carried out in momentum space where the
discretized Fock-space expansion of $\varphi_0(x)$ becomes

$$\varphi_0(x^-)|_{x^+=0} = \sum_{n=1,2,..} \frac{1}{\sqrt{4\pi n}}[a_n e^{ik_n^+ x^-} + a_n^+ e^{-ik_n^+ x^-}],$$

$$[a_n, a_m^+] = \delta_{nm}. \tag{4.9}$$

Then the Haag series for the zero mode takes the form :

$$\begin{aligned} \Omega &= \phi_0 + \sum_{n=1}^{\infty} C_n^{(1,1)} a_n^+ a_n + \sum_{n,m=1}^{\infty} (C_{n,m}^{(2,1)} a_n^+ a_m^+ a_{n+m} + h.c.) \\ &+ \sum_{n,m,l=1}^{\infty} (C_{n,m,l}^{(3,1)} a_n^+ a_m^+ a_l^+ a_{n+m+l} + h.c.) \\ &+ \sum_{n,m,l=1}^{\infty} (C_{n,m,l}^{(2,2)} a_n^+ a_m^+ a_l a_{n+m-l} + h.c.) + ..., \end{aligned} \tag{4.10}$$

where we have added a C-number part ϕ_0.
An analogous expansion exists for the particle sector field $\varphi(x)$. Its explicit form

is not given here, since its form is the same as in the conventional treatment, except for the suppression of all zero momentum componants. The form (4.10) implies apparently $< 0|\phi|0 >= \phi_0$.

The task is the determination of all coefficients of the expansions of Ω and $\varphi(x)$; in principle it can be accomplished by taking matrix elements between all relevant Fock sectors, i.e. the vacuum sector, the 1-particle sector, the 2-particle sector etc. It is a unique feature of LCQFT that this can actually be done because the action of the Fock-space operators a_n^+, a_m on all conceavable states is known due to the triviality of the vacuum. This is at variance with ETQFT where the existence of a nontrivial ground state would render this procedure impossible. Though the solution of the coupled, nonlinear equations for the expansion coefficients is highly nontrivial, the equations to be solved can be written down which is not the case in the ET-approach.

To make what follows as simple as possible we replace $\varphi(x)$ by $\varphi_0(x)$ and take into account only the first two terms on the r.h.s. of eq. (4.10). In this approximation the zero mode Ω becomes

$$\Omega = \phi_0 + \sum_{n=1}^{\infty} C_n a_n^+ a_n, \qquad (4.11)$$

where we have made the change of notation $C_n^{(1,1)} \to C_n$. The action of the operator valued part of Ω on a single particle state $|k >= a_k^+|0 >$ is tantamount to the action of a momentum dependent single particle field. Therefore the form (4.11) defines a mean field approximation. The task is now to find ϕ_0 and the C_n 's, $n = 1, 2, ...$ .They are easily obtained by replacing $\varphi(x)$ by $\varphi_0(x)$ on the r.h.s. of eq.(4.8) and evaluating the matrix elements $< 0|\theta|0 >$ and $< n|\theta|n >, n = 1, 2, 3, ...$, leading to the equations

$$< 0|\theta|0 >= \mu^2 \phi_0 + \frac{\lambda}{6}\phi_0^3 + \frac{\lambda}{24\pi} \sum_{m=1}^{\infty} \frac{C_m}{m} = 0, \qquad (4.12)$$

$$< n|\theta|n >= \frac{\lambda}{6}C_n^3 + \frac{\lambda\phi_0}{2}C_n^2 + (\mu^2 + \frac{\lambda\phi_0^2}{2} + \frac{\lambda}{4\pi n})C_n + \frac{\lambda\phi_0}{4\pi n} = 0, \qquad (4.13)$$

where μ^2 is a tadpole-renormalized mass :

$$\mu^2 = m^2 + \frac{\lambda}{2} < 0|\varphi_0^2(x)|0 >= m^2 + \frac{\lambda}{8\pi} \sum_{m=1}^{\infty} \frac{1}{m}.$$

Dividing both equations by μ^2 one sees that λ and μ^2 come only in the ratio $\frac{\lambda}{\mu^2}$; we therefore introduce a dimensionless coupling constant $g = \frac{\lambda}{4\pi\mu^2}$. For any $n = 1, 2, 3, ...$ eq. (4.13) can be solved for $C_n = C_n(g, \phi_0)$ which in turn can be injected into the sum $\sum_{m=1}^{\infty} C_m/m$ which is convergent since asymptotically

$C_m \sim \frac{1}{m}$. Thus eq. (4.12) becomes an equation for the determination of $\phi_0 = \phi_0(g)$:

$$\phi_0 + \frac{2\pi}{3} g\phi_0^3 + \frac{g}{6} \sum_{m=1}^{\infty} \frac{C_m(g, \phi_0)}{m} = 0.$$

Whereas for arbitrary ϕ_0 the solution must be obtained numerically, in the vicinity of the phase transition - where ϕ_0 and all C_n are small - the equations can be linearized and an analytical approach becomes possible.

The results are the following :

(1) At $g = 3.18$ there is a second order phase transition from $\phi_0 = 0$ to $\phi_0 \neq 0$.

(2) The critical exponents are $\beta = 1/2, \gamma = 1$ and $\delta = 3$ and correspond to the mean-field result of the conventional treatment.

(3) There exist two degenerate solutions in the following sense : For a given ϕ_0 there is always a sign - conjugate solution $-\phi_0$; the two possible zero-modes $\Omega_\pm$ lead to two different Hamiltonians $H_\pm = H(\varphi + \Omega_\pm)$ which have the same spectrum. Therefore symmetry breaking takes place on the level of the Hamiltonian, not on the level of the groundstate which is unique and always trivial. Choosing one of the two possible solutions is equivalent to choosing one of the two possible zero-modes or associated Hamiltonians. Therefore in LCQFT the symmetry of the world is the symmetry of the Hamiltonian.

5. Chiral symmetry breaking

Due to the fact that in LCQFT the information on nonperturbative physics resides in field operator zero-modes the issue of chiral transformations and chiral symmetry breaking is conceptually and technically rather different from the ET case. The essential features can best be exposed with of a simple model which allows a comparison of the LC-and ET formulation.

The model in question is the **O(2)** nonlinear σ-model defined by the Lagrangian density

$$\mathcal{L} = \frac{1}{2}(\partial_\mu \sigma)^2 + \frac{1}{2}(\partial_\mu \pi)^2 + V(\sigma^2 + \pi^2) \;;$$

$$V = \frac{1}{2}\mu^2(\sigma^2 + \pi^2) + \frac{\lambda}{4}(\sigma^2 + \pi^2)^2, \qquad (5.1)$$

which is invariant with respect to rotations in the (σ, π)-plane.

In the ET-formulation the symmetry features of interest for a comparison with the LC-case are :

(1) There is a conserved current

$$j_\mu = (\partial_\mu \sigma)\pi - (\partial_\mu \pi)\sigma, \qquad (5.2)$$

and a conserved charge

$$Q = \int d^3x j_0(x). \tag{5.3}$$

(2) Q satisfies the commutation relations

$$[Q, \pi] = i\sigma \quad ; \quad [Q, \sigma] = -i\pi \quad ; \quad [Q, H] = 0, \tag{5.4}$$

i.e. it is the generator of the **O**(2)-symmetry.

(3) The classical potential energy density of the constant fields σ_c, π_c has its minimum for

$$(a) \qquad \sigma_c = \pi_c = 0 \qquad \text{if} \quad \mu^2 > 0, \tag{5.5}$$

$$(b) \quad \sigma_c^2 + \pi_c^2 = -\frac{\mu^2}{\lambda} \qquad \text{if} \quad \mu^2 < 0. \tag{5.6}$$

In case (a) the vacuum is unique and $Q|vac >= 0$. After quantization the σ- and π-fields are massive.

In case (b), where infinitely many degenerate vacuum solutions exist, the vacuum is not annihilated by Q, i.e. $Q|vac > \neq 0$, whereas $[H, Q] = 0$ still holds. If quantization is based on the broken-symmetry solution $\sigma_c = \sqrt{\frac{-\mu^2}{\lambda}}$, $\pi_c = 0$, the σ-field is massive, whereas the pion becomes a Goldstone boson $(m_\pi = 0)$.

(4) Defining (for case b)) shifted fields $\sigma = \sigma_c + \sigma'$, $\pi = \pi' + \pi_c$, the Hamiltonian $H'(\sigma', \pi', \sigma_c)$ and the charge :

$$Q' = \int dx^3 [(\partial \sigma_0')\pi' - (\sigma_0 \pi')(\sigma' + \sigma_c)] \tag{5.7}$$

expressed in terms of $\sigma'\pi'$, and σ_c, trivially satisfy $[H', Q'] = 0$. Writing the charge in terms of the fluctuating fields σ' and π' only i.e. :

$$Q'' = \int dx^3 [(\partial_0 \sigma')\pi' - (\partial_0 \pi')\sigma'], \tag{5.8}$$

leads to

$$[Q'', H'] \neq 0. \tag{5.9}$$

In the LC-formulation the explicit form of $\mathcal{L}(x)$ becomes :

$$\mathcal{L} = (\partial_+ \sigma)(\partial_- \sigma) + (\partial_+ \pi)(\partial_- \pi) - \frac{1}{2}(\partial_\perp \sigma)^2 - \frac{1}{2}(\partial_\perp \pi)^2 - V(\sigma^2 + \pi^2). \tag{5.10}$$

The conjugate momenta $\pi_\sigma = \partial_- \sigma, \pi_\pi = \partial_- \pi$ are spatial derivatives, i.e. dependent variables, i.e. the system is constrained. The equations of motion are:

$$\partial_+ \partial_- \sigma - \partial_\perp^2 \sigma + \frac{\delta V}{\delta \sigma} = 0 \tag{5.11}$$

$$\partial_+\partial_-\pi - \partial_\perp^2\pi + \frac{\delta V}{\delta\pi} = 0. \tag{5.12}$$

As in paragraph 4 we impose periodic boundary conditions in the x^--direction and decompose the fields into zero-modes :

$$\sigma_0 = \frac{1}{2L}\int_{-L}^{+L} dx^-\sigma(x) \quad ; \quad \pi_0 = \frac{1}{2L}\int_{-L}^{+L} dx^- \ \pi(x),$$

and normal modes $\varphi_\sigma, \varphi_\pi$:

$$\sigma = \sigma_0 + \varphi_\sigma \quad ; \quad \pi = \pi_0 + \varphi_\pi.$$

Projecting (5.11) and (5.13) onto the vacuum sector leads to two constraints:

$$\theta_\sigma = \frac{1}{2L}\int_{-L}^{+L} dx_-[(\mu_\sigma^2 - \partial_\perp^2)\sigma + \lambda(\sigma^3 + \frac{1}{2}\sigma\pi^2 + \frac{1}{2}\pi^2\sigma)] = 0 \tag{5.13}$$

$$\theta_\pi = \frac{1}{2L}\int_{-L}^{+L} dx^-[(-\partial_\perp^2)\pi + \lambda(\pi^3 + \frac{1}{2}\pi\sigma^2 + \frac{1}{2}\sigma^2\pi)] = 0, \tag{5.14}$$

which yield the zero-modes in terms of the normal modes. The signature of symmetry breaking is the existence of solutions allowing for nonzero C-number parts of the zero-modes. With the knowledge of the zero-modes the Hamiltonian can be determined as :

$$H^{LC} = \int d^2x_\perp \int_{-L}^{+L} dx^-[\frac{1}{2}(\partial_\perp\varphi_\sigma)^2 + \frac{1}{2}(\partial_\perp\varphi_\pi)^2 + V((\sigma_0+\varphi_\sigma)^2 + (\pi_0+\varphi_\pi)^2)]$$
$$\tag{5.15}$$

The symmetry features of the LC-formulation are the following.

(1) There is a conserved current

$$j_\mu = (\partial_\mu\sigma)\pi - (\partial_\mu\pi)\sigma, \tag{5.16}$$

and a conserved charge

$$Q^{LC} = \int dx_\perp^2 \int_{-L}^{+L} dx^-[(\partial_-\sigma)\pi - (\partial_-\pi)\sigma] \tag{5.17}$$

which due to the boundary conditions becomes

$$Q^{LC} = \int dx_\perp^2 \int_{-L}^{+L} dx^-[(\partial_-\varphi_\sigma)\varphi_\pi - (\partial_-\varphi_\pi)\varphi_\sigma] \tag{5.18}$$

i.e. Q^{LC} depends only on the normal modes of the fields, even in the presence of zero-modes. In this sense Q^{LC} corresponds to Q'' of eq. (5.8).

(2) The commutation relations are :

$$[Q^{LC}, \varphi_\sigma] = -i\varphi_\pi \quad ; \quad [Q^{LC}, \varphi_\pi] = i\varphi_\sigma, \tag{5.19}$$

$$[Q^{LC}, \sigma_0] = -i\pi_0 \quad , \quad [Q^{LC}, \pi_0] = i\sigma_0,$$

$$[Q^{LC}, H^{LC}] = 0 \quad \text{for} \quad \mu^2 > 0 \quad , \quad \sigma_0 = \pi_0 = 0, \tag{5.20}$$

$$[Q^{LC}, \sigma_0] \neq -i\pi_0 \quad , \quad [Q^{LC}, \pi_0] \neq i\sigma_0,$$

$$[Q^{LC}, H^{LC}] \neq 0 \quad \text{for} \quad \mu^2 < 0 \quad , \quad <\sigma_0>^2 = -\frac{\mu^2}{\lambda} \tag{5.21}$$

in analogy to eq. (5.9).

(3) For $\mu^2 > 0$ and $\mu^2 < 0$ one always has $Q^{LC}|0> = 0$ (the vacuum is trivial and Q^{LC} depends only on normal operators). With $<\sigma_0> = -\sqrt{-\frac{\mu^2}{\lambda}}$, $<\pi_0> = 0$ the σ-field becomes massive while for the pion field $m_\pi = 0$.

The results of eqs. (5.20) and (5.21) are obtained from a perturbative expansion of the operator-valued parts of the zero-modes σ_0 and π_0. Comparison of the two cases shows that the symmetry properties of the zero-modes are changed by the spontaneous symmetry breaking. The results on the spectrum of the theory, i.e. on the particle spectrum, can be obtained only after a (perturbative) solution for the zero-modes has been accomplished. If this is done, the spectrum reflects the symmetry properties of the zero-modes. To give an explicit example we show the Hamiltonian which is obtained for the broken phase with the first order solution (in λ) of the constraints :

$$H_{(1)}^{LC} = \int dx_\perp^2 \int_{-L}^{+L} [\frac{1}{2}(\partial_\perp \varphi_\sigma)^2 + \frac{1}{2}(\partial_\perp \varphi_\pi)^2 + \frac{m_\sigma^2}{2}\varphi_\sigma^2 + \frac{\lambda}{4}(\varphi_\sigma^2 + \varphi_\pi^2)^2]$$

$$+\lambda. <\sigma_0> \int dx_\perp^2 \int_{-L}^{+L} dx^- \varphi_\sigma(\varphi_\sigma^2 + \varphi_\pi^2) + H_{classical}$$

Clearly the pion mass term has disappeared, so there is a pionic Goldstone boson.

To summarize one can say that in analogy to the scalar ϕ^4-field theory discussed in the preceeding section LCQFT handles nontrivial properties of the theory on the level of zero-modes of field operators which are determined from constraints.

As to the fermionic case the LC-version of chiral symmetry is still in its infancy. It is mainly due to the already mentionned conceptual problems connected with the fermionic vacuum. What is true both for fermionic and bosonic theories is the property $Q^{LC}|0> = 0$. Also the existence of nonzero vacuum expectation values of composite operators like $< 0|\bar{\psi}\psi|0 >$ has been verified.

Furthermore the prototype of a theory wich exhibits dynamical mass generation via chiral symmetry breaking - the Nambu-Jona-Lasinio model - has been solved on the light-cone in the mean-field approximation, where only a C-number valued zero-mode of $\bar{\psi}\psi$ shows up [15]. Some aspects of the Gross-Neveu model have also been investigated [16]. A very fundamental approach to chiral symmetry breaking in the framework of QCD - based on renormalization group techniques - comes from the Wilson group at Ohio State University [17].

6. Gauge theories on the light-cone

A gauge theory on light-cone is at the start a Lagrangian field theory with redundant degrees of freedom which change under gauge rotations. The canonical quantization of such a theory is notoriously difficult because the Lagrangian contains too many degrees of freedom which must be eliminated by gauge fixing procedures. In general it is globally impossible to do so due to the existence of Gribov horizons in the space of gauge potentials which limit the domains inside which the gauge can be fixed. Already in the ET-form the quantization is very difficult due to the presence of constraints.

Different strategies have been developed to attack the problem :

a) The lattice formulation where one sums over all possible gauge configurations.

b) One first fixes the gauge and tries to solve the constraints on the classical level, using either the Dirac-Bergman [1] or the Fadeev-Jackiw method [18] and then quantizes. Apparently a problem arises in this approach, if one is confronted with noncanonical brackets.

c) One first quantizes (preferably in cartesian coordinates) and then fixes the gauge.

In what follows only the pure gauge theory (Yang-Mills-theory) is discussed, since the inclusion of fermionic matter fields would introduce additional severe problems.

The light-front Yang-Mills Lagrangian is linear in the velocities and therefore singular in the sense of Dirac. For what follows the Fadeev-Jackiw method [18] is adapted ; periodic boundary conditions are choosen for all spatial directions in a finite volume : $-L \leq (x^-, x_i) \leq +L$; $i = 1, 2$.

The Lagrangian density is written in the form

$$\mathcal{L}_{YM}(x) = \Pi_-^a \dot{A}_-^a + \Pi_i^a \dot{A}_i^a - \frac{1}{2}(\Pi_-^a \Pi_-^a + B_-^a B_-^a)$$

$$+ A_+^a (D_-^{ab}\Pi_-^b + D_i^{ab}\Pi_i^b) - H(x) + A_+^a G^a. \qquad (6.1)$$

Here $H(x)$ is the Hamiltonian density, G is the Gauss-law operator :

$$G^a = D_i^{ab}[A_i]E^{ib} = 0, \qquad (6.2)$$

D is the matrix operator of the covariant derivative.

The points stand for the"time" derivative $\frac{\partial}{\partial x^+}$. Moreover Π^a_- and Π^a_i are the field momenta :

$$\Pi^a_- = E^a_- = \dot{A}^a_- - D^{ab}_- A^b_+ \quad ; \quad \Pi^a_i = E^a_i = \partial_- A^a_i - D^{ab}_i A^b_-. \tag{6.3}$$

The chromo-electric and chromo-magnetic fields expressed through the field tensor $F^{ik,a}$ are :

$$E^{ia} = F^{i+,a} \quad ; \quad B^{ia} = \frac{1}{2}\epsilon^{ijk} F^{jk,a}. \tag{6.4}$$

The $2 \times 3 = 6$ momenta E^a_i are dependent quantities since only spatial derivatives show up in their definition. Using (6.3) in eq. (6.1) makes the Lagrangian highly non-canonical. This is a typical LC-feature.

The application of the Fadeev-Jackiw method leads to very non-canonical elementary brackets which moreover are field dependent. Up to now the only feasable strategy to proceed seems first to fix the gauge and then quantize. It is interesting to note that every thing becomes completely canonical in the light-come gauge where $A^a_- = 0$. This gauge choice is the basis of practically all perturbative QCD-calculations which have been quite success ful in the past. Physically the light-cone gauge amounts to suppress field zero modes which are probably irrelevant in the perturbative domain but are responsible on the other hand for nonperturbative effects. In order to take that into account a modified gauge choice has been suggested for $SU(2)$ by Franke et al [19] in which $A^i_- = 0$ for $i = 1, 2$; $A^3_- = a$; $\partial_- a = 0$, i.e. a is a zero-mode (independant of x^-).

In this case the elementary brackets become diagonal, but dependent on the zero mode operator a. A closed form for the Hamiltonian can be obtained [19] which contains a-dependent Jacobians and render the kinetic energy quite complicated. Moreover at this level the theory still contains a LC-typical second class constraint quite analogous to the constraint appearing in the scalar ϕ^4-theory. It is related to the eventual presence of zero-modes (x^--independent) of the gauge potentials $A^3_i, i = 1, 2$. Due to the difficulty to solve the constraint in question the physical implications and consequences of this constraint are not yet clear.

As far as the zero-mode a and its interplay with the transverse gluons A_i is concerned a rather simple picture emerges [20, 21] : The total Hilbert space is a direct product of the Fock space $\mathcal{R}_a$ associated with the transverse gluons and Hilbert space $\mathcal{R}_a$ of functionals $\Psi\{a\}$ depending on the zero-mode a. Though the expansion coefficients of the transverse gluon fields operators are a-dependent, this causes no problem, since they act multiplicatively in the a-representation. In this representation the vacuum is trivial in space-time ; its nontrivial character comes from the zero-mode degrees of freedom which are believed to be connected with non-trivial topology.

From the above discussions emerges another important results : The problems of zero-modes and gauge fixing are apparently very intimately interwinded:

The zero-modes are therefore gauge dependent objects which makes their interpretation more difficult than for scalar fields.

References

[1]   Dirac P.A.M., Rev of Modern Physics 216(1949) 393.

[2]   Susskind L., *Phys. Rev.* 165 (1968) 1535.

[3]   Weinberg S., *Phys. Rev.* 150 (1966) 1313.

[4]   Bardacki K. and Halpern M.B., *Phys. Rev.* 176 (1968) 1686.

[5]   Chang S.J. and Ma S.K., *Phys. Rev.* 180 (1969) 1506.

[6]   Domokos G., in : "Lectures in Theoretical Physics" Vol. XIV, 1971, A.O. Barut and W.E. Brittin eds. Colorado University Press Boulder (1972).

[7]   Dirac P.A.M., Canad. J. Math 2 (1950) 1.

[8]   Dirac P.A.M., Lectures on Quantum Mechanics Benjamin, New-York (1964).
Heinzl T., Krusche S. and Werner E., *Nucl. Phys.* **A532** (1991) 429.

[9]   Heinzl T., Krusche S. and Werner E., *Phys. Lett.***B256** (1991) 55.

[10]   Heinzl T., Krusche S. and Werner E., *Phys. Lett.* **B275** (1992) 410.

[11]   Heinzl T., Krusche S. and Werner E., *Nucl. Phys.* **A532** (1991) 429.

[12]   Robertson D.G., *Phys. Rev.* **D47** (1993) 2549.

[13]   Pinsky S.S., Van de Sande B. and Bender C.M., *Phys. Rev.* **D48** (1993) 816. Pinsky S.S. and Van de Sande B., *Phys. Rev.* **D49** (1994) 2001. Pinsky S. S., Van de Sande B. and Hiller J.R., *Phys. Rev.* **D51** (1995) 726.

[14]   Heinzl T., Stern C., Werner E. and Zellermann B., Preprint TPR-95-20, to appear in Z. Phys. C.

[15]   Dietmaier C., Heinzl T., Schaden M. and Werner E., Z. Phys. **A333** (1989) 215.

[16]   Pesando I., Mod. *Phys. Lett.* **A10** (1995) 525.

[17]   Wilson K. et al, *Phys. Rev.* **D49** (1994) 6720.

[18]   Fadeev L.D. and Jackiw R., *Phys. Rev. Lett.* 60 (1988) 1692.

[19]   Franke V.A., Novozhilov Yu.V. and Prokhvatilov E.V., Lett. Math. Phys. 5(1891) 239, 431.

[20]   Pause T., Diploma Thesis, Regensburg, 1995.

[21]   Heinzl T., Nucl. Phys. B (Proc. Suppl.) 39 (1995) 217.

CHAPTER I

Effective Hamiltonian and Renormalization Group

Renormalization of Hamiltonians

Stanisław D. Głazek

Institute of Theoretical Physics, Warsaw University
ul. Hoża 69, 00-681 Warsaw, Poland

A matrix model of an asymptotically free theory with a bound state is solved using a perturbative similarity renormalization group for hamiltonians. An effective hamiltonian with a small width, calculated including the first three terms in the perturbative expansion, is projected on a small set of effective basis states. The resulting small hamiltonian matrix is diagonalized and the exact bound state energy is obtained with accuracy of order 10%. Then, a brief description and an elementary illustration are given for a related light-front Fock space operator method which aims at carrying out analogous steps for hamiltonians of QCD and other theories.

1. INTRODUCTION

This lecture has two aims. The first aim is to show a simple example of a new kind of calculation of effective hamiltonians, based on the perturbative similarity renormalization group [1, 2]. The second aim is to show how one can generalize the simple example and start systematic perturbative calculations for quantum field theoretic hamiltonians in the light-front Fock space.

Although the methods we present are quite general, the main motivation came from QCD. QCD is asymptotically free and its perturbative running coupling constant grows at small momentum transfers beyond limits. This rise invalidates usual perturbative expansions in the region of scales where the bound states are formed.

Ref.[3] outlined a light-front hamiltonian approach to this problem in QCD, using the perturbative similarity renormalization group. Independently, Wegner [4] proposed a flow equation for hamiltonians in solid state physics. He introduced an explicit expression for the generator of the similarity transformation which leads to a Gaussian similarity factor of a uniform width.

Wilson and I have solved numerically a simple matrix model to gain quantitative experience with the similarity scheme using Wegner's equation. [5] We also made perturbative studies. [6] This lecture is based on those works in the part describing the model. The remaining part contains an outline of how one can attempt to make similar steps for light-front hamiltonians in quantum field theory using creation and annihilation operators.[7]

2. MODEL

Consider a quantum theory which is characterized by a large range of energy scales as measured by certain H_0. QCD has this feature. It extends in energies from ∞ (asymptotic freedom) down to the infrared energy region. We represent the theory by a model with a hamiltonian $H = H_0 + H_I$ acting in a space spanned by a finite discrete set of nondegenerate eigenstates of the hamiltonian H_0,

$$H_0|i> = E_i|i> . \tag{2.1}$$

Matrix elements of the interaction are assumed to be

$$< i|H_I|j > = -g\sqrt{E_i E_j} . \tag{2.2}$$

g is a dimensionless coupling constant.

We choose $E_i = 2^i$ and $M \leq i \leq N$. M is large and negative and N is large and positive. We use $M = -21$ and $N = 16$ in our numerical example. Let the energy equal 1 correspond to 1 GeV. Then, the ultraviolet cutoff corresponds to 65 TeV and the infrared cutoff corresponds to 0.5 eV.

The same model can be alternatively derived by discretization of the 2-dimensional Schrödinger equation with a potential of the form a coupling constant times a δ-function. [8]

For $g > 1/38$, the hamiltonian matrix has one negative eigenvalue and 37 positive eigenvalues. g is adjusted to obtain the negative eigenvalue equal -1 GeV; $g \sim 0.06$. This eigenvalue corresponds to the s-wave bound state energy in the 2-dimensional Schrödinger equation.

We calculate effective hamiltonians, $\mathcal{H} \equiv \mathcal{H}(\lambda)$, using the similarity renormalization group equations in the differential form. The effective hamiltonians are parametrized by their energy width λ. The notion of the hamiltonian width will become clear shortly. We use Wegner's flow equation [4]

$$\frac{d\mathcal{H}}{d\lambda^2} = -\frac{1}{\lambda^4}[[\mathcal{D}, \mathcal{H}], \mathcal{H}] , \tag{2.5}$$

with the initial condition $\mathcal{H}(\infty) = H$. The matrix $\mathcal{D}$ is the diagonal part of $\mathcal{H}$ with elements $\mathcal{D}_{mn} = \mathcal{H}_{mm}\delta_{mn}$. Thus, $\mathcal{H}(\lambda)$ is a unitary transform of H and both have the same spectrum (see Wegner's lecture in this volume).

Equation (2.5) can be approximately solved for a small g keeping only terms order 1 and g. One obtains

$$\mathcal{H}_{mn} \;=\; E_m\,\delta_{mn} \;-\; g\sqrt{E_m E_n}\,\exp\left[-(E_m - E_n)^2/\lambda^2\right]. \qquad (2.6)$$

Here, $\mathcal{D}_{mm} = (1 - g)E_m$. The Gaussian factor of width λ is the similarity function. This explains the notion of the hamiltonian width. Ref. [5] demonstrated that the Wegner flow equation has a renormalization group interpretation. Including terms order g^2, we let g depend on λ and we introduce $\tilde{g}(\lambda) \equiv \tilde{g}$. It follows from equations satisfied by the matrix elements $\mathcal{H}_{mn}$ with the indices m and n close to M that, neglecting small energies,

$$d\tilde{g}/d\lambda \;=\; -\tilde{g}^2\,\frac{d}{d\lambda}\sum_{\ell}\exp\left[-2E_\ell^2/\lambda^2\right], \qquad (2.7)$$

and $\tilde{g}(\infty) = g$. Analytic integration of Eq. (2.7) in the model gives, approximately,

$$\tilde{g}_a(\lambda) \;=\; (1.45\log\lambda - 0.9)^{-1}. \qquad (2.8)$$

$\tilde{g}_a(\lambda)$ grows when λ gets smaller and it exhibits the asymptotic freedom behavior: it is smaller for more violent interactions (i.e. of wider range in energy).

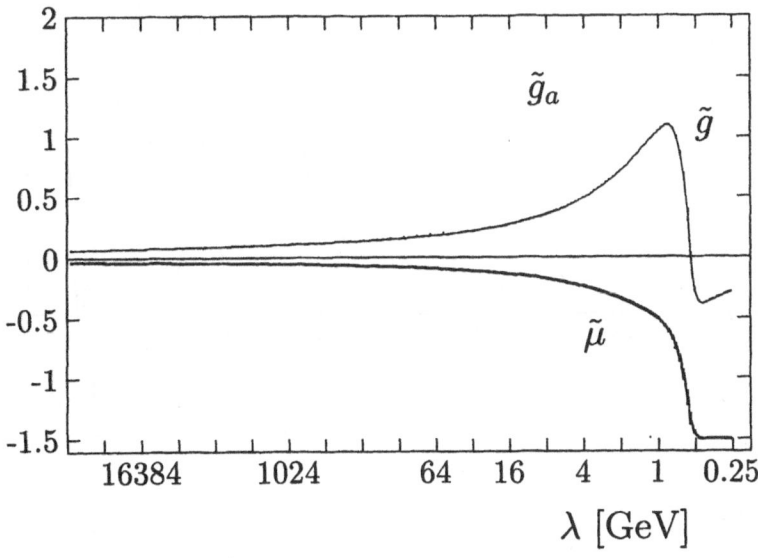

Fig. 1. — The approximate running coupling $\tilde{g}_a(\lambda)$ from Eq. (2.8) and the exact running coupling $\tilde{g}(\lambda)$, plotted as functions of the effective hamiltonian width λ. The matrix element $\tilde{\mu}(\lambda) = \mathcal{H}_{-1,-1}(\lambda) - 0.5$ GeV is also plotted to show the width range where the bound state eigenvalue appears on the diagonal.

$\tilde{g}_a(\lambda)$ blows up to infinity for $\lambda \sim 1.9$ GeV. In this approximation, matrix elements of $\mathcal{H}$ for $E_m \sim E_n \ll \lambda$ can be written as

$$\mathcal{H}_{mn}(\lambda) = E_m \delta_{mn} - \tilde{g}_a(\lambda) \sqrt{E_m E_n} \exp\left[-[E_m - E_n]^2/\lambda^2\right] + corrections \,.$$
(2.9)

Now, $\mathcal{D}_{mm}(\lambda) = [1 - \tilde{g}_a(\lambda)]E_m$. The energy order of low energy states is reversed when $\tilde{g}_a(\lambda)$ grows above 1.

The exact running coupling, $\tilde{g}(\lambda)$, is defined by writing $\mathcal{H}_{M,M+1}(\lambda) = -\tilde{g}(\lambda)$ $\sqrt{E_M E_{M+1}}$. Eq. (2.9) shows that $\tilde{g}_a = \tilde{g}$ for large λ. To find $\tilde{g}$ for all values of λ, we solved Eq. (2.5) numerically. Fig. 1. shows that the approximate solution blows up in the flow before the effective hamiltonian width is reduced to the scale where the bound state is formed. That scale, order 1 GeV, equals λ at which the bound state eigenvalue appears on the diagonal. The diagonal matrix element is also shown in Fig. 1.

The key feature, visible in Fig. 1, is that the exact effective coupling con-

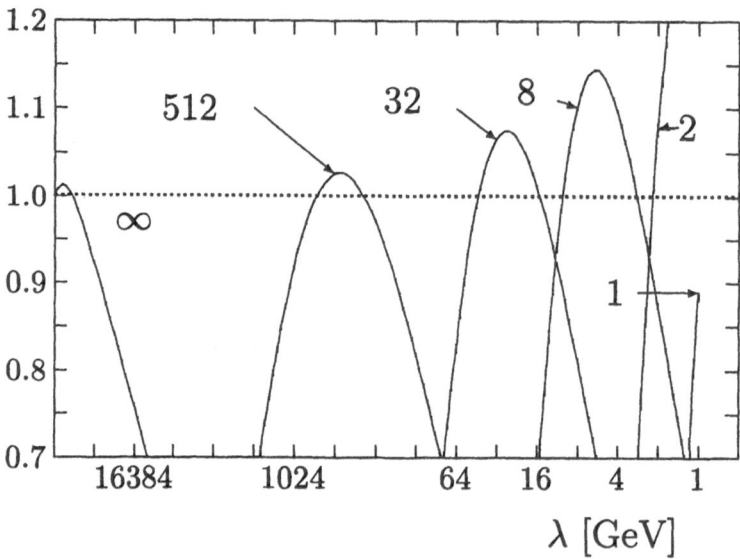

Fig. 2. — The accuracy of the bound state eigenvalues obtained from effective hamiltonians whose renormalization group flow with the width λ is calculated expanding in powers of the effective coupling constant $\tilde{g}(\lambda_0)$ and including terms order 1, $\tilde{g}(\lambda_0)$ and $\tilde{g}^2(\lambda_0)$. The accuracy is given as ratio of the bound state eigenvalue obtained by diagonalization of the effective hamiltonian of width λ to the exact value, -1 GeV. The curves correspond to the indicated values of λ_0 (in units of GeV). The result of expansion in the initial coupling g is denoted by ∞. The arrows show points where $\lambda = \lambda_0$.

window/whole	$\tilde{n} = 2$	$\tilde{n} = 1$	$\tilde{n} = 0$
$\tilde{m} = -8$	0.993	0.993	0.961
$\tilde{m} = -5$	0.940	0.940	0.908

Table I. — Ratio of the bound state eigenvalue of the small window hamiltonian with indices limited by $\tilde{m}$ and $\tilde{n}$, to the eigenvalue of the whole effective hamiltonian at $\lambda = 1$ GeV calculated using expansion up to second power in the running coupling $\tilde{g}(1\text{GeV})$. 0.993 corresponds to the absolute accuracy of the bound state eigenvalue equal 12% and 0.908 to 19% (see the text).

stant does not grow unlimitedly. The similarity renormalization group for hamiltonians provides a new option for investigating bound state dynamics in asymptotically free theories. The question is how far down in λ we can reach using perturbation theory instead of the exact solution. The answer is: down to 1 GeV in second order with 10% accuracy. This is illustrated in Fig. 2.

The remaining question of how small the space of states can be on which one can project the narrow effective hamiltonian and reproduce the bound state eigenvalue by diagonalization of the projected matrix, is answered in *Table I*. The eigenvalue of the whole $\mathcal{H}(\lambda = 1\text{GeV})$ is equal -0.8902 GeV instead of -1 GeV. Window matrices with energy range order 1 GeV reproduce the same result with accuracy given in Table 1. This is encouraging to pursue a similar approach to QCD.

3. FOCK SPACE METHOD

The model study shows that the perturbative similarity renormalization group allows a calculation of a small width effective hamiltonian, which can be projected on a small space of states. The small hamiltonian can be solved exactly and the bound state eigenvalue of the full theory is obtained with 10% accuracy. The question is how to repeat these steps in quantum field theory.

The method we propose [7] is based on the idea that one can unitarily transform the creation and annihilation operators, i.e.

$$a_\lambda^\dagger = U_\lambda a_\infty^\dagger U_\lambda^\dagger, \qquad (3.1)$$

and the same for a's. $a_\infty^\dagger$ and a_∞ appear in the initial hamiltonian H. We call them "bare". $a_\lambda^\dagger$ and a_λ appear in the effective hamiltonian $\mathcal{H}_\lambda$. They create and annihilate effective particles. In a way, U_λ is analogous to the Melosh transformation in the case of quarks. However, we are building the transformation using the similarity renormalization group idea, the transformation is fully dynamical and it can be applied to other particles than quarks, too.

The effective hamiltonians satisfy the equation

$$\frac{d}{d\lambda}\mathcal{H}_\lambda = [\mathcal{H}_\lambda, \mathcal{T}_\lambda], \qquad (3.2)$$

where $T_\lambda = U_\lambda^\dagger dU_\lambda/d\lambda$. The unitary transformation generator T is constructed so that the effective hamiltonians have width λ in the relative momentum transfer,

$$\mathcal{H}_\lambda = F_\lambda[\mathcal{G}_\lambda]. \tag{3.3}$$

The operation F_λ on the interaction terms $\mathcal{G}_\lambda$, inserts the similarity factors, f_λ. They are most easy to think about as form factors in the interaction vertices. The smaller is λ the softer are the interactions and the effective particles get more dressed.

Following the general idea of the similarity scheme [2], one can find the equation satisfied by the vertex operators in the effective hamiltonians [7], i.e.

$$\frac{d}{d\lambda}\mathcal{G}_\lambda = \left[f_\lambda \mathcal{G}_{2\lambda}, \left\{ \frac{d}{d\lambda}(1 - f_\lambda)\mathcal{G}_{2\lambda} \right\}_{\mathcal{G}_{1\lambda}} \right]. \tag{3.4}$$

$\mathcal{G}_\lambda = \mathcal{G}_{1\lambda} + \mathcal{G}_{2\lambda}$, $\mathcal{G}_{1\lambda}$ is the $a^\dagger a$ part of the hamiltonian and $\mathcal{G}_{2\lambda}$ is the remaining part which changes momenta of the individual particles. The curly bracket with subscript $\mathcal{G}_{1\lambda}$ denotes the similarity energy denominator factor.

An elementary example illustrates how it works in Yukawa theory which is defined by the following initial hamiltonian

$$H_Y = \int dx^- d^2 x^\perp \left[\bar{\psi}_m \gamma^+ \frac{-\partial^{\perp 2} + m^2}{2i\partial^+}\psi_m + \frac{1}{2}\phi(-\partial^{\perp 2} + \mu^2)\phi \right.$$

$$\left. + g\bar{\psi}_m\psi_m\phi + g^2\bar{\psi}_m\phi\frac{\gamma^+}{2i\partial^+}\phi\psi_m \right]_{x^+=0}. \tag{3.5}$$

The one particle energy is obtained in the form,

$$\mathcal{G}_{1\,meson\,\lambda} = \int [k]\frac{k^{\perp 2} + \mu_\lambda^2}{k^+}a_k^\dagger a_k. \tag{3.6}$$

In second order perturbation theory in the coupling constant g, Eq. (3.4) implies

$$\frac{d\mu_\lambda^2}{d\lambda} = g^2 \int [x\kappa]\frac{df^2(z_\lambda^2)}{d\lambda}\frac{8(x - \frac{1}{2})^2\mathcal{M}^2}{\mathcal{M}^2 - \mu^2}r_\epsilon(x, \kappa). \tag{3.7}$$

Here $\mathcal{M}^2 = (\kappa^2 + m^2)/x(1 - x)$ and $r_\epsilon(x, \kappa)$ denotes the regularization factor which is an analog of the number N in the matrix model. The similarity function $f^2(z_\lambda^2)$ can be made as simple as, for example, $\theta(\lambda^2 + 3\mu^2 - \mathcal{M}^2)$. In this case, integration of Eq. (3.7) gives the following effective meson mass term

$$\mu_\lambda^2 = \mu_1^2 + \frac{\alpha}{48\pi}\left[\lambda^2 - \lambda_1^2 + (\mu^2 - 6m^2)\log\frac{\lambda^2}{\lambda_1^2} \right] + \mu_{conv}^2(\lambda, \lambda_1) + o(g^4). \tag{3.8}$$

$\mu_{conv}^2(\lambda, \lambda_1)$ denotes a finite term which has a limit when $\lambda \to \infty$. It equals 0 for $\lambda = \lambda_1$. μ_1 is the effective meson mass in the hamiltonian $\mathcal{H}(\lambda_1)$. In the second order calculation, it is equal to the physical meson mass if $\lambda_1^2 \leq 4m^2 - 3\mu^2$.

The reason I show this example is that one can do similar calculations for other terms in the effective hamiltonians. [7] For example, in second order perturbation theory, effective interactions between quarks are partly similar to the results obtained by Perry and his collaborators. [9] [10] [11]

The questions how many orders of pertrubation theory are required in the calculation of the effective hamiltonian for constituent quarks and gluons in QCD and how large must be the subspace of the light-front Fock space to diagonalize the effective hamiltonian of QCD, require much more work to answer than in the matrix model.

Acknowledgments

I am grateful to Pierre Grange for organizing the Les Houches workshop on new light-front computational methods and to Robert Perry for organizing the session on effective hamiltonians and renormalization issues, and inviting me to speak. I am most indebted to Ken Wilson for discussions during my stay at The Ohio State University as a Fulbright Scholar in the academic year 1995/1996. It is my pleasure to thank Robert Perry for helpful comments and I would like to express my gratitude and thank him and Billy Jones, Martina Brisudová and Brent Allen for discussions and hospitality extended to me at OSU. I have also discussed the subjects of my talk with Tomek Masłowski and Marek Więckowski. Research described in this paper has been supported in part by Maria Skłodowska-Curie Foundation under Grant No. MEN/NSF-94-190.

References

[1] Głazek St.D., Wilson K.G., *Phys. Rev. D* **48** (1993) 5863.
[2] Głazek St.D., Wilson K.G., *Phys. Rev. D* **49** (1994) 4214.
[3] Wilson K.G. et al, *Phys. Rev. D* **49** (1994) 6720.
[4] Wegner F. , *Ann. Physik* **3** (1994) 77.
[5] Wilson K.G., Głazek St.D., in "Computational Physics: Proceedings of the Ninth Physics Physics Summer School at the Australian National University"; Gardner and C.M. Savage Eds. World Scientific, Singapore, 1997.
[6] Głazek St.D. and Wilson K.G. , "Asymptotic Freedom and Bound States in Hamiltonian Dynamics", in preparation.
[7] Głazek St.D., "Renormalization of Hamiltonians in the Light-Front Fock Space", Warsaw University Report No. IFT/2/1997.
[8] E.g. see Jackiw R., in "M. A. B. Bég Memorial Volume"; Ali and P. Hoodbhoy Eds (World Scientific, Singapore, 1991) p. 25.
[9] Perry R.J. , "A Simple Confinement Mechanism for Light-Front Quantum Chromodynamics"; in "Theory of Hadrons and Light-Front QCD", Ed. Głazek Ed. , World Scientific, Singapore, 1995, p. 56, and references therein. In particular, see the works by Perry and collaborators.
[10] Brisudová M. and Perry R.J. , *Phys. Rev. D* **54** (1996) 1831.
[11] Brisudová M., Perry R.J. and Wilson K.G., *Phys. Rev. Lett.* **78** (1997) 1227.

Spin Glasses and the Renormalization Group

Giorgio Parisi[1]

([1]) *Dipartimento di Fisica, Università La Sapienza*
and INFN Sezione di Roma
Piazzale Aldo Moro, Roma 00187, Italy

---

1. SPIN GLASSES

In this talk I will present some models of spin glasses an I will show that they are a gauge theory of a rather peculiar type. I will study here a very simple model of spin glasses[1, 2]. I consider a material where there are three kinds of atoms: M, A and B. M is a magnetic atom (it has a non zero magnetic moment) while A and B are magnetically inert.

I suppose that at low temperature the system crystallises in such a way that the M-atoms stay on a regular lattice and the A and B atoms stay on the links of the same lattice. The position of the magnetically inert atoms is supposed to be random. In other words we consider an M A_x B_{100-x} alloy; the case $x = 50$ corresponds to an equal proportion of A and B.

Let us assume that the magnetic interaction among the M-atoms is of the nearest neighbour type and that it is mediated by the non magnetic atoms. The interaction among two M-atoms is ferromagnetic if the link is occupied by an A atom, while it is antiferromagnetic if the link is occupied by a B atom. We also assume for simplicity that the strengths of the ferromagnetic and of the antiferromagnetic interactions are equal.

Usually the magnetic interaction is relevant only at temperatures much lower than the melting temperature and it may be neglected during the formation of the alloy. If the temperature is decreased fast enough the position of the atoms in not influenced by the magnetic interaction. We can describe this situation by saying that we are in presence of a *quenched* disorder.

If we assume that the spin are Ising variables, the corresponding Hamiltonian, in presence of a magnetic field h, is

$$H_U[\sigma] \equiv -\sum_{i,k} \sigma_i U_{i,k} \sigma_k - h \sum_i \sigma_i . \tag{1}$$

The variables σ are defined on the sites of the lattice and they take the values ± 1. The variables $U_{i,k}$ are defined on the links of the lattice, i.e. when i and k are nearest neighbours; they also take the values ± 1.

The variables U are random independent variables. For each choice of the U we can define a statistical expectation value:

$$\langle g(\sigma) \rangle_U = \frac{\sum_\sigma \exp(-\beta H_U[\sigma]) g(\sigma)}{\sum_\sigma \exp(-\beta H_U[\sigma])}. \tag{2}$$

We are interested in computing the statistical expectation values averaged over the probability distribution of the samples, in other words the quantity

$$\langle g(\sigma) \rangle \equiv \overline{\langle g(\sigma) \rangle_U} \equiv \int dP(U) \langle g(\sigma) \rangle_U , \tag{3}$$

where we denote by an horizontal bar the average over the U and $P(U)$ is the probability distribution of the variables U.

In the infinite volume limit the expectation value of intensive quantities does not depend on the realisation of the couplings U (i.e. all the samples have essentially the same properties) and the sample to sample fluctuations vanish in this limit.

A very interesting quantity to evaluate is the magnetic susceptibility. A naive computation would give the following formula

$$\chi = \beta(1 - \overline{\langle \sigma_i \rangle_U^2}) = \beta(1 - \overline{m(i)_U^2}) = \beta(1 - q_{EA}) , \tag{4}$$

where $m(i)_U \equiv \langle \sigma_i \rangle_U$ is the site dependent spontaneous magnetisation and $q_{EA} \equiv \overline{m(i)_U^2}$ is the so called Edward Anderson order parameter[1]. We shall see later how this formula for the susceptibility is modified by a more sophisticated treatment. Physical intuition tell us that at high temperature at zero magnetic field there is no spontaneous magnetization and consequently $q_{EA} = 0$. At low temperature each sample should develop its own spontaneous magnetization, and consequently $q_{EA} \neq 0$ at low temperature. The non vanishing of q_{EA} should there fore mark the spin glass transition. Before studying this model further it is convenient to analyze its symmetries and in particular the consequences of gauge invariance.

2. GAUGE INVARIANCE

It easy to see that (at zero magnetic field) the Hamiltonian introduced in the previous section is invariant with respect to the local gauge transformation[3]:

$$\sigma_i \to -\sigma_i ; \qquad U_{i,k} \to U_{i,k} . \tag{5}$$

The set of all possible realizations of the system at zero magnetic field is gauge invariant under the gauge group Z_2. The couplings and the spins play respectively the role of the gauge connection and of the matter field. At non-zero magnetic field the gauge invariance is explicitly broken. The Hamiltonian at zero magnetic field is (apart from a constant) the square of the covariant lattice-gradient of the σ variables. It can be written as

$$\sum_i \sum_\mu (\sigma_i - U_{i,i+\mu}\sigma_{i+\mu})^2 \ . \tag{6}$$

The relevant quantities are gauge invariant. If two realizations of the couplings U and U' differ by a gauge transformation, their thermodynamic properties are the same. It is important to concentrate our attention on gauge invariant quantities.

Let us study in more details how the thermodynamical quantities depend on the choice of the couplings. A quantity which is often used to characterise the gauge fields is the Wilson loop. We can associate to each closed circuit on the lattice the ordered product of all the links of the circuit. We thus define:

$$W(C) \equiv \prod_{(i,k)\in C} U_{i,k} \ . \tag{7}$$

This quantity may take the values ± 1. If $W(C) = -1$, the loop is said to be *frustrated* [3]: along that loop it is not possible to find a configuration of spins such that

$$U_{i,k}\sigma_i\sigma_k = 1 \quad \forall \ (i,k) \in C. \tag{8}$$

If there are frustrated loops (i.e. if there is no gauge transformation which brings all couplings U to 1) it is not possible to find a configuration of the spins such that all terms in the Hamiltonian are positive. Some defects (i.e. links for which the contribution to the energy is negative) must be present.

At low temperature the equilibrium probability distribution is concentrated on those spin configurations which have the minimal energy. It is interesting to study also those configurations which are local minima of the Hamiltonian in the sense that the Hamiltonian increases when we flip a spin. These local minima are very important in the dynamics outside equilibrium because at low temperatures the system may be trapped for a very long time in these minima.

If we study the structure of local and global minima with great care, we discover that frustration implies the presence of defects which can be put in many ways on the lattice. The ground state is degenerate. The number of local and not global minima is also large.

This phenomenon is well known in gauge field. On the lattice the choice of the Landau gauge corresponds to find the maximum of

$$\sum_{i,k} \mathrm{Tr} g_i^* U_{i,k} g_k, \tag{9}$$

where g is the gauge transform which brings the gauge fields in the Landau gauge[4]. This problem is equivalent (in the our case) to find the minimum of the Hamiltonian eq. (1). Gribov ambiguity tell us that in the general case there are many possibility of choosing the gauge. This result implies the existence of many minima of the Hamiltonian eq. (1)(¹)

3. THE REPLICA METHOD

We face now the problem of evaluating the quantities which appear in equation (2). The first proposal would be to sum over the variables U and to remain with an effective interaction for the variables σ. This approach clashes with fact that both the numerator and the denominator of eq. (2) depend on the variables U and the sum over the U is not easy.

This difficulty may be avoided by introducing n identical copies (or replicas) of the same system [1]. We define

$$\langle g(\sigma)\rangle_n = \frac{\sum_\sigma \sum_U \exp(-\beta H_n[\sigma, U])g(\sigma^1)}{\sum_\sigma \sum_U \exp(-\beta H_n[\sigma, U])}, \tag{10}$$

where the spins σ_i^a carry an other index (a) which ranges from one to n. The new Hamiltonian is the sum of n identical Hamiltonians

$$H_n(\sigma, U) = \sum_{a=1,n} H_U[\sigma^a]. \tag{11}$$

It easy to check that

$$\langle g(\sigma)\rangle_n = \frac{\sum_U \langle g(\sigma)\rangle_U (Z_U)^n}{\sum_U (Z_U)^n}, \tag{12}$$

where the U-dependent partition function is defined as

$$Z_U = \sum_\sigma \exp(-\beta H_U[\sigma]). \tag{13}$$

We finally find that

$$\langle g(\sigma)\rangle = \langle g(\sigma)\rangle_n|_{n=0} . \tag{14}$$

In this way one finds that properties of the matter fields averaged over the disordered (quenched) gauge fields can be computed by considering the gauge fields interacting with n copies of the matter field and computing the expectation values in the limit $n \to 0$. The argument is familiar to those who work in

---

(¹) In the continuum Gribov ambiguity is normally present in non Abelian gauge theories. It is also present in Abelian theories when we allow configurations which are singular in the continuum limit, e.g. like magnetic monopoles.

numerical simulation of latttice gauge theories, where n plays the role of the number of quark flavours in the sea[2].

The average of the U fields can be done and one remains with an effective interaction for the σ variables. This effective interaction must be written in terms of the gauge invariant combinations. In this case the most appropriate variables are

$$Q_i^{ab} = \sigma_i^a \sigma_i^b. \tag{15}$$

The diagonal terms of the matrix Q are identically equal to 1, so that we can consider only the off-diagonal term. All computations must be done for generic values of n and we must send n to zero at the end.

As usually we can expand the effective interaction in powers of Q, being careful to preserve the various symmetries of the problem, i.e.:

- The group of permutations of the n replicas $S(n)$.

- The spin reversal symmetry for each replica, i.e. n times the direct product of the Z_2 global group, where each Z_2 group acts on a different replica[3].

In the continuum limit in D dimensions the simplest form for the effective free energy, in which all important terms are contained, is the following :

$$F[Q] = \int d^D x \left(\frac{1}{2} \sum_\mu \mathrm{Tr}(\partial_\mu Q(x))^2 + W(Q(x)) \right), \tag{16}$$

where the function $W(Q)$ is given by

$$W(Q) = \tau \mathrm{Tr}(Q^2) + g_3 \mathrm{Tr}(Q^3) + g_4 \mathrm{Tr}(Q^4) + y \sum_{ab} Q_{ab}^4 . \tag{17}$$

The usual strategy consists in computing the minimum of F and constructing the usual perturbative expansion for the small fluctuations around this minimum (the so called loop expansion). As we shall see even the first step is not very simple.

Let us assume that the $F[Q]$ is minimized by a function $Q(x)$ which does not depend from x. We can this set $Q(x) = Q$ and look for the minimum of $W(Q)$.

For $\tau > 0$ we find that there is only a minimum at $Q = 0$. When τ is negative we easily find that there is a stationary points at

$$Q_{ab} = q, \quad \forall \, a \, b. \tag{18}$$

---

[2] It is evident why I proposed the name *quenched* for the approximation of neglecting quark loops in QCD.

[3] This symmetry is present only at zero magnetic field.

This point is invariant under the replica group $S(n)$. In order to decide if this stationary point is a minimum of W, we have to compute the small fluctuations around this point. The Hessian

$$\mathcal{H}_{ab,cd} = \frac{\partial W}{\partial Q_{ab} Q_{cd}} , \qquad (19)$$

must have non negative eigenvalues (in a field theory language there should be no negative squared masses).

If we consider the case where $y = 0$, the $S(n)$ symmetry is promoted to an $O(n)$ symmetry (i.e. there is an accidental symmetry). This $O(n)$ symmetry is spontaneously broken by the choice in eq. (18). Consequently the Hessian has zero eigenvalues (i.e. there are Goldstone Bosons).

If y is different from zero, the $O(n)$ symmetry is explicitly broken and the Goldstone Bosons acquire a mass squared proportional to y. A detail computation shows that in the general case y is negative and the Goldstone Bosons acquire a negative mass squared (i.e. the Hessian has negative eigenvalues).

This instability implies that the proposed stationary point is not a local minimum and one has to look for a minimum which will be no more invariant under the S_n group. Such a minimum can be constructed [5, 6]; we have to introduce an infinite sequence of steps breaking the $S(n)$ symmetry. After a rather long computation one finally finds a matrix Q to which one can associate a function $q(u)$, where u belongs to the interval $0 - 1$.

There is a simple physical interpretation of this symmetry breaking. The presence of frustration implies that for each realization of the couplings there are different equilibrium states with different local magnetisation (i.e. different vacua). We indicate by w^α the probability of finding the system in the state labelled by α and by m_i^α the local magnetization (i.e. the expectation value of σ_i in the state α). The overlap among different states may be defined as

$$q^{\alpha\gamma} = \frac{\sum_{i=1,N} m_i^\alpha m_i^\gamma}{N}. \qquad (20)$$

After some computations one finds that

$$\overline{\sum_{\alpha\,\gamma} w^\alpha w^\gamma f(q^{\alpha\gamma})} \equiv \int dq P(q) f(q) = \int_0^1 du f(q(u)). \qquad (21)$$

The probability of having two states with given overlap q is thus controlled by the function $q(u)$.

This approach is able to explain in a qualitative and sometimes a quantitative way many of the properties of real spin glass. When the replica symmetry is broken, a change in the external magnetic field changes the structure of equilibrium states and one must be quite carefully in the definition of the susceptibility. One finds that there are two susceptibilities:

- The linear response susceptibility which quantifies the response to a small variation of the magnetic field measured on a short time scale such that no global rearrangement of spins is possible. This susceptibility is given by $\chi_{LR} = \beta(1 - q_{EA})$.

- The thermodynamic susceptibility which quantifies the response to a small variation of the magnetic field on a very long time scale such that global rearrangements of spins are possible. This susceptibility is larger[4] and it is given by $\chi_{eq} = \beta(1 - \int du q(x))$.

The difference between these two susceptibilities is one of the most characteristic phenomena experimentally observed in spin glasses and it is well explained by this theory [2].

4. RENOMALIZATION GROUP RESULTS

At this stage of the theory, where fluctuations are neglected, is not clear which of the theoretical predictions does survive when the effect of the fluctuations is included. In order to have more precise theoretical predictions for real three dimensional systems it is necessary to go beyond the mean field approximation. This has been the subject of intensive studies.

If we approach the critical temperature from above, the situation is apparently quite clear. The critical exponents are the same as in mean field theory in dimensions greater than 6. They can be computed in an expansion in $\epsilon = 6 - D$ and the are know up to the order ϵ^3.

The real problem arise in the region below the critical temperature. Usually the structure of the Goldstone modes (if any) and of their interaction determines the lower critical dimensions. Unfortunately the computation of the corrections to the mean field theory are rather involved. The correlation functions at zero loops are rather complicated [7] and the one loop corrections to the correlation functions have not yet fully computed. Also the simplest task, i.e. the precise identification of the Goldstone mode and of their decoupling at zero external momentum has not been completely understood [8].

Numerical simulations in 4 and 3 dimensions [9] strongly suggest the correctness of the broken replica picture. No anomaly is observed and most of the theoretical predictions have been successfully tested.

When we will master the loops corrections, we will be hopefully able to set up a renormalization group study of the properties of the system in the low temperature region and in particular to compute the value of the lowest critical dimension, i.e. the dimension where the structure of replica symmetry breaking scheme is no more consistent and the predictions of the mean field theory do not apply anymore. A direct computation of the lower critical dimension done using a domain wall argument [10] gives 2.5 and it would be very interesting to test this results using a different approach.

---

[4] One finds that $q_{EA} = \max(q(u))$.

References

[1] Edwards S.F. and Anderson P.W., *J. Phys.* **F 5** (1975) 965.
[2] Binder K. and Young A.P., *Rev. Mod. Phys.* **58** (1986) 801.
[3] Toulouse G., *Comm. on Phys.* **2** (1977) 115.
[4] Marinari E., Parrinello C. and Ricci R. *Nucl. Phys.* **B36** (1991) 487.
[5] Mézard M., Parisi G. and Virasoro M. A., Spin Glass Theory and Beyond (World Scientific, Singapore, 1987) p. 1.
[6] Parisi G., Field Theory Disorder and Simulations (World Scientific, Singapore, 1992) p. 151.
[7] De Dominicis C. and Kondor I., *Europhys. Lett.* **2** (1986) 617.
[8] Ferrero M. and Parisi G., *On infrared divergences in spin glasses*, cond-mat/9511049.
[9] Marinari E., Parisi G. and Ruiz-Lorenz J.J. *J. Phys.* **A 26** (1993) 6711.
[10] Franz S., Parisi G. and Virasoro M.A., *J. Phys.* **A 31** (1995) 5991.

Hamiltonian Flow in Condensed Matter Physics

F. Wegner[1]

([1]) *Institut für Theoretische Physik, Ruprecht-Karls-Universität*
Philosophenweg 19, D-69120 Heidelberg
Germany

A recently developed method to diagonalize or block-diagonalize Hamiltonians by means of an appropriate continuous unitary transformation is reviewed. Two applications in condensed matter physics are given as examples: (i) the interaction of an n-orbital model of fermions in the limit of large n is brought to block-diagonal form, and (ii) the generation of the effective attractive two-electron interaction due to the elimination of electron-phonon interaction is given. The advantage of this method in particular in comparison to conventional perturbation theory is pointed out.

1. INTRODUCTION

In this contribution I describe a method for diagonalizing Hamiltonians, which has recently been developed [1] in Heidelberg and applied to models in solid-state physics. The applications I will give will be in condensed matter physics. The reason why I present them in this workshop on light-front physics is that a similar scheme has been developed in Ohio in the group around Ken Wilson [2, 3] under the name of similarity renormalization. It will be explained in other lectures in this volume.

In most cases the diagonalization of a many-body Hamiltonian is a difficult task. In order to approach this problem we introduce a continuous transformation (as function of a flow-parameter l), which is constructed so that the off-diagonal matrix elements decay continuously. The aim is finally to obtain the diagonalized Hamiltonian or a block-diagonal Hamiltonian. In the next section the scheme will be explained.

This scheme has two important inherent features:

(i) The matrix elements between degenerate states or nearly degenerate states will decay only slowly, and the convergency may be subtle. But this has to be expected as we know from small energy-denominators in perturbation theory. Of course this is related to infrared physics and singularities and worthwhile to be investigated. As we will see later the present method yields smoother results than conventional perturbation theory.

(ii) This scheme will in general generate complicated interactions. Starting from a problem with two-particle interactions the transformation will generate three-particle, four-particle, etc. interactions. In the original paper [1] an n-orbital model was considered in the limit $n \to \infty$. Although in this limit these many-particle interactions will be generated, it turns out that the one-particle energies are independent of l and the equations for the two-particle interaction are closed in themselves, so that an explicit calculation can be performed to a large extent. Kehrein and Mielke observed that for systems with impurities (Anderson impurity model [4, 5] and spin-boson model [6, 7]) it is sufficient to keep only rather simple contributions to the Hamiltonian in order to obtain good results.

Here I will briefly explain, how the method works for the n-orbital model (sect. 2) and for the elimination of the electron-phonon interaction [8]. This problem has already been attacked in the fifties by Fröhlich [9], a procedure that can be found in all textbooks of theoretical solid-state physics. His effective interaction has an energy denominator, which can vanish and gives rise to both attraction and repulsion of the electron pairs. With the present procedure we obtain in a simple way an attractive interaction between all pairs which yields very good agreement [10, 11] with more sophisticated methods [12, 13]. The permanent adjustment of the infinitesimal unitary transformation to the Hamiltonian yields a smoother effective interaction than conventional perturbation theory.

2. FLOW EQUATIONS

The basic idea of the flow equations [1] is, that the initial Hamiltonian $H = H(0)$ undergoes a continuous unitary transformation

$$H(l) = U(l)HU^\dagger(l) \tag{1}$$

(U being unitary). Differentiation yields

$$\frac{dH(l)}{dl} = [\eta(l), H(l)] \tag{2}$$

with the generator

$$\eta(l) = \frac{dU(l)}{dl}U^\dagger(l) = -\eta^\dagger(l). \tag{3}$$

This generator should be chosen the way, that the off-diagonal matrix elements decay. One can easily see, that for a finite matrix

$$\eta = [H_d, H] \tag{4}$$

is a good choice, where H_d is the diagonal part of the Hamiltonian. If the diagonal matrix elements are denoted by ϵ, then

$$\eta_{k,n} = (\epsilon_k - \epsilon_n) H_{k,n}. \tag{5}$$

A simple calculation yields

$$\frac{dH_{k,n}}{dl} = \sum_m (\epsilon_k + \epsilon_n - 2\epsilon_m) H_{k,m} H_{m,n} \tag{6}$$

and

$$\frac{d}{dl} \sum_{k,n,k \neq n} H_{k,n} H_{n,k} = -\frac{d}{dl} \sum_k \epsilon_k^2 = -2 \sum_{k,n} (\epsilon_k - \epsilon_n)^2 H_{k,n} H_{n,k}. \tag{7}$$

Thus the sum of the squares of the off-diagonal matrix elements is indeed negative or zero. The procedure comes to an end, when all off-diagonal matrix elements vanish. It may be, however, that off-diagonal matrix elements survive, if the corresponding one particle energies are degenerate. We will see, however, that since the diagonal matrix elements themselves vary as a function of l, it may happen that even the off-diagonal matrix elements between asymptotically degenerate states vanish. In this formulation the flow parameter l has the dimension $1/\text{energy}^2$. Indeed in many cases it turns out, that the uncertainty of the basis states decays under the flow like $1/\sqrt{l}$.

3. n-ORBITAL MODEL

3.1. The model and its flow equations

As a first example we would like to apply this method to an interacting electronic system [1]. We consider an n-orbital model, that is, the electrons carry a momentum and in addition a quantum number (flavor) s, which runs from 1 to n. Finally we are interested in the limit n going to ∞.
The Hamiltonian will be expressed in terms of the operators

$$N_{p,q} = \frac{1}{n} \sum_s c_{p,s}^\dagger c_{q,s} \tag{8}$$

with creation and annihilation operators $c^\dagger$ and c and momenta p and q. All operators will be expressed as normal-ordered polynomials of the $N_{p,q}$. Thus $N_{p,p} = n_p + : N_{p,p} :$, where n_p is the occupation number for the groundstate of the free system $n_p = \theta(k_F - |p|)$. One can convince oneself that the commutator between two operators yields in leading order in $1/n$

$$[: A :, : B :] = [A, B]_1 + [A, B]_2 + O(n^{-2}),\tag{9}$$

$$n[A, B]_1 = \sum_{p,q,r} : \frac{\partial A}{\partial N_{p,q}} \frac{\partial B}{\partial N_{q,r}} N_{p,r} : - \sum_{p,q,r} : \frac{\partial A}{\partial N_{p,q}} \frac{\partial B}{\partial N_{r,p}} N_{r,q} :,\tag{10}$$

$$n[A, B]_2 = \sum_{p,q} : \frac{\partial A}{\partial N_{p,q}} \frac{\partial B}{\partial N_{q,p}} : (n_p - n_q).\tag{11}$$

Let us consider a Hamiltonian

$$H = H_1 + H_2 + ...,\tag{12}$$

$$H_1 = n \sum_q \epsilon_q : N_{q,q} :,\tag{13}$$

$$H_2 = \frac{n}{2\Omega} \sum_{\delta,Q,K} v_{\delta,Q,K} : N_{Q+\delta/2,Q-\delta/2} N_{K-\delta/2,K+\delta/2} :,\tag{14}$$

where δ is the momentum transfer and Ω the volume of the system and H_k contains the k-particle interaction. Similarly one has

$$\eta = \eta_2 + ...,\tag{15}$$

$$\eta_2 = \frac{n}{2\Omega} \sum_{\delta,Q,K} \eta_{\delta,Q,K} : N_{Q+\delta/2,Q-\delta/2} N_{K-\delta/2,K+\delta/2} : .\tag{16}$$

In leading order in $1/n$, that is in the limit n going to ∞ one obtains

$$\eta_2 = [H_1, H_2 - H_{2d}]_1 + [H_{2d}, H_2]_2,\tag{17}$$

$$\frac{\partial H_1}{\partial l} = 0,\tag{18}$$

$$\frac{\partial H_2}{\partial l} = [\eta_2, H_1]_1 + [\eta_2, H_2]_2.\tag{19}$$

Thus H_1 is constant in leading order in $1/n$. Although more-particle contributions H_k with $k > 2$ will be generated (which I did not write down), they do not couple back into the equation for H_2. Thus the equations for H_2 are closed in themselves.

3.2. Solution

In a first attempt I have used the flow equations literally, that is as the diagonal part H_d I chose the contributions diagonal in momentum representation. It turns out that in the thermodynamic limit, that is for Ω approaching ∞ the diagonal part reduces to H_1 and the diagonal part of H_2 becomes negligible. The solutions of the corresponding equations do not yield a converging behavior. We assume that it is not allowed to neglect many particle interactions in H_d.

Therefore I made a different choice for the diagonal part. I considered all terms to be diagonal, which conserve the number of quasiparticles, that is the number of electrons above and the number of holes below the Fermi-edge. Then H_0, which is a constant in addition to the Hamiltonian already written down, becomes the energy of (hopefully) the groundstate, H_1 contains the energies of the one-quasiparticle excitations. H_2 contains the interaction between two quasiparticles, etc. Thus to obtain the two-quasiparticle states, it is sufficient to diagonalize a two-particle problem, which is of course much easier than to solve the N-particle problem.

With this in mind and

$$s_{\delta,k} = n_{k-\delta/2} - n_{k+\delta/2} \tag{20}$$

the terms $v_{\delta,Q,K} : N_{Q+\delta/2,Q-\delta/2} N_{K-\delta/2,K+\delta/2}$ belong to H_d, if $s_{\delta,Q} = s_{\delta,K}$.

With

$$\epsilon_k = k^2/2 \tag{21}$$

one obtains

$$
\begin{aligned}
\eta_{\delta,Q,K} &= (Q-K)\delta v_{\delta,Q,K}(1 - \delta_{s_{\delta,Q},s_{\delta,K}}) \\
&+ \frac{1}{\Omega}\sum_P s_{\delta,P}(\delta_{s_{\delta,P},s_{\delta,Q}} - \delta_{s_{\delta,P},s_{\delta,K}})v_{\delta,Q,P}v_{\delta,P,K},
\end{aligned}
\tag{22}
$$

$$\frac{\partial v_{\delta,Q,K}}{\partial l} = (K-Q)\delta \eta_{\delta,Q,K} + \frac{1}{\Omega}\sum_P s_{\delta,P}(\eta_{\delta,Q,P}v_{\delta,P,K} - v_{\delta,Q,P}\eta_{\delta,P,K}). \tag{23}$$

One observes that only matrix elements with the same momentum transfer δ are connected to each other. Further inspection shows that even the equations for matrix elements $v_{\delta,Q,K}$ for which $s_{\delta,Q}$ and $s_{\delta,K}$ are different from zero close into themselves.

For small momentum transfer there are only small regions of K and Q around k_F, in which s is different from zero. We assume, that in this limit v does not depend essentially on $|K|$ and $|Q|$, but only on the direction of these vectors.

Let us now restrict to the one-dimensional case. If the momenta lie in the interval of size δ, in which s equals $+1$ and -1, then I write instead of the momenta $+$ and $-$, resp. The equations for $v_{\pm,\pm}$ are closed,

$$\frac{dv_{+,+}}{dl} = \frac{dv_{-,-}}{dl} = -\frac{\delta^2}{\pi^2}Av_{+,-}v_{-,+}, \tag{24}$$

$$\frac{dv_{+,-}}{dl} = -\frac{\delta^2}{\pi^2} A^2 v_{+,-},$$
(25)

$$A = 2\pi k_F + \frac{1}{2}(v_{+,+} + v_{-,-}).$$
(26)

One immediately sees that $v_{+,+} - v_{-,-}$ and $B = A^2 - v_{+,-}v_{-,+}$ are constants and the solutions can be given in the form

$$A^2(l) = \frac{BA^2(0)}{A^2(0) - v_{+,-}(0)v_{-,+}(0)\exp(-\gamma l)}$$
(27)

$$v_{+,-}(l)v_{-,+}(l) = \frac{Bv_{+,-}(0)v_{-,+}(0)}{A^2(0) - v_{+,-}(0)v_{-,+}(0)\exp(-\gamma l)}$$
(28)

$$\gamma = \frac{2\delta^2 B}{\pi^2}.$$
(29)

Apparently the $v_{\pm,\pm}$ are smooth functions of l. For positive B the off-diagonal matrix elements $v_{+,-}$ and $v_{-,+}$ vanish as l goes to ∞. If B is negative, then the diagonal matrix elements $v_{+,+}$ and $v_{-,-}$ vanish, whereas the off-diagonal matrix elements approach a finite limit. In this latter case, however, the system is unstable.

Next one considers the off-diagonal matrix element $v_{\pm,K}$ and obtains

$$\frac{\partial v_{+,K}}{\partial l} = -\frac{\delta^2}{4\pi^2}(A + c_K)^2 v_{+,K} - \frac{\delta^2}{4\pi^2}(3A - c_K)v_{+,-}v_{-,K}$$
(30)

$$c_K = \frac{1}{2}(v_{+,+} - v_{-,-}) - 2\pi K.$$
(31)

Together with a similar equation for $v_{-,K}$ this is a system of coupled linear differential equations. Since for large l $v_{+,-}$ tends to zero, the equations become decoupled and it is the negative prefactor in front of the $v_{+,K}$ which guarantees asymptotically an exponential decay. There is one exception: if $A + c_K$ vanishes, which happens for a special value $K_{c,\pm}$, then $v_{K,\pm}$ does not decay.

Finally one writes down the flow equation for $v_{Q,K}$. These couplings tend to a finite value unless both Q and K are at K_c. In a different model we will see below that a change of the diagonal matrix elements as a function of l may be sufficient, so that matrix elements even at such resonances go to zero.

3.3.   Further results

Starting from this transformation one can determine expectation values and correlation functions. In order to do this, the operators have to be subject to the same unitary transformations as the Hamiltonian, that is the same flow equation is applied to operators. After transformation of the operators one can evaluate them in the $l = \infty$-basis. This has been done for the average occupation number in the n-orbital model [1], which yields a correction in order $1/n$. The result is compatible with that for the Luttinger model [14], which can easily be generalized to n orbitals. In the Luttinger model one finds a power law behavior of the occupation number both above and below Fermi-edge. The occupation number itself is a continuous function at the edge. Similar calculations in dimensions $d > 1$ [15] indicate, that there is a jump of the occupation number at least for small interactions in agreement with the idea of a Landau liquid.

4.   ELIMINATION OF THE ELECTRON-PHONON COUPLING

4.1.   The effective electron-electron interaction

This section is to a large extent based on the diploma thesis of Peter Lenz and on ref. [8]. Our aim is to calculate the effective electron-electron interaction responsible for the superconductivity. The Hamiltonian of an electronic system coupled to phonons consists of three contributions

$$H = H_0 + H_{e-ph} + H_{e-e}. \tag{32}$$

Here H_0 is the free part

$$H_0 = \sum_q \omega_q a_q^\dagger a_q + \sum_k \epsilon_k : c_k^\dagger c_k :, \tag{33}$$

where ω_q and ϵ_k are the phonon and electron energies, resp. The electron-phonon interaction is given by

$$H_{e-ph} = \sum_{k,q} M_{k,q} a_{-q}^\dagger c_{k+q}^\dagger c_k + h.c. \tag{34}$$

and the electron-electron interaction by

$$H_{e-e} = \sum_{k,k',q} V_{k,k',q} : c_{k+q}^\dagger c_{k'-q}^\dagger c_{k'} c_k : \tag{35}$$

Here k and k' include the z-component of the spin s and s', resp. In all contributions of the Hamiltonian we have used normal ordering. All terms which after normal ordering are not of these types will be neglected. Here we proceed similarly as before, since we consider those terms to be diagonal which conserve the number of particles. This applies to H_0 and H_{e-e}, whereas H_{e-ph} does not conserve the number of phonons and thus is considered to be off-diagonal. Furthermore we will assume the electron-electron interaction to be small and only the second order contribution from H_{e-ph} to H_{e-e} will be calculated. Thus we choose

$$\eta = [H_0, H_{e-ph}] = \sum_{k,q} M_{k,q} \alpha_{k,q} a_{-q}^\dagger c_{k+q}^\dagger c_k - h.c. \tag{36}$$

with the energy difference

$$\alpha_{k,q} = \epsilon_{k+q} - \epsilon_k + \omega_q. \tag{37}$$

This generator yields several contributions to $dH/dl = [\eta, H]$. The contribution to the change of M results from $[\eta, H_0]$

$$\frac{\partial M_{k,q}(l)}{\partial l} = -\alpha_{k,q}^2 M_{k,q}(l) \tag{38}$$

with the solution

$$M_{k,q}(l) = M_q \exp(-\alpha_{k,q}^2 l), \tag{39}$$

where M_q is the initial electron-phonon coupling. The contribution to the electron-phonon coupling is obtained from $[\eta, H_{e-ph}]$. The terms which describe the interaction between the electron pairs (with zero momentum) obey

$$\frac{\partial V_{k,-k,q}(l)}{\partial l} = -(\alpha_{k,q} + \alpha_{-k-q,-q}) M_{k,q}(l) M_{-k-q,q}(l) \tag{40}$$

with the solution

$$V_{k,-k,q}(\infty) = V_{k,-k,q}(0) - M_q^2 \frac{\omega_q}{\omega_q^2 + (\epsilon_{k+q} - \epsilon_k)^2}. \tag{41}$$

Several remarks are in order, since Fröhlich's result differs from this one by a minus-sign between the two squares in the denominator:

(i) The interaction from the electron-phonon coupling is attractive for all values of k and q.

(ii) Mielke has also obtained an effective attractive interaction [10] without pole by means of Glazek and Wilson's similarity transformation [2]. The critical temperature determined from this interaction yields values very similar [10] to those determined with the method by MacMillan and Dynes [13] based on the Eliashberg theory [12] and close to the experimental values [11].

(iii) Similar sums of two squares in the denominator appear in the matrix-elements $v_{K,Q}$ of the n-orbital model [1] and in a revised treatment of the Schrieffer-Wolf transformation with the present scheme [5].

(iv) For on-shell matrix elements V, that is for those which obey $\epsilon_{k+q} + \epsilon_{k'-q} = \epsilon_k + \epsilon_{k'}$, Fröhlich's result and ours coincide.

(v) We observe that the permanent adjustment of $\eta(l)$ to the current $H(l)$ yields smoother interactions than conventional perturbation theory.

(vi) Perturbation theory for Hamiltonians is not uniquely defined.

4.2. Comparison with Fröhlich's treatment

Let us now compare Fröhlich's and our treatment. Fröhlich introduces a transformation

$$
\begin{aligned}
H^{Fr} &= e^{-S} H e^S = H + [H, S] + \frac{1}{2}[[H, S], S] + \dots \\
&= H_0 + H_{e-ph} + [H_0, S] + [H_{e-ph}, S] + \frac{1}{2}[[H_0, S], S] + \dots \tag{42}
\end{aligned}
$$

He assumes S to be of the order of the electron-phonon coupling and requires the contribution in first order, that is $H_{e-ph} + [H_0, S]$, to vanish. This yields

$$S^{Fr} = -\sum_{k,q} M_q \left(\frac{a_{-q}^\dagger}{\alpha_{k,q}} - \frac{a_q}{\alpha_{k+q,q}} \right) c_{k+q}^\dagger c_k. \tag{43}$$

Our treatment yields

$$\exp(-S^{LW}) = T_l \exp\left(\int dl \eta(l) \right) \tag{44}$$

where T_l is an ordering of l, since η's with different argument do not commute. An expansion in powers of η yields

$$S^{LW} = -\int_0^\infty dl \eta(l) - \frac{1}{2} \int_0^\infty dl \int_0^l dl' [\eta(l), \eta(l')] + \dots \tag{45}$$

The first term agrees with Fröhlich's S^{Fr}. The second term yields the difference due to the permanent adjustment of η to H.

4.3. Asymptotics of $\omega_q(l)$

By now we have not considered a variation of the one-particle energies with l. For $\omega(l)$ one has the flow equation

$$\frac{\partial \omega_q(l)}{\partial l} = 2 \sum_k M_{k,q}^2(l) \alpha_{k,q}(l)(n_{k+q} - n_k), \tag{46}$$

$$M_{k,q}(l) = M_q \exp\left(-\int_0^l dl' \alpha_{k,q}^2(l') \right). \tag{47}$$

This yields a nonlinear integro-differential equation for $\omega_q(l)$. Comparable equations were obtained before by Kehrein, Mielke and Neu [6] for the tunneling frequency, $\Delta(l) = \Delta(\infty) + 1/(2\sqrt{l})$. Lenz found that the general asymptotic behavior obeys $\omega(l) = \omega(\infty) + c(l)/\sqrt{l}$ for the electron-phonon system [8], where $c(l)$ is a periodic function in $\ln(l)$ and the average of c^2 equals $1/4$. This decay to the asymptotic value is sufficiently slow, so that even the off-diagonal matrix elements M, for which $\alpha(\infty) = 0$, vanish.

5.   CONCLUDING REMARKS

In two applications, the n-orbital model and the electron-phonon coupling in superconductivity, it has been shown, how the idea of flow equations for Hamiltonians can be applied to condensed matter physics in order to obtain effective block-diagonal interactions. As already mentioned, this scheme has also been applied to two impurity models, the Anderson impurity model and the spin-boson model. A further application is the elimination of the coupling between states with different double-occupancy in the Hubbard model [16].

An interesting subject only briefly mentioned for the operator of occupancy in the n-orbital model is the question, what happens to operators under the flow equations. It has been investigated for the spin-boson model [6, 7]. The interesting observation is that the spin-operator completely transforms into a linear combination of boson creation and annihilation operators. From this we learn that quasi-particles may be quite different from bare particles. In this connection also the question of dissipation can be approached [17, 18].

References

[1] Wegner F., *Ann. Phys. (Leipzig)* **3** (1994) 77.

[2] Glazek S.D. and Wilson K.G., *Phys. Rev.* **D48** (1993) 5863.

[3] Wilson K.G., Walhout T.S., Harindranath A., Zhang W.M., Perry R.J. and Glazek S.D., *Phys. Rev.* **D49** (1994) 6720.

[4] Kehrein S.K. and Mielke A., *J. Phys.* **A27** (1994) 4259,5705.

[5] Kehrein S.K. and Mielke A., *Ann. Phys. (NY)* **252** (1996) 1.

[6] Kehrein S., Mielke A. and Neu P., *Z. Phys.* **B99** (1996) 269.

[7] Kehrein S.K. and Mielke A., *Phys. Lett.* **A219** (1996) 313.

[8] Lenz P. and Wegner F., *Nucl. Phys.* **B482** (1996) 693.

[9] Fröhlich H., *Proc. Roy. Soc. London* **A215** (1952) 291.

[10] Mielke A., *Ann. Physik (Leipzig)* **6** (1997) 215.

[11] Mielke A., Calculating critical temperatures of superconductivity from a renormalized Hamiltonian, preprint, 1997.

[12] Eliashberg G.M., *Zh. Eksp. Teor. Fiz.* **28** (1960) 966; **29** (1960) 1437 [*Sov. Physics JETP* **11**, 696; **12**, 1000].

[13] MacMillan W., *Phys. Rev.* **167** (1968) 331.

[14] Luttinger J.M., *J. Math. Phys.* **4** (1963) 1154.

[15] Kabel A. and Wegner F., to appear in *Z. Physik* **B** (1997).

[16] Stein J., Flow equations and the strong-coupling expansion for the Hubbard Model, to appear in *J. Stat. Phys.* (1996).

[17] Kehrein S.K. and Mielke A., *Ann. Phys. (Leipzig)* **6** (1997) 90.

[18] Kehrein S.K. and Mielke A., Diagonalization of system plus environment Hamiltonians, preprint cond-mat/9701123.

CHAPTER II

Quantization of Constrained Systems

LECTURE 4

Coherent States and
Constrained Systems

John R. Klauder

Departments of Physics and Mathematics
University of Florida
Gainesville, Fl 32611

---

1. INTRODUCTION

The quantization of systems with constraints is important in many applications
that include lightfront quantization schemes along with familiar gauge theories.
Principal techniques for the quantization of such systems involve conventional
operator techniques [1], path integral techniques in terms of the original phase
space variables [2], extended operator techniques involving ghost variables in
addition to the original variables and extended path integral techniques also
including ghost fields [3]. In most popular techniques Gribov ambiguities of
one or another kind appear that render the approach difficult if not outright
impossible [4]. Gribov difficulties were first established for functional integral
techniques, but it has recently been shown that even the popular BRST approach is not immune to Gribov ambiguities [5].

To address this unsatisfactory situation it would seem desirable to develop
an alternative approach to the quantization of systems with constraints which
can work with just the original phase space variables plus needed Lagrange
multipliers without the introduction of auxiliary variables such as ghosts, and
furthermore, can inherently avoid any associated Gribov ambiguities. There
have been some doubts expressed whether such a goal is actually attainable.

After all, it is argued—correctly, in fact—that Gribov ambiguities generally arise from obstructions, e.g., topological, and it is unclear how such general properties are to be overcome [6]. Quite simply, the answer is that the obstructions will not be overcome; instead, they will not arise in the first place!

In the present brief account we sketch the underlying philosophy directing the reader to [7] for additional details and applications. We limit our discussion to bosonic degrees of freedom.

1.1. General Principles

First, consider the phase space of a dynamical system without any constraints whatsoever. We assume that the phase space of the unconstrained system is flat and admits a standard quantization of its canonical variables either in an operator form or in an equivalent path integral form. Second, we introduce a set of constraints, which, for the sake of discussion, we choose as a closed set of first-class constraints; extensions to treat more general constraints are presented in [7]. In the presence of constraints the original set of variables is no longer composed solely of physical variables but now contains unphysical variables as well. While such variables cause little concern from a classical standpoint, they are viewed as unwelcome from a quantum standpoint inasmuch as one wants to quantize only physical variables. Thus it is deemed necessary to eliminate the unphysical variables leaving only the true physical degrees of freedom. Quantization of the true degrees of freedom is supposed to proceed as in the initial step. In the general case, however, such a quantization of the remaining degrees of freedom is not possible because the reduced phase space has changed its structure significantly and one or another obstruction has arisen where none existed before. An obstruction precludes the existence of self-adjoint (observable!) canonical operators satisfying the canonical commutation relations. In path integral treatments, such obstructions arise from the introduction of delta functionals that enforce the classical constraints and the concomitant need to introduce subsidiary delta functionals to select a compatible dynamical gauge. Evidence for an obstruction is the impossibility to introduce a compatible gauge condition, a situation which is often rendered concrete in the vanishing or even sign changing of the requisite Faddeev-Popov determinant. With an unsuitable choice of gauge fixing fermion in the BRST scheme the Gribov ambiguity can appear as the breakdown of unitarity and a general dependence of physical quantities on the gauge condition. These are fundamental problems that seem difficult to overcome.

The main source of difficulty with the scenario in the preceding paragraph arises from the "need" to pass to the true set of classical physical variables since it is in this step that the obstruction appears. If we impose the constraints on the original phase space and divide by the gauge group so that every point in the quotient space corresponds to a physically distinct situation, then we have effectively arrived at the *moduli space* for the system at hand. Quantization of the degrees of freedom that characterize the moduli space is generally impossible because of some obstruction that has arisen in taking the quotient. A main feature of our approach is that we generally do *not* form a moduli space, and if we do reduce the initial classical phase space we do so in such a way as to remain compatible with the quantization procedure.

In part, these positive features are made possible by the use of coherent states and associated coherent-state techniques. Limitations of space only permit us to sketch how coherent states may be useful; for additional details see [7].

1.2. Coherent States

Canonical quantization is consistent only for Cartesian phase space coordinates [8], and we assume that our original and unconstrained set of classical dynamical variables fulfill that condition. Then, for each classical coordinate q^j and momentum p_j, $1 \leq j \leq J$, we may introduce associated irreducible, self-adjoint, canonical operators Q^j and P_j which satisfy, in units where $\hbar = 1$, the canonical commutation relations $[Q^j, P_k] = i\delta^j_k \mathbb{1}$, with all other commutators vanishing. With $|0\rangle \in \mathsf{H}$ a suitable normalized state—typically the ground state of a (unit-frequency) harmonic oscillator (but not always!)—we introduce the canonical coherent states as

$$|p, q\rangle \equiv \exp(-i\Sigma_j \, q^j P_j) \exp(i\Sigma_j \, p_j Q^j) \, |0\rangle \qquad (1)$$

for all $(p, q) \in \mathbb{R}^{2J}$, where $p = \{p_j\}$ and $q = \{q^j\}$. These states admit a resolution of unity in the form

$$\mathbb{1} = \int |p, q\rangle\langle p, q| \, d\mu(p, q) \,, \qquad d\mu(p, q) \equiv d^J p \, d^J q/(2\pi)^J \qquad (2)$$

integrated over $\mathbb{R}^{2J}$.

The unit operator resides in the Hilbert space H of the unconstrained system. We may conveniently represent this Hilbert space as follows. We first introduce a *reproducing kernel* $\langle p'', q'' | p', q' \rangle$ as the overlap matrix element between any two coherent states. This expression is a bounded, continuous function that characterizes a reproducing kernel Hilbert space representation of H appropriate to the unconstrained system as follows. A dense set of vectors in the associated functional Hilbert space is given by vectors of the form

$$\psi(p, q) \equiv \langle p, q | \psi \rangle = \sum_{l=1}^{L} \alpha_l \, \langle p, q | p_{(l)}, q_{(l)} \rangle \,, \qquad (3)$$

for arbitrary sets $\{\alpha_l\}$ and $\{p_{(l)}, q_{(l)}\}$ with $L < \infty$. The inner product of two such vectors is given by

$$(\psi, \phi) \equiv \sum_{l,m=1}^{L,M} \alpha_l^* \, \beta_m \, \langle p_{(l)}, q_{(l)} | \bar{p}_{(m)}, \bar{q}_{(m)} \rangle \qquad (4)$$

$$= \int \psi(p, q)^* \phi(p, q) \, d\mu(p, q) \,, \qquad (5)$$

where ϕ is a second function defined in a manner analogous to ψ. A general vector in the functional Hilbert space is defined by a Cauchy sequence of such vectors, and all such vectors are given by bounded, continuous functions. The first form of the inner product applies in general only to vectors in the dense set, while the second form of the inner product holds for arbitrary vectors in the Hilbert space.

1.3.   Constraints

Now suppose we introduce constraints into the quantum theory. In particular, we assume that $\mathbb{E}$ denotes a nonzero projection operator onto the constraint subspace, i.e., the subspace on which the quantum constraints are satisfied, which is also called the physical Hilbert space H_{phys}. Hence, if $|\psi\rangle \in \mathsf{H}$ denotes a general vector in the original (unconstrained) Hilbert space, the vector $\mathbb{E}|\psi\rangle \in \mathsf{H}_{phys}$ represents its component within the physical subspace. As a Hilbert space, the physical subspace also admits a functional representation by means of a reproducing kernel which may be taken as $\langle p'', q''|\mathbb{E}|p', q'\rangle$. In the same manner as before it follows that a dense set of vectors in $\mathsf{H}_{phys} \equiv \mathbb{E}\mathsf{H}$ is given by functions of the form

$$\psi(p, q) \equiv \langle p, q|\mathbb{E}|\psi\rangle = \sum_{l=1}^{L} \alpha_l \langle p, q|\mathbb{E}|p_{(l)}, q_{(l)}\rangle , \tag{6}$$

for arbitrary sets $\{\alpha_l\}$ and $\{p_{(l)}, q_{(l)}\}$ with $L < \infty$. The inner product of two such vectors is given by

$$(\psi, \phi) \equiv \sum_{l,m=1}^{L,M} \alpha_l^* \beta_m \langle p_{(l)}, q_{(l)}|\mathbb{E}|\overline{p}_{(m)}, \overline{q}_{(m)}\rangle \tag{7}$$

$$= \int \psi(p, q)^* \phi(p, q) \, d\mu(p, q) . \tag{8}$$

Again, a general vector in the functional Hilbert space is defined by means of a Cauchy sequence, and all such vectors are given by bounded, continuous functions. Note well, in the case illustrated, that although $\mathbb{E}\mathsf{H} \subset \mathsf{H}$, the functional representations of the unconstrained and the constrained Hilbert spaces are *identical*, namely as functions of $(p, q) \in \mathbb{R}^{2J}$, and the form of the inner product is *identical* in the two cases. This situation holds even if H_{phys} is one dimensional!

1.4.   Dynamics

Suppose further that the Hamiltonian operator $\mathcal{H}$ respects the first-class character of the constraints. It follows in that case that $[\mathbb{E}, \mathcal{H}] = 0$ or stated otherwise that

$$e^{-i\mathcal{H}t}\mathbb{E} = \mathbb{E}e^{-i\mathcal{H}t}\mathbb{E} = \mathbb{E}e^{-i(\mathbb{E}\mathcal{H}\mathbb{E})t}\mathbb{E} . \tag{9}$$

Note that any observable $\mathcal{O}$—$\mathcal{H}$ included—satisfies $[\mathbb{E}, \mathcal{O}] = 0$, and relations similar to (9) follow with $\mathcal{H}$ replaced by $\mathcal{O}$. Dynamics in the physical subspace is then fully determined by the propagator on H_{phys}, which is given in the relevant functional representation by

$$\langle p'', q'' | e^{-i\mathcal{H}t} \mathbb{E} | p', q' \rangle . \tag{10}$$

In (10) we have defined a fully gauge invariant propagator without having to reduce the number or even the domain of the original classical variables nor change the original form of the inner product on the functional Hilbert space representation. Compare, e.g., [9, 3, 10].

The foregoing scenario has assumed that the appropriate H_{phys} is given by means of a projection operator $\mathbb{E}$ acting on the original Hilbert space. This assumption holds true whenever the set of quantum constraints admits zero as a common point in their discrete spectrum; in that case $\mathbb{E}$ defines the subspace where the constraints all vanish. However, this situation may not always hold true, but even when zero lies in the continuous spectrum for some or all of the constraints, a suitable result may generally be given by matrix elements of a sequence of rescaled projection operators, say $c_\delta \, \mathbb{E}_\delta$, $c_\delta > 0$, as $\delta \to 0$. Specifically, we consider the limit of a sequence of reproducing kernels $c_\delta \, \langle p'', q'' | \mathbb{E}_\delta | p', q' \rangle$, which—if the limit is a nonvanishing continuous function—defines a new reproducing kernel, and thereby a new reproducing kernel Hilbert space, within which the appropriate constraints are fulfilled. In such a limit certain variables may cease to be relevant and as a consequence the local integral representation of the inner product may require modification. On the other hand, the definition of the inner product by sums involving the reproducing kernel will generally hold. A simple example should help clarify the situation in the present case.

Let the expression

$$\langle p'', q'' | \mathbb{E}_\delta | p', q' \rangle \equiv$$
$$\pi^{-J/2} \int_{-\delta}^{\delta} \cdots \int_{-\delta}^{\delta} \exp[-\tfrac{1}{2}(k - p'')^2 + ik \cdot (q'' - q') - \tfrac{1}{2}(k - p')^2] \, d^J k \tag{11}$$

denote a reproducing kernel for any $\delta > 0$. (If $\delta = \infty$ the result is the usual canonical coherent state overlap and characterizes the unconstrained Hilbert space.) In the present case it follows that $\mathbb{E}_\delta \equiv \prod_{j=1}^{J} \mathbb{E}(-\delta < P_j < \delta)$. When $\delta \to 0$, then (11) vanishes. However, if we first multiply by δ^{-J}—or more conveniently by $\pi^{J/2}(2\delta)^{-J}$—before taking the limit, the result becomes

$$\lim_{\delta \to 0} \pi^{J/2}(2\delta)^{-J} \langle p'', q'' | \mathbb{E}_\delta | p', q' \rangle = \exp(-\tfrac{1}{2} p''^2) \exp(-\tfrac{1}{2} p'^2) , \tag{12}$$

which is continuous and therefore defines a reproducing kernel for some Hilbert space. Note that the classical variables q'' and q' have disappeared. In the present example, the physical Hilbert space is *one* dimensional, and the inner product may be given either by a sum as in (7) involving the p variables alone or by a local integral representation now using the measure $\pi^{-J/2} d^J p$. This example illustrates the case where the constraints are $P_j = 0$ for all j, a situation where zero lies in the continuous spectrum. Nevertheless, the reduction of the functional representation has taken place in a fashion in which, for more complex situations, "obstructions" would have no consequence. Let us briefly see how these ideas appear in use.

2. PATH INTEGRAL REALIZATION

A classical action (summation implied)

$$I = \int [p_j \dot{q}^j - H(p,q) - \lambda^a \phi_a(p,q)] \, dt \, , \tag{13}$$

$1 \leq j \leq J$, $1 \leq a \leq A \leq 2J$, is said to describe a system with closed first-class constraints provided $\{p_j, q^j\}$ are dynamical variables, $\{\lambda^a\}$ are Lagrange multipliers, and

$$\{\phi_a(p,q), \phi_b(p,q)\} = c_{ab}{}^c \, \phi_c(p,q) \, , \tag{14}$$

$$\{\phi_a(p,q), H(p,q)\} = h_a{}^b \, \phi_b(p,q) \, , \tag{15}$$

where $\{\cdot, \cdot\}$ denotes the Poisson bracket.

In the quantum theory we assume that

$$[\Phi_a(P,Q), \Phi_b(P,Q)] = i c_{ab}{}^c \, \Phi_c(P,Q) \, , \tag{16}$$

$$[\Phi_a(P,Q), \mathcal{H}(P,Q)] = i h_a{}^b \, \Phi_b(P,Q) \, , \tag{17}$$

where Φ_a and $\mathcal{H}$ denote self-adjoint constraint and Hamiltonian operators, respectively. In [7] a coherent-state path integral was defined (by a suitable lattice limit) in such a way that

$$\mathcal{M} \int \exp\{i \int_0^T [p_j \dot{q}^j - H(p,q) - \lambda^a \phi_a(p,q)] \, dt\} \, \mathcal{D}p \, \mathcal{D}q \, \mathcal{D}C(\lambda)$$

$$= \langle p'', q'' | e^{-i\mathcal{H}T} \mathbb{E} | p', q' \rangle \, . \tag{18}$$

In this expression $\mathcal{M}$ is a formal normalization and $C(\lambda)$, $\int \mathcal{D}C(\lambda) = 1$, is a measure on the Lagrange multipliers designed to introduce (at least) one factor of a suitable projection operator $\mathbb{E}$. Because the set of self-adjoint operators $\{\Phi_a\}$ forms a closed Lie algebra, we may define $\exp(-i\xi^a \Phi_a)$, for appropriate sets $\{\xi^a\}$ of real parameters, as unitary group operators. For a *compact group* with a normalized Haar measure $\delta\xi$, $\int \delta\xi = 1$, then

$$\mathbb{E} = \int e^{-i\xi^a \Phi_a} \, \delta\xi = \mathbb{E}(\Phi_a = 0) \, . \tag{19}$$

For a *noncompact group*, some generators of which have continuous spectra, then, for example, we choose a weight function $f(\xi)$ such that

$$\mathbb{E} = \int e^{-i\xi^a \Phi_a} \, f(\xi) \, \delta\xi = \mathbb{E}(\Sigma \Phi_a^2 \leq \delta^2) \, , \tag{20}$$

for $0 < \delta \ll 1$. As discussed in [7], the properly defined path integral with a suitable integration measure for the Lagrange multipliers—essentially different than in the Faddeev treatment [2]—automatically leads to gauge invariant results for compact groups, and after a suitable δ-limiting process (illustrated above) to gauge invariant results for noncompact groups.

Finally, observe that no delta functionals of the classical constraints accompanied by delta functionals of the subsidiary conditions are introduced, and thus no Faddeev-Popov determinant is required. Specifically, in our approach, no effort is made to first pass to the moduli space, and hence no associated obstruction to the quantization can arise. In point of fact, quantization is carried out for the *un*constained system and any constraints are imposed only after quantization. This approach is identical to that of Dirac [1], and thus our results agree with those obtained by his operator techniques.

Acknowledgements

Thanks are expressed to J. Govaerts and S.V. Shabanov for discussions related to the present formulation of constrained system quantization.

References

[1] Dirac P.A.M., Lectures on Quantum Mechanics (Belfer Graduate School of Science, Yeshiva University, New York, 1964).
[2] Faddeev L.D., *Theor. Math. Phys.* **1** (1970) 1.
[3] Henneaux M. and Teitelboim C., Quantization of Gauge Systems (Princeton University Press, Princeton, 1992).
[4] Gribov V.N., *Nucl. Phys.* **B139** (1978) 1; Singer I.M., *Commun. Math. Phys.* **60** (1978) 7; see also Govaerts J., Hamiltonian Quantisation and Constrained Dynamics, Leuven Notes in Mathematical and Theoretical Physics, Vol. 4, Series B: Theoretical Particle Physics (Leuven University Press, 1991).
[5] Scholtz F.G. and Shabanov S.V., "Gribov vs BRST", hep-th/9701051; see also Govaerts J., *Int. J. Mod. Phys.* A **4** (1989) 173; ibid. 4487.
[6] Roepstorff G., in Path Integrals: Dubna '96 (JINR Publishing Dept., Dubna), p. 119.
[7] Klauder J.R., *Ann. Phys.* **254** (1997) 419.
[8] Dirac P.A.M., The Principles of Quantum Mechanics (Clarendon Press, Oxford, 1976) Fourth Edition, p. 114.
[9] Prokhorov L.V. and Shabanov S.V., *Sov. Phys. Uspekhi* **34** (1991) 108.
[10] Govaerts J., *J. Phys. A: Math. Gen.* **30** (1997) 603.

Unified Description and Canonical Reduction to Dirac's Observables of the Four Interactions

L. Lusanna

Sezione INFN di Firenze
L.go E. Fermi 2 (Arcetrii)
50125 Firenze, Italy

1. Systems with Constraints

The standard $SU(3) \times SU(2) \times U(1)$ model of elementary particles, all its extensions with or without supersymmatry, all variants of string theory, all formulations of general relativity are described by singular Lagrangians. Therefore, their Hamiltonian formulation needs Dirac's theory of 1st and 2nd class constraints[1] determining the submanifold of phase space relevant for dynamics: this means that the basic mathematical structure behind our description of the four interactions is presymplectic geometry (namely the theory of submanifolds of phase space with a closed degenerate 2-form; strictly speaking only 1st class constraints are associated with presymplectic manifolds: the 2nd class ones complicate the structure). For a system with 1st and 2nd class constraints the physical description becomes clear in coordinates adapted to the presymplectic submanifold. Locally in phase space, an adapted Darboux chart can always be found by means of Shanmugadhasan's canonical transformations [2] [strictly speaking their existence is proved only for finite-dimensional systems, but they underlie the existence of the Faddeev-Popov measure for the path integral]. The new canonical basis has: i) as many new momenta as 1st class constraints (Abelianization of 1st class constraints); ii) their conjugate canonical variables (Abelianized gauge variables); iii) as many pairs of canonical variables as pairs of 2nd class constraints (standard form, adapted to the chosen Abelianization, of the irrelevant variables); iv) pairs of canonically conjugate Dirac's observables (canonical basis of physical variables adapted to the chosen Abelianization; they give a trivialization of the BRST construction of observables). Putting equal to zero the Abelianized gauge variables one defines a local gauge of the model. If a system with constraints admits one (or more) global Shanmugadhasan canonical transformations, one obtains one (or more) privileged global gauges in which the physical Dirac observables are globally defined and globally separated from the gauge and the irrelevant degrees of freedom [for systems with a compact configuration space this is impossible]. These privileged gauges (when they exist) can be called generalized Coulomb gauges.

To find them the main problem is to discover how the original canonical variables depend upon the gauge variables. This can be achieved by solving (if possible) the so-called multitemporal equations [see Refs.[3]a), b)]: by considering each of the original 1st class constraints as the Hamiltonian for the evolution in a suitable parameter (called a generalized time), these equations are the associated Hamilton equations. If one succeeds in solving these equations (which are formally integrable due the 1st class nature of the constraints), one finds the the suitable parameters are just the Abelianized gauge variables and, then, one can construct the conjugate Abelianized 1st class constraints and the standard form of the 2nd class ones. Every set of suitable generalized times gives rise to a different generalized Coulomb gauge. Let us remark that in certain cases it is possible to find some special global Shanmugadhasan canonical transformations such that the effective new Hamiltonian of the system is automatically the sum of the physical Hamiltonian (depending only on Dirac's observables) and a gauge Hamiltonian (depending only on the gauge variables and the 1st class constraints): in these cases there is a decoupling of the gauge degrees of freedom without the need to add gauge-fixings (i.e. without putting to zero the Abelianized gauge variables).

Therefore, given a system with constraints without a compact configuration space, one has to investigate whether there is any obstruction to the existence of global Shanmugadhasan canonical transformations, namely whether one can do a global canonical reduction. Possible obstructions can arise when the configuration (and then also the phase) space and/or related mathematical structures like fiber bundles are not topologically trivial: usually they may be present when certain cohomological groups of the classical manifolds are not trivial (these groups are at the basis of the possible existence of anomalies in the quantization of the system). A related problem in gauge theories is the possible existence of Gribov-type ambiguities (nontrivial stability groups of gauge transformations for gauge potentials and/or field strengths): since they imply the nonexistence of global gauges, they destroy the possibility of a global decoupling of physical and gauge degrees of freedom. These ambiguities imply that both the configuration space and the phase space constraint manifold are in general stratified manifolds (i.e. disjoint union of manifolds) possibly with singularities. Moreover, for constrained systems defined in Minkowski spacetime the constraint manifold is always a stratified manifold, because one always assumes that the kinematical Poincaré group is globally implemented for isolated systems. This implies that the ten Poincaré generators must be finite (this is a first restriction on the boundary conditions of the fields present in the isolated system to allow use of group theory) and that the constraint manifold is the disjoint union of manifolds, each one of which contains all the system configurations belonging to the same type of Poincaré orbit (spacelike orbits should be absent at the classical level not to have causality problems). The main stratum, dense in the constraint manifold, will be the one associated with nonzero spin (i.e. with nonzero Pauli-Lubanski Casimir) timelike orbits. The existence of this Poincaré stratification raises the general question whether there could exist generalized Coulomb gauges with some kind of manifest covariance under Lorentz transformations and with some kind of universal breaking of manifest Lorentz covariance (unavoidable when one eliminates all the gauge degrees of freedom).

Moreover, one would like to have of the results obtained in Minkowski spacetime M^4 described in a form which can be extended to incorporate general relativity.

Given this general setting for constrained systems, a research program started trying to get a description only in terms of Dirac's observables and with an explicit control on covariance of (to start with) the $SU(3) \times SU(2) \times U(1)$ standard model of elementary particles coupled to tetrad gravity (more natural than metric gravity for the coupling to fermion fields). See Refs.[3] for the genesis and the developments of this program, which is well defined only for isolated systems [to recover theories with external fields one should consider special limits of some parameter of some subsystem].

In the next Sections a review of the results will be given.

2. Noncovariant Generalized Coulomb Gauges

Dirac[4] found the Coulomb gauge physical Hamiltonian of the isolated system formed by a fermion field plus the electromagnetic field [see the second paper in Ref.[5] for the case of Grassmann-valued fermion fields], which contains the coupling of physical fermions with the radiation field and the nonlocal Coulomb self-energy of the fermion field:

$$\int d^3x d^3y \, (\check{\psi}^\dagger \check{\psi})(\vec{x}, x^o) \frac{1}{4\pi|\vec{x} - \vec{y}|} (\check{\psi}^\dagger \check{\psi})(\vec{y}, x^o)$$

The Dirac observables are the transverse vector potential $\vec{A}_\perp(\vec{x}, x^o)$, the transverse electric field $\vec{E}_\perp(\vec{x}, x^o)$ and physical fermion fields dressed with a Coulomb cloud, $\check{\psi}(\vec{x}, x^o) = e^{i\eta_{em}(\vec{x}, x^o)}\psi(\vec{x}, x^o)$, $\eta_{em} = -\frac{1}{\Delta}\vec{\partial} \cdot \vec{A}$.

Extending this approach, the generalized Coulomb gauge of the following isolated systems has been found [see Ref.[3] c) for other systems like the Nambu string, relativistic two-body systems with action-at-a-distance interactions and for nonrelativistic Newton mechanics reformulated with 1st class constraints]:

a) Yang-Mills theory with Grassmann-valued fermion fields[5] in the case of a trivial principal bundle over a fixed-x^o R^3 slice of Minkowski spacetime with suitable Hamiltonian-oriented boundary conditions; this excludes monopole solutions and, since R^3 is not compactified, one has only winding number and no instanton number. After a discussion of the Hamiltonian formulation of Yang-Mills theory, of its group of gauge transformations and of the Gribov ambiguity, the theory has been studied in suitable weigthed Sobolev spaces where the Gribov ambiguity is absent. The global Dirac observables are the transverse quantities $\vec{A}_{a\perp}(\vec{x}, x^o)$, $\vec{E}_{a\perp}(\vec{x}, x^o)$ and fermion fields dressed with Yang-Mills (gluonic) clouds. The nonlocal and nonpolynomial (due to classical Wilson lines along flat geodesics) physical Hamiltonian has been obtained: it is nonlocal but without any kind of singularities, it has the correct Abelian limit if the structure constants are turned off, and it contains the explicit realization of the abstract Mitter-Viallet metric.

b) The Abelian and non-Abelian SU(2) Higgs models with fermion fields[6, 7], where the symplectic decoupling is a refinement of the concept of unitary gauge. There is an ambiguity in the solutions of the Gauss law constraints, which reflects the existence of disjoint sectors of solutions of the Euler-Lagrange equations of Higgs models. The physical Hamiltonian and Lagrangian of the Higgs phase have been found; the self-energy turns out to be local and contains a local four-fermion interaction.

c) The standard SU(3)xSU(2)xU(1) model of elementary particles[8] with Grassmann - valued fermion fields. The final reduced Hamiltonian contains nonlocal self-energies for the electromagnetic and color interactions, but "local ones" for the weak interactions implying the nonperturbative emergence of 4-fermions interactions. To obtain a nonlocal self-energy with a Yukawa kernel for the massive Z and $W^\pm$ bosons one has to reformulate the model on spacelike hypersurfaces and make a modification of the Lagrangian.

3.   Wigner-Covariant Rest-Frame Instant Form

The next problem is how to covariantize these results. Again the starting point was given by Dirac[1] with his reformulation of classical field theory on spacelike hypersurfaces foliating Minkowski spacetime. In this way one gets parametrized field theory with a covariant 3+1 splitting of flat spacetime and already in a form suited to the coupling to general relativity in its ADM canonical formulation (see also Ref.[9] , where a theoretical study of this problem is done in curved spacetimes) The price is that one has to add as new configuration variables the points $z^\mu(\tau, \vec{\sigma})$ of the spacelike hypersurface Σ_τ [the only ones carrying Lorentz indices; the scalar parameter τ labels the leaves of the foliation and $\vec{\sigma}$ are curvilinear coordinates on Σ_τ] and then to define the fields on Σ_τ so that they know the hypersurface Σ_τ of τ-simultaneity [for a Klein-Gordon field $\phi(x)$ this new field is $\tilde{\phi}(\tau, \vec{\sigma}) = \phi(z(\tau, \vec{\sigma}))$]. Then, besides a Lorentz-scalar form of the constraints of the given system, from the Lagrangian rewritten on the hypersurface [function of z^μ through the induced metric $g_{\check{A}\check{B}} = z^\mu_{\check{A}} \eta_{\mu\nu} z^\nu_{\check{B}}$, $z^\mu_{\check{A}} = \partial z^\mu / \partial \sigma^{\check{A}}$, $\sigma^{\check{A}} = (\tau, \sigma^{\check{r}})$] one gets 4 further first class constraints $\mathcal{H}_\mu(\tau, \vec{\sigma}) \approx 0$ implying the independence of the description from the choice of the spacelike hypersufaces. Being in special relativity, it is convenient to restrict ourselves to arbitrary spacelike hyperplanes $z^\mu(\tau, \vec{\sigma}) = x^\mu_s(\tau) + b^\mu_{\check{r}}(\tau)\sigma^{\check{r}}$. Since they are described by only 10 variables [an origin $x^\mu_s(\tau)$ and 3 orthogonal spacelike unit vectors generating the fixed constant timelike unit normal to the hyperplane], we remain only with 10 first class constraints determining the 10 variables conjugate to the hyperplane [they are a 4-momentum p^μ_s and the 6 independent degrees of freedom hidden in a spin tensor $S^{\mu\nu}_s$] in terms of the variables of the system.

If we now restrict ourselves to timelike ($p^2_s > 0$) 4-momenta, we can restrict the description to the so-called Wigner hyperplanes orthogonal to p^μ_s itself. To get this result, we must boost at rest all the variables with Lorentz indices by using the standard Wigner boost $L^\mu{}_\nu(p_s, \overset{\circ}{p}_s)$ for timelike Poincaré orbits, and then add the gauge-fixings $b^\mu_{\check{r}}(\tau) - L^\mu{}_{\check{r}}(p_s, \overset{\circ}{p}_s) \approx 0$. Since these gauge-fixings depend on p^μ_s, the final canonical variables, apart p^μ_s itself, are of 3 types: i) there is a non-covariant center-of-mass variable $\tilde{x}^\mu(\tau)$ [the classical basis of the Newton-Wigner position operator]; ii) all the 3-vector variables become Wigner spin 1 3-vectors [boosts in M^4 induce Wigner rotations on them]; iii) all the other variables are Lorentz scalars. Only the 4 1st class constraints determining p^μ_s are left. One obtains in this way a new kind of instant form of the dynamics (see Ref.[10]), the Wigner-covariant 1-time rest-frame instant form[11] with a universal breaking of Lorentz covariance. It is the special relativistic generalization of the nonrelativistic separation of the center of mass from the relative motion [$H = \frac{\vec{P}^2}{2M} + H_{rel}$]. The role of the center of mass is taken by the Wigner hyperplane, identified by the point $\tilde{x}^\mu(\tau)$ and by its normal p^μ_s. The 4 first class constraints can be put in the following form: i) the vanishing of the total (Wigner spin 1) 3-momentum of the system $\vec{p}[system] \approx 0$, saying that the Wigner hyperplane $\Sigma_W(\tau)$ is the intrinsic rest frame [instead, $\vec{p}_s$ is left arbitrary, since it reflects the orientation of the Wigner hyperplane with respect to arbitrary reference frames in M^4]; ii) $\pm \sqrt{p^2_s} - M[system] \approx 0$, saying that the invariant mass M of the system replaces the nonrelativistic Hamiltonian H_{rel} for the relative degrees of freedom, after the addition of the gauge-fixing $T_s - \tau \approx 0$ [identifying the time parameter τ with the Lorentz scalar time of the center of mass in the rest frame; M generates the evolution in this time]. When one is able, as in the case of N free particles [11], to find the (Wigner spin 1) 3-vector $\vec{\eta}(\tau)$ conjugate to $\vec{p}[system](\approx 0)$, the gauge-fixing $\vec{\eta} \approx 0$ eliminates the gauge variables describing the 3-dimensional intrinsic center of mass inside the Wigner hyperplane [$\vec{\eta} \approx 0$ forces it to coincide with $x^\mu_s(\tau) = z^\mu(\tau, \vec{\sigma} = \vec{\eta} = 0)$ and breaks the translation invariance $\vec{\sigma} \mapsto \vec{\sigma} + \vec{a}$], so that we remain only with Newtonian-like degrees of freedom with rotational covariance: i) a 3-coordinate (not Lorentz covariant) $\vec{z}_s = \sqrt{p^2_s}(\tilde{x}_s - \frac{\vec{p}_s}{p^o_s}\tilde{x}^o)$ and its conjugate momentum $\vec{k}_s = \vec{p}_s/\sqrt{p_s^2}$ for the absolute center of mass in M^4; ii) a set of relative conjugate pairs of variables with Wigner covariance inside the Wigner hyperplane .

The systems till now analyzed to get their rest-frame generalized Coulomb gauges are:

a) The system of N scalar particles with Grassmann electric charges plus the electromagnetic field [11]. The starting configuration variables are a 3-vector $\vec{\eta}_i(\tau)$ for each particle $[x_i^\mu(\tau) = z^\mu(\tau, \vec{\eta}_i(\tau))]$ and the electromagnetic gauge potentials $A_{\check{A}}(\tau, \vec{\sigma}) = \frac{\partial z^\mu(\tau, \vec{\sigma})}{\partial \sigma^{\check{A}}} A_\mu(z(\tau, \vec{\sigma}))$, which know implicitely the embedding of Σ_τ into M^4. One has to choose the sign of the energy of each particle, because there are not mass-shell constraints (like $p_i^2 - m_i^2 \approx 0$) among the constraints of this formulation, due to the fact that one has only 3 degrees of freedom for particle, determining the intersection of a timelike trajectory and of the spacelike hypersurface Σ_τ. The final Dirac's observables are: i) the transverse radiation field variables; ii) the particle canonical variables $\vec{\eta}_i(\tau)$, $\check{\kappa}_i(\tau)$, dressed with a Coulomb cloud. The physical Hamiltonian contains the Coulomb potentials extracted from field theory and there is a regularization of the Coulomb self-energies due to the Grassmann character of the electric charges Q_i $[Q_i^2 = 0]$. In Ref.[12] there is the study of the Lienard-Wiechert potentials and of Abraham-Lorentz-Dirac equations in this rest-frame Coulomb gauge and also scalar electrodynamics is reformulated in it. Also the rest-frame 1-time relativistic statistical mechanics is developed [11].

b) The system of N scalar particles with Grassmann-valued color charges plus the color SU(3) Yang-Mills field[13]: it gives the pseudoclassical descrption of the relativistic scalar-quark model, deduced from the classical QCD Lagrangian and with the color field present. The physical invariant mass of the system is given in terms of the Dirac observables. From the reduced Hamilton equations the second order equations of motion both for the reduced transverse color field and the particles are extracted. Then, one studies the N=2 (meson) case. A special form of the requirement of having only color singlets, suited for a field-independent quark model, produces a "pseudoclassical asymptotic freedom" and a regularization of the quark self-energy. With these results one can covariantize the bosonic part of the standard model given in Ref.[8].

c) It is in an advanced stage the description of Dirac and chiral fields and of spinning particles on spacelike hypersurfaces[14]. After its completion, the rest-frame form of the full standard $SU(3) \times SU(2) \times U(1)$ model can be achieved.

Finally, to eliminate the three 1st class constraints $\vec{p}[system] \approx 0$ by finding their natural gauge-fixings, when fields are present, one needs to find a rest-frame canonical basis of center-of-mass and relative variables for fields (in analogy to particles). Such a basis has already been found for a real Klein-Gordon field[15]. This kind of basis will allow, after quantization, to find the asymptotic states of the covariant Tomonaga-Schwinger formulation of quantum field theory on spacelike hypersurfaces: these states are needed for the theory of quantum bound states [since Fock states do not constitute a Cauchy problem for the field equations, because an in (or out) particle can be in the absolute future of another one due to the tensor product nature of these asymptotic states, bound state equations like the Bethe-Salpeter one have spurious solutions which are excitations in relative energies, the variables conjugate to relative times (which are gauge variables[11])].

4. Ultraviolet Cutoff

As said in Ref.[12, 13], the quantization of these rest-frame models has to overcome two problems. On the particle side, the complication is the quantization of the square roots associated with the relativistic kinetic energy terms. On the field side (all physical Hamiltonian are nonlocal and, with the exception of the Abelian case, nonpolynomial), the obstacle is the absence (notwithstanding there is no no-go theorem) of a complete regularization and renormalization procedure of electrodynamics (to start with) in the Coulomb gauge: see Ref.[16] (and its bibliography) for the existing results for QED.

However, as shown in Refs.[11, 5] [see their bibliography for the relevent references referring to all the quantities introduced in this Section], the rest-frame instant form of dynamics automatically gives a physical ultraviolet cutoff in the spirit of Dirac and Yukawa: it is the Møller radius[17] $\rho = \sqrt{-W^2}c/P^2 = |\vec{S}|c/\sqrt{P^2}$ ($W^2 = -P^2\vec{S}^2$ is the Pauli-Lubanski Casimir), namely the classical intrinsic radius of the worldtube, around the covariant noncanonical Fokker-Price center of inertia Y^μ, inside which the noncovariance of the canonical center of mass $\tilde{x}^\mu$ is concentrated. At the quantum level ρ becomes the Compton wavelength of the isolated system multiplied its spin eigenvalue $\sqrt{s(s+1)}$, $\rho \mapsto \hat{\rho} = \sqrt{s(s+1)}\hbar/M = \sqrt{s(s+1)}\lambda_M$ with $M = \sqrt{P^2}$ the invariant mass and $\lambda_M = \hbar/M$ its Compton wavelength. Therefore, the criticism to classical relativistic physics, based on quantum pair production, concerns the testing of distances where, due to the Lorentz signature of spacetime, one has intrinsic classical covariance problems: it is impossible to localize the canonical center of mass $\tilde{x}^\mu$ (also named Pryce center of mass and having the same covariance of the Newton-Wigner position operator) in a frame independent way.

Since ρ describes a nontestable classical short distance region [there is a conceptual connection with the aspect of Mach's principle according to which only relative motions are measurable], it sounds reasonable [13] that for a confined system of effective radius $r_o = 1/\Lambda_{QCD}$ (the fundamental scale of QCD) one has $\rho \le r_o^2 M = M/\Lambda_{QCD}^2$ [this ensures the mass-spin relation $|\vec{S}| = \alpha'_s M^2 + \alpha_o$ of phenomenological Regge trajectories]. Let us note that in string theory[18] the relevant dimensional quantity is the tension $T_s = 1/2\pi\alpha'_s$ (the energy per unit length), which, at the quantum level, determines a minimal length

$$L_s = \sqrt{\hbar/T_s} = \sqrt{2\pi\hbar\alpha'_s} \overset{\hbar=1}{=} \sqrt{2\pi\alpha'_s}$$

[for a classical string one has $|\vec{S}| \le \alpha'_s M^2$; a QCD string has $2\pi\alpha'_s \le r_o^2 = \Lambda_{QCD}^{-2}$].

Let us remember [11] that ρ is also a remnant in flat Minkowski spacetime of the energy conditions of general relativity: since the Møller noncanonical, noncovariant center of energy has its noncovariance localized inside the same worldtube with radius ρ (it was discovered in this way) [17], it turns out that an extended relativistic system with the material radius smaller of its intrinsic radius ρ has: i) the peripheral rotation velocity can exceed the velocity of light; ii) its classical energy density cannot be positive definite everywhere in every frame. Now, the real relevant point is that this ultraviolet cutoff determined by ρ exists also in Einstein's general relativity (which is not power counting renormalizable) in the case of asymptotically flat spacetimes, taking into account the Poincaré Casimirs of its asymptotic ADM Poincaré charges (when supertranslations are eliminated with suitable boundary conditions; let us remark that Einstein and Wheeler use closed universes because they don't want to introduce boundary conditions, but in this way they loose Poincaré charges and the possibility to make contact with particle physics).

By comparison, in string cosmology[18], at the quantum level, the string tension

$$T_{cs} = 1/2\pi\alpha'_{cs} = L_{cs}^2/\hbar$$

gives rise to a minimal length $L_{cs} \stackrel{\hbar=1}{=} \sqrt{2\pi\alpha'_{cs}} \geq L_P$ [$L_P = 1.6\,10^{-33}cm$ is the Planck length] and is determined by the vacuum expectation value of the background metric of the vacuum (if the ground state is flat Minkowski spacetime), while the grand unified coupling constant α_{GUT} (replacing α_s of QCD) is determined by the vacuum expectation value of the background dilaton field. This minimal length $L_{cs} \geq L_P$ (suppressing the gravitational corrections) could be a lower bound for the Møller radius of an asymptotically flat universe. The upper bound on ρ (namely a physical infrared cutoff) could be the Hubble distance $cH_o^{-1} \approx 10^{28}cm$ considered as an effective radius of the universe. Therefore, it seems reasonable that our physical ultraviolet cutoff ρ is meaningful in the range $L_P \leq L_{cs} < \rho < cH_o^{-1}$.

Moreover, the extended Heisenberg relations of string theory[18], i.e. $\Delta x = \frac{\hbar}{\Delta p} + \frac{\Delta p}{T_{cs}} = \frac{\hbar}{\Delta p} + \frac{\hbar\Delta p}{L_{cs}^2}$ implying the lower bound $\Delta x > L_{cs} = \sqrt{\hbar/T_{cs}}$ due to the $y + 1/y$ structure, have a counterpart in the quantization of the Møller radius[11]: if we ask that, also at the quantum level, one cannot test the inside of the worldtube, we must ask $\Delta x > \hat{\rho}$ which is the lower bound implied by the modified uncertainty relation $\Delta x = \frac{\hbar}{\Delta p} + \frac{\hbar\Delta p}{\hat{\rho}^2}$. This would imply that the center-of-mass canonical noncovariant (Pryce) 3-coordinate $\vec{z} = \sqrt{P^2}(\vec{\tilde{x}} - \frac{\vec{P}}{P_o}\tilde{x}^o)$ [11] cannot become a self-adjoint operator. See Hegerfeldt's theorems (quoted in Refs.[5, 11]) and his interpretation pointing at the impossibility of a good localization of relativistic particles (experimentally one determines only a worldtube in spacetime emerging from the interaction region). Since the eigenfunctions of the canonical center-of-mass operator are playing the role of the wave function of the universe, one could also say that the center-of-mass variable has not to be quantized, because it lies on the classical macroscopic side of Copenhagen's interpretation and, moreover, because, in the spirit of Mach's principle that only relative motions can be observed, no one can observe it. On the other hand, if one rejects the canonical noncovariant center of mass in favor of the covariant noncanonical Fokker-Pryce center of inertia Y^μ, $\{Y^\mu, Y^\nu\} \neq 0$, one could invoke the philosophy of quantum groups to quantize Y^μ to get some kind of quantum plane for the center-of-mass description.

5. Tetrad Gravity

The next step of the program is the search of Dirac's observables for classical tetrad gravity in globally hyperbolic asymptotically flat spacetimes $M^4 = \Sigma \times R$ with Σ diffeomorphic to R^3, so to have the asymptotic Poincaré charges and the same ultraviolet cutoff ρ as for the other interactions.

In Ref.[19] there is a new formulation of tetrad gravity avoiding the use of Schwinger's time gauge condition and, with the technology developed for Yang-Mills theory[5], 13 of its 14 1st class constraints have been Abelianized [the Abelianization of the 6 constraints generating space-diffeomorphisms and Lorentz rotations has been done in 3-orthogonal coordinates on Σ so that the 3-metric is diagonal]. The last constraint (the superHamiltonian one) becomes an integral equation for the momentum conjugate to the conformal factor of the 3-metric. See Ref.[3] c) for an expanded summary of the results and of the still open problems.

Further problems are how to deparametrize the theory[20], so to reexpress it in the form of parametrized field theories on spacelike hypersurfaces in Minkowski spacetime. This is an extremely important point, because, if we add N scalar particles to tetrad gravity (whose reduction to Dirac's observables should define the N-body problem in general relativity), the deparametrization should be the bridge to the previously quoted theory on spacelike hypersurfaces in Minkowski spacetime[11, 12, 13] in the limit of zero curvature. A new formulation of the N-body problem would be relevant to try to understand the energy balance in the emission of gravitational waves from systems like binaries. If it will be possible to find the Dirac observables for the particles, one will understand how to extract from the field theory the covariantization of Newton potential [one expects one scalar and one vector (gravitomagnetism) potential] and a mayor problem will be how to face the expected singularities of the mass-self-energies.

Finally one should couple tetrad gravity to the electromagnetic field, to fermion fields and then to the standard model, trying to make to reduction to Dirac's observables in all these cases.

References

[1] Dirac P.A.M. and Can.J.Math. **2**, 129 (1950); "Lectures on Quantum Mechanics", Belfer Graduate School of Science, Monographs Series (Yeshiva University, New York, N.Y., 1964).

[2] Shanmugadhasan S., J.Math.Phys. **14**, 677 (1973). Lusanna L., Int.J.Mod.Phys. **A8**, 4193 (1993). Chaichian M., Louis Martinez D. and

Lusanna L., Ann.Phys.(N.Y.)**232**, 40 (1994). Lusanna L., Phys.Rep. **185**, 1 (1990); Riv. Nuovo Cimento **14**, n.3, 1 (1991); J.Math.Phys. **31**, 2126 (1990); J.Math.Phys. **31**, 428 (1990).

[3] Lusanna L., a) "Classical Observables of Gauge Theories from the Multitemporal Approach", talk given at the Conference 'Mathematical Aspects of Classical Field Theory', Seattle 1991, in Contemporary Mathematics **132**, 531 (1992); b) "Hamiltonian Constraints and Dirac's Observables: from Relativistic Particles towards Field Theory and General Relativity", talk at the Workshop "Geometry of Constrained Dynamical Systems", Newton Institute, Cambridge, 1994, (J.M.Charap, Ed.), Cambridge Univ.Press, Cambridge, 1995. c) "Solving Gauss' Laws and Searching Dirac Observables for the Four Interactions", talk at the "Second Conf. on Constrained Dynamics and Quantum Gravity", S.Margherita Ligure 1996 (HEP-TH/9702114).

[4] Dirac P.A.M., Can.J.Phys. **33**, 650 (1955).

[5] Lusanna L., Int.J.Mod.Phys. **A10**, 3531 and 3675 (1995).

[6] Lusanna L. and Valtancoli P., "Dirac's Observables for the Higgs model: I) the Abelian Case", to appear in Int.J.Mod.Phys. A (HEP-TH/9606078).

[7] Lusanna L. and Valtancoli P., "Dirac's Observables for the Higgs model: II) the non-Abelian SU(2) Case", to appear in Int.J.Mod.Phys. A (HEP-TH/9606079).

[8] Lusanna L. and Valtancoli P., "Dirac's Observables for the SU(3)xSU(2)xU(1) Standard Model", Firenze Univ.preprint, May 1997.

[9] Kuchar K., J.Math.Phys. **17**, 777, 792, 801 (1976); **18**, 1589 (1977).

[10] Dirac P.A.M., *Rev.Mod.Phys.* **21** (1949) 392.

[11] Lusanna L., Int.J.Mod.Phys. **A12**, 645 (1997).

[12] Alba D. and Lusanna D., "The Lienard-Wiechert Potential of Charged Scalar Particles and théir Relation to Scalar Electrodynamics in the Rest-Frame Instant Form", Firenze Univ.preprint, May 1997.

[13] Alba D. and Lusanna L., "The Classical Relativistic Quark Model in the Rest-Frame Wigner-Covariant Coulomb Gauge", Firenze Univ.preprint,May 1997.

[14] Bigazzi F., DePietri R. and Lusanna L., in preparation.

[15] Longhi G. and Materassi M., "Collective and Relative Variables for a Classical Relativistic Field", in preparation.

[16] Leibbrandt G., "Non-Covariant Gauges", ch.9 (World Scientific, Singapore, 1994).

[17] Møller C., Ann.Inst.H.Poincaré **11**, 251 (1949); "The Theory of Relativity" (Oxford Univ.Press, Oxford, 1957).

[18] Veneziano G., "Quantum Strings and the Constants of Nature", in "The Challenging Questions", ed.A.Zichichi, the Subnuclear Series n.27 (Plenum Press, New York, 1990).

[19] Lusanna L. and Russo S., "Dirac's Observables for Tetrad Gravity", in preparation.

[20] Isham C.J. and Kuchar K., Ann.Phys.(N.Y.) **164**, 288 and 316 (1984). Kuchar K., Found.Phys. **16**, 193 (1986).

Time Evolution in General Gauge Theories

R. Marnelius[1]

([1]) *Institute of Theoretical Physics, Chalmers University of Technology, Göteborg University, S-412 96 Göteborg, Sweden*

INTRODUCTION

In this talk I will discuss some properties of time evolutions in general gauge theories within a BRST quantization [1]. More precisely I will discuss the choices of Hamiltonians within the Hamiltonian framework set up by Batalin, Fradkin and Vilkovisky which is called the BFV formulation [2] (for a review see *e.g.* [3]). This I will do from the point of view of an operator formulation for inner product solutions within the BFV scheme which I have been developing during some years [4]-[8]. This formalism turns out to yield more information about quantum properties than just an effective BRST invariant Lagrangian or Hamiltonian formulation. In fact, an effective Hamiltonian is more difficult to extract within this scheme, but the procedure provides for a deeper understanding of the standard BFV prescriptions. These results will be briefly reviewed. As a particular example of a natural consequence I will at the end show that QED is coBRST invariant. However, let me first review the standard BFV formulation.

STANDARD BFV-BRST

Within the BFV formulation Hamiltonians of general gauge theories are assumed to have the form

$$H = H_0 + \int v_i \theta_i, \tag{1}$$

where v_i are Lagrange multipliers and θ_i constraint variables. (Repeated indices are summed over and integrals are over space coordinates.) H_0 and θ_i satisfy the super Poisson bracket conditions

$$\{H_0, \theta_i\} = C_{ij}\theta_j, \quad \{\theta_i, \theta_j\} = C_{ijk}\theta_k. \tag{2}$$

where C_{ij} and C_{ijk} may be functions on the phase space. In the corresponding BRST quantization BFV introduces the following additional degrees of freedom:

- π_i – conjugate momenta to the Lagrange multipliers v_i. (They are additional abelian constraint variables.)
- $\mathcal{C}_i$, $\mathcal{P}_i$ – ghosts and their conjugate momenta.
- $\bar{\mathcal{C}}_i$, $\bar{\mathcal{P}}_i$ – antighosts and their conjugate momenta.

Their Grassmann parities and ghost numbers are

$$\varepsilon(\mathcal{C}_i) = \varepsilon(\bar{\mathcal{C}}_i) = \varepsilon(\theta_i) + 1, \quad gh(\mathcal{C}_i) = 1 = -gh(\bar{\mathcal{C}}_i). \tag{3}$$

In this extended phase space the BRST charge is given by

$$Q = \int \left(\mathcal{C}_i\theta_i + \ldots + \bar{\mathcal{P}}_i\pi_i \right), \tag{4}$$

where the dots indicates terms determined by the super Poisson bracket condition $\{Q, Q\} = 0$. The Hamiltonian for the effective BRST invariant theory is defined to be

$$H_{BRST} = H_0' + i\{Q, \psi\}, \quad \{H_{BRST}, Q\} = 0, \tag{5}$$

where

$$H_0' = H_0 + \text{ghost dependent terms}, \quad \{H_0', Q\} = 0, \tag{6}$$

and where in turn ψ is an odd gauge fixing fermion which usually is chosen such that

$$H_{BRST} = H + \int (\ )\pi_i + \text{ghost dependent terms}. \tag{7}$$

Such a Hamiltonian leads to a BRST invariant effective Lagrangian of the standard form

$$\mathcal{L}_{BRST} = \mathcal{L} + \mathcal{L}_{gf} + \mathcal{L}_{gh}. \tag{8}$$

The general allowed form for ψ as prescribed by BFV is

$$\psi = \int \left(\mathcal{P}_i v_i + \bar{\mathcal{C}}_i \chi_i \right), \tag{9}$$

where χ_i are gauge fixing variables to θ_i. (The matrix $\{\chi_i, \theta_j\}$ is required to be invertible.)

One may observe that neither H_0' nor ψ are uniquely determined. For instance, in an abelian gauge theory H_{BRST} is invariant under the transformations

$$H_0 \longrightarrow H_0 + x_i \theta_i, \quad \psi \longrightarrow \psi + \mathcal{P}_i x_i \tag{10}$$

for any BRST invariant variable x_i.

OPERATOR QUANTIZATION ON INNER PRODUCT SPACES

Case 1: $H_0' = 0$

This case includes all reparametrization invariant theories, such as particles, strings, and gravity. The operator quantization proceeds here as follows: Quantize all degrees of freedom and construct an extended inner product state space V. Physics is then what is contained in the subspace $V_{ph} \subset V$ defined by $QV_{ph} = 0$. V_{ph} is degenerate since the zero norm states QV is contained in V_{ph}. The nondegenerate inner product space is therefore $V_s = V_{ph}/QV$, the states of BRST singlets. V_s is an inner product space if V is an inner product space. An important concept in this connection is the coBRST charge $^*Q$ [9]. It is defined by

$$^*Q = \eta Q \eta, \tag{11}$$

where η is an hermitian metric operator such that $\eta^2 = 1$ and $\langle u|\eta|u \rangle \geq 0$ $\forall |u\rangle \in V$. Thus, η maps V onto a Hilbert space and $^*Q$ is just the hermitian conjugate of Q in this Hilbert space. We have $^*Q^2 = 0$. In terms of the coBRST charge the BRST singlets $|s\rangle \in V_s$ are determined by

$$Q|s\rangle = \ ^*Q|s\rangle = 0 \tag{12}$$

or equivalently

$$\Delta|s\rangle = 0, \quad \Delta \equiv [Q, \ ^*Q]_+. \tag{13}$$

Now it is usually very difficult to find the appropriate inner product space V. (There are even cases which allow for several different choices.) Fortunately, there is a possibility to construct formal operator expressions for the singlets $|s\rangle$ without prescribing V. In fact, such expressions will at the end tell you the appropriate prescription for V [8]. Since this formalism is not yet completely rigorously proved, I will present the main ingredients as a set of proposals:

Proposal 1: If $Q = \delta + \delta^\dagger$, where δ, $\delta^\dagger$ are independent nilpotent operators each containing effectively half the constraints of Q, then the solutions of $\delta|ph\rangle = \delta^\dagger|ph\rangle = 0$ are formally inner product solutions what concerns the unphysical degrees of freedom.

Proposal 2: Q for any gauge theory in BFV form may be decomposed as $Q = \delta + \delta^\dagger$, where δ, $\delta^\dagger$ are independent and each containing effectively half the constraints of Q and such that $\delta^2 = 0$ and $[\delta, \delta^\dagger]_+ = 0$.

Proposal 3: The formal solutions of $\delta|ph\rangle = \delta^\dagger|ph\rangle = 0$ have up to zero norm states the general form $|ph\rangle = e^{[Q,\psi]_+}|\phi\rangle$ where ψ is an odd gauge fixing fermion of the form (9), and where $|\phi\rangle$ satisfies simple hermitian conditions.

Proposal 1 has been shown to be valid in all investigated cases. Proposal 2 has been proved for general Lie group theories [4]. Concerning proposal 3 the following may be said: Formal inner product solutions of the form $|ph\rangle = e^{[Q,\psi]_+}|\phi\rangle$ exist for any gauge theory if Q is in BFV form. In fact, in [6] it was shown that the BRST singlets $|s\rangle$ locally may be written as $|s\rangle = e^{[Q,\psi]_+}|\phi_s\rangle$ where $|\phi_s\rangle$ is a ghost and gauge fixed $|\phi\rangle$). There it was shown that $D_i|s\rangle = 0$ where D_i are a complete set of BRST doublets ($C,[Q,C]_\pm$-pairs) satisfying $[D_i, D_j]_\pm = c_{ijk}D_k$, and that $[D_i, D_j^\dagger]_\pm$ is an invertible matrix operator. ($\langle\phi|\phi\rangle$ is undefined while $\langle\phi|e^{[Q,\psi]_+}|\phi\rangle$ is well defined and independent of ψ.)

What is the form of the coBRST charge within the BFV formalism? It turns out that the coBRST charge has the form of an allowed gauge fixing fermion [7]. However, one may notice that $|s\rangle = e^{[Q,{}^*Q]_+}|\phi_s\rangle = e^\triangle|\phi_s\rangle$ does not satisfy ${}^*Q|s\rangle = 0$. Only $|s'\rangle = U|s\rangle$ does, where U is a unitary operator[7].

Case 2: $H_0' \neq 0$

This case may always be transformed to case 1 ($H_0' = 0$) by making the gauge theory reparametrization invariant. The transformed theory may then be treated as before. This is the fundamental approach here. *Thus, nontrivial time evolution does not cause any basic problems.* However, problems do occur whenever one tries to find inner product solutions without going to the corresponding reparametrization invariant theory. A simple and natural prescription for inner product solutions in the latter case is first to construct inner product solutions as in case 1 ignoring H_0' and then to require $|ph,t\rangle$ or $|s,t\rangle$ to be determined by a Schrödinger equation with the Hamiltonian H_0'. From the corresponding reparametrization invariant theory one finds that this is possible provided the gauge fixing conditions satisfy some weak conditions like

$$[Q,[H_0',\psi]_-]_+ = 0, \quad [\psi,[H_0',\psi]_-]_+ = 0. \tag{14}$$

That the Hamiltonian must be H_0' follows from the fact that the BRST charge in the corresponding reparametrization invariant theory is

$$\tilde{Q} = Q + \mathcal{C}(\pi + H_0') + \bar{\mathcal{P}}\pi_v, \tag{15}$$

where $\mathcal{C}$ and $\bar{\mathcal{P}}$ are new ghost variables, π_v is the conjugate momentum to a new Lagrange multiplier, and π is the conjugate momentum to a dynamical time variable. Thus, since $\pi + H_0'$ is a new constraint variable it is easily understood that $(\pi + H_0')|\ \rangle = 0$ is a natural equation, and this is the Schrödinger equation with H_0' as Hamiltonian. It turns out that the BRST singlets satisfy this Schrödinger equation strictly under weak conditions like (14). However, the problem with this procedure is that ψ and H_0' are not uniquely given. It is well known that one by unitary transformations may change the constraint variables in the BRST charge and H_0' is part of a constraint in the reparametrization invariant theory (15). On the other hand, this freedom may be used to find a H_0' and ψ's satisfying conditions (14). Whether or not this is possible in general is unclear though.

According to the procedure above we have in the case when H_0' has no explicit time dependence

$$|ph, t\rangle = e^{-iH_0't}|ph\rangle. \tag{16}$$

Since the first condition in (14) implies $[H_0', [Q, \psi]_+]_- = 0$ by means of the Jacobi identities, (16) combined with proposal 3 implies

$$|ph, t\rangle = e^{-iH_0't + [Q, \psi]_+}|\phi\rangle. \tag{17}$$

This leads to

$$\langle ph, t'|ph, t\rangle = \int d\omega' d\omega \ \phi'^*(\omega'^*)\phi(\omega)\langle \omega', t'|\omega^*, t\rangle, \tag{18}$$

where ω denotes all coordinates of the original BRST invariant theory. (Due to the indefinite metric state space not all hermitian operators have real eigenvalues.) After the replacement $\psi \to (t' - t)\psi/2$, which is possible for any finite $t' - t \neq 0$, one may derive the path integral representation [5, 1]:

$$\langle \omega', t'|\omega^*, t\rangle = \int D\omega D\pi_\omega \ \exp\left\{ i \int_t^{t'} (\pi_\omega \dot{\omega} - H_{BRST}) \right\}, \tag{19}$$

where

$$H_{BRST} = H_0' + i\{Q, \psi\} \tag{20}$$

is the effective Hamiltonian function in agreement with the BFV prescription (5). Notice, however, that H_{BRST} is not real in general. Often a real effective Hamiltonian requires us to choose an imaginary time t [5].

One may notice that the second condition in (14) is satisfied if $\psi^2 = 0$. If $[H_0', \psi]_- = 0$ and $\psi^2 = 0$ then the effective Hamiltonian H_{BRST} in the path integral (19) satisfies $\{H_{BRST}, \psi\} = 0$ which implies that ψ generates a new nilpotent symmetry. That this may be realized is shown in the following example.

EXAMPLE: QED

Consider for simplicity the Lagrangian density for a free electromagnetic field
(the metric is time-like)

$$\mathcal{L} = -\frac{1}{4}F^{\mu\nu}F_{\mu\nu}, \quad F_{\mu\nu} \equiv \partial_\mu A_\nu - \partial_\nu A_\mu. \tag{21}$$

The canonical momenta to A_μ are

$$E^\mu = \frac{\partial \mathcal{L}}{\partial \dot{A}_\mu} = F^{\mu 0}. \tag{22}$$

$E^0 = 0$ is a primary constraint. The Hamiltonian density, which is equal to
the canonical energy density T^{00}, is given by

$$\mathcal{H} = -\frac{1}{2}E^i E_i + E^i \partial_i A^0 + \frac{1}{4}F^{ij}F_{ij}. \tag{23}$$

The Hamiltonian equations of motion are generated by

$$H \equiv \int d^3x \left(\mathcal{H}(x) + \dot{A}^0 E^0 \right), \tag{24}$$

where $\dot{A}^0$ is an arbitrary function which represents the gauge freedom. (The
Lorentz condition $\partial_\mu A^\mu = 0$ demands $\dot{A}^0 = -\partial_i A^i$.) Since

$$\dot{E}^0(x) = \{E^0(x), H\} = \partial_i E^i(x) \tag{25}$$

consistency requires the secondary constraint (Gauss' law) $\partial_i E^i = 0$.

The standard Faddeev-Popov Lagrangian for QED is

$$\mathcal{L} = -\frac{1}{4}F^{\mu\nu}F_{\mu\nu} - \frac{1}{2\alpha}(\partial_\mu A^\mu)^2 - i\partial_\mu \bar{C}\partial^\mu C, \tag{26}$$

where α is a real parameter. The corresponding Hamiltonian within the BFV
scheme is given by (20) where the standard choice is

$$\mathcal{H}_0 = -\frac{1}{2}E^i E_i + \frac{1}{4}F^{ij}F_{ij}, \quad Q = \int d^3x(C\partial_i E^i - \bar{P}E^0). \tag{27}$$

(We have a minus sign in Q since $\pi_v = -E^0$.) The gauge fixing fermion is

$$\psi = \int d^3x(\mathcal{P}v + \bar{C}\chi), \tag{28}$$

where $v = -A^0$ and $\chi = \partial_i A^i + E^0/2\alpha$. However, this ψ is neither conserved
nor nilpotent, and it does not satisfy the conditions (14).

Now there is another option for $\mathcal{H}_0$, namely

$$\mathcal{H}_0 = -\frac{1}{2}E^i E_i + \frac{1}{2\nabla^2}(\partial_i E^i)^2 + \frac{1}{4}F^{ij}F_{ij}, \quad \nabla^2 \equiv -\partial_i\partial^i. \qquad (29)$$

This choice determines the Lagrange multiplier v to be

$$v \equiv -A^0 - \frac{1}{2\nabla^2}\partial_i E^i. \qquad (30)$$

Exactly the same effective Hamiltonian as before is now obtained by the formula (20) with $\mathcal{H}_0$ given by (29) and ψ given by (28) now with v as in (30) and $\chi = \partial_i A^i + E^0/2\alpha$. In distinction to the previous construction $\mathcal{H}_0$ in (29) satisfies the strong condition $\{\psi, \mathcal{H}_0\} = 0$. For $\alpha = 1$ (the Feynman gauge) ψ is furthermore nilpotent and may be identified with a coBRST charge. In this case we have $\{\psi, H_{BRST}\} = 0$ and ψ generates a symmetry transformation. It is

$$rA^0 = \frac{1}{2}\bar{C}, \quad rA^i = -\frac{1}{2\nabla^2}\partial^i\dot{\bar{C}},$$

$$rC = \frac{1}{2}i(A^0 - \frac{1}{\nabla^2}\partial_i\dot{A}^i), \quad r\bar{C} = 0. \qquad (31)$$

(Like the BRST transformation it is only nilpotent on-shell.) Of course, also the bosonic charge, $i\{Q, \psi\}$, is conserved and generates a symmetry transformation. The symmetry transformation (31) was also given in [10]. (I am thankful to Joaquim Gomis for pointing out this reference to me.)

The corresponding construction for Yang-Mills theories is more difficult. The standard BRST fixed Lagrangian does not allow for a conserved coBRST charge due to the Gribov ambiguities [11]. (χ contains the Coulomb gauge $\partial_i A_a^i$.)

References

[1] Marnelius R., *Time evolution in general gauge theories on inner product spaces*, Nucl. Phys. **B** (in press)

[2] Batalin I. A. and Vilkovisky G. A, Phys. Lett. **B69** (1977) 309.
    Fradkin E. S. and Fradkina T. E., Phys. Lett. **B72** (1978) 343.
    Batalin I. A. and Fradkin E. S., Phys. Lett. **B122** (1983) 157.

[3] Batalin I. A. and Fradkin E. S., Riv. Nuovo Cim. **9** (1986) 1.

[4] Marnelius R., Nucl. Phys. **B395** (1993) 647.
    Nucl. Phys. **B412** (1994) 817.

[5] Marnelius R., Phys. Lett. **B318** (1993) 92.

[6] Batalin I. and Marnelius R., Nucl. Phys. **B442** (1995) 669.

[7] Fülöp G. and Marnelius R., Nucl. Phys. **B456** (1995) 442.

[8] Marnelius R., Nucl. Phys. **B418** (1994) 353.

[9] Nishijima K., Nucl. Phys. **B238** (1984) 601.
    Spiegelglas M., Nucl. Phys. **B283** (1987) 205.
    Kalau W., van Holten J.W., Nucl. Phys. **B361** (1991) 233.

[10] Lavelle M. and McMullan D., Phys.Rev.Lett. **71** (1993) 3758.

[11] Gribov V. N., Nucl. Phys. **B139** (1978) 1.

CHAPTER III

Results in 3 + 1 Dimensions

Developing Transport Theory to Study the Chiral Phase Transition

S.P. Klevansky, A. Ogura, P. Rehberg and J. Hüfner

*Institut für Theoretische Physik, Philosophenweg 19,
D-69120 Heidelberg, Germany*

1. INTRODUCTION

One of the fundamental questions that is open to both theory and experiment is how to observe the chiral phase transition. Indications that a phase transition from a chirally symmetric to a chirally broken phase have been obtained a long time ago: lattice gauge simulations of quantum chromodynamics (QCD), which are equilibrium calculations that can include only temperature but not finite density, show a phase transition at a finite critical temperature T_c [1]. Unfortunately, due to the difficulties inherent in a confining theory such as QCD, it is impossible to observe this transition directly experimentally, let alone obtain experimental values for the usual quantities that are associated with critical phenomena, such as critical exponents.

The experimental search for clues to a phase transition that has been conceived, is to proceed via heavy ion collisions, in which both high temperatures and densities are to be reached. Much effort has been expended in this direction and will continue to be so during the early course of the next century, particularly at RHIC and CERN. However, from the theoretical side, there is at present no single conclusive observable that might indicate the occurrence of the chiral phase transition. In addition, the simple equilibrium picture of the static lattice simulations is inadequate for describing heavy ion collisions. Thus we may speculate as to whether a concrete signal of this phase transition can be obtained from a non-equilibrium formulation of the problem or transport theory, that is based on an underlying chiral Lagrangian. Here one may surmise whether phenomena such as critical scattering, i.e. divergence of the scattering cross-sections at the phase transition, may leave an obvious signal, or simply the fact that the dynamically generated quark mass is a function of space and time may leave some visible remnant of a phase transition.

Thus our goal is to construct a consistent transport theory that is based on the Nambu–Jona-Lasinio (NJL) model, which to order $1/N_c$ also includes the mesonic degrees of freedom and the hadronization of quarks into these degrees of freedom. In this report, we discuss primarily the first term in the $1/N_c$ expansion in the collision, explicitly relating this to Feynman diagrams for quark-quark and quark-antiquark scattering (for details, see Ref.[2]), leaving the mesonic sector to a future publication [3].

2.   GENERAL TRANSPORT THEORY FOR FERMIONS.

Non-equilibrium phenomena are completely described by the Schwinger - Keldysh formalism for Green functions [4]. Many conventions exist in the literature. For our purposes, it is simplest to use the convention of Landau [5]. In this, the designations $+$ and $-$ attributed to the closed time path that is shown in Fig.1, and the fermionic Green functions are defined as

$$
\begin{aligned}
iS^c(x,y) &= \langle T\psi(x)\bar{\psi}(y)\rangle = iS^{--}(x,y) \\
iS^a(x,y) &= \langle \tilde{T}\psi(x)\bar{\psi}(y)\rangle = iS^{++}(x,y) \\
iS^>(x,y) &= \langle \psi(x)\bar{\psi}(y)\rangle = iS^{+-}(x,y) \\
iS^<(x,y) &= -\langle \bar{\psi}(y)\psi(x)\rangle = iS^{-+}(x,y).
\end{aligned}
\tag{1}
$$

In a standard fashion, one constructs equations of motion for the matrix of Green functions and one moves to relative and centre of mass variables, $u = x - y$ and $X = (x + y)/2$. A Fourier transform with respect to the relative coordinate, or Wigner transform,

$$
S(X,p) = \int d^4 u\, e^{ip \cdot u} S\left(X + \frac{u}{2}, X - \frac{u}{2}\right)
\tag{2}
$$

is then performed. Of particular interest is the equation of motion for $S^{-+} = S^<$. Regarding this, together with the equation of motion for the adjoint function $S^{<\dagger}$, one arrives at the so-called transport and constraint equations by adding and subtracting these. One finds

$$
\frac{i\hbar}{2}\{\gamma^\mu, \frac{\partial S^{-+}}{\partial X^\mu}\} + [\not{p}, S^{-+}(X,p)] = I_-
\tag{3}
$$

and

$$
\frac{i\hbar}{2}[\gamma^\mu, \frac{\partial S^{-+}}{\partial X^\mu}] + \{\not{p} - m_0, S^{-+}(X,p)\} = I_+,
\tag{4}
$$

respectively. Here

$$I_{\mp} = I_{\text{coll}} + I_{\mp}^A + I_{\mp}^R, \tag{5}$$

where the collision term is

$$
\begin{aligned}
I_{\text{coll}} &= \Sigma^{-+}(X,p)\hat{\Lambda}S^{+-}(X,p) - \Sigma^{+-}(X,p)\hat{\Lambda}S^{-+}(X,p) \\
&= I_{\text{coll}}^{\text{gain}} - I_{\text{coll}}^{\text{loss}},
\end{aligned} \tag{6}
$$

and terms containing retarded (R) and advanced (A) components are contained in

$$I_{\mp}^R = -\Sigma^{-+}(X,p)\hat{\Lambda}S^R(X,p) \pm S^R(X,p)\hat{\Lambda}\Sigma^{-+}(X,p) \tag{7}$$

and

$$I_{\mp}^A = \Sigma^A(X,p)\hat{\Lambda}S^{-+}(X,p) \mp S^{-+}(X,p)\hat{\Lambda}\Sigma^A(X,p). \tag{8}$$

In these equations, Λ is given as

$$\hat{\Lambda} = \exp\left(-\frac{i\hbar}{2}\left(\frac{\overleftarrow{\partial}}{\partial X^\mu}\frac{\overrightarrow{\partial}}{\partial p_\mu} - \frac{\overleftarrow{\partial}}{\partial p_\mu}\frac{\overrightarrow{\partial}}{\partial X^\mu}\right)\right). \tag{9}$$

Note that the equations (3) and (4) are general equations that are generic for *any* relativistic fermionic theory [6].

3. APPROXIMATION SCHEMES: APPLICATION TO THE NJL MODEL

A double expansion in inverse powers of the number of colors $1/N_c$ and $\hbar$ is performed. We explain this briefly in the context of the model.

The simplest form of the NJL Lagrangian

$$\mathcal{L} = \overline{\psi}\left(i\hbar\,\slashed{\partial} - m_0\right)\psi + G\left[\left(\overline{\psi}\psi\right)^2 + \left(\overline{\psi}i\gamma^5\psi\right)^2\right], \tag{10}$$

is considered. Here G is a coupling strength and m_0 is the current quark mass. The color degree of freedom is implicit. As this is a strong coupling theory, an expansion in G is inadmissable, and the usual technique involves an expansion in $1/N_c$. In this ordering, the lowest order Feynman graph is a Hartree term, which describes the mean field experienced by the quarks, while higher orders introduce collisions. We examine these two cases individually. In order to make connection with the classically known transport equations, we consider an expansion of the transport and constraint equations in powers of $\hbar$.

In the transport and constraint equations, Eqs.(3) and (4), the right hand sides I_- and I_+, respectively depend on $\hbar$ in two ways: (i) The derivative operator $\hat{\Lambda}$, Eq.(9), contains $\hbar$ to all orders. Its expansion in $\hbar$ generates the various terms of the so-called gradient expansion, and (ii) the self-energies Σ^R and Σ^{-+} may have overall factors in $\hbar$. While to order $O((1/N_c)^0)$, $\Sigma^R = m$, and is directly related to the mass, which we treat as a quantity that has a direct classical interpretation, we find $\Sigma^{-+} \propto \hbar d\sigma/d\Omega$, and is only related via a factor $\hbar$ to a quantity with direct physical interpretation. Note that while we use the overall factors in $\hbar^0$ or $\hbar^1$ in Σ^R or Σ^{-+} in our $\hbar$ classification, we do not expand the quantities m or $d\sigma/d\Omega$ in powers of $\hbar$. We will detail this strategy in the calculation of the mean field term and collision integrals.

3.1.   The Hartree Approximation

The leading term in the $1/N_c$ expansion corresponds to the Hartree approximation. The Hartree self-energy can be evaluated, given a form for the Green function. Here we assume that

$$
S_H^{-+}(X,p) = 2\pi i \frac{1}{2E_p}[\delta(p_0 - E_p)\sum_s u_s(p)\bar{u}_s(p)f_q(X,p)
$$
$$
+\delta(p_0 + E_p)\sum_s v_s(-p)\bar{v}_s(-p)\bar{f}_{\bar{q}}(X,-p) \qquad (11)
$$

with $E_p^2 = p^2 + m^2(X)$ and $f_{q,\bar{q}}$ the (spin independent) quark and antiquark distribution functions. This is the quasiparticle assumption. Similar expressions can be written down for the other Green functions of Eq.(1). Inserting the quasiparticle ansäzte into the Wigner transformed expression for the self-energy in the Hartree approximation, $-i\Sigma_H^{--} = 2i\hbar tri S_H^{--}(x,x)$ leads to the non-equilibrium gap equation

$$
m(X) = m_0 + 4GN_c m(X) \int \frac{d^3p}{(2\pi\hbar)^3} \frac{1}{E_p(X)}[1 - f_q(X,p) - f_{\bar{q}}(X,p)]. \qquad (12)
$$

Some manipulations are now necessary for the evaluation of (3) and (4). The evaluation of $I_\pm$ is made according to the $\hbar$ expansion. Then, on performing a spinor trace and integrating over positive or negative energies, leads one to the Vlasov equation

$$
p^\mu \partial_\mu f_{q,\bar{q}}(X,\vec{p}) + m(X)\partial_\mu m(X)\partial_p^\mu f_{q,\bar{q}}(X,\vec{p}) = 0, \qquad (13)
$$

where $p^0 = E_p(X)$, and the constraint equation

$$
(p^2 - m^2(X))f_{q,\bar{q}} = 0. \qquad (14)
$$

to leading order in $\hbar$. Note that the constraint equation validates the quasiparticle ansatz, and that the equations Eq.(13) and (14) can be solved self-consistently.

3.2.   The Collision Term

There are several graphs which contribute to the next order in the $1/N_c$ expansion. One of these is displayed in Fig. 1(a). The accompanying graph of Fig. 1(b) is of higher order in $1/N_c$, and is not required on this basis, but, as

$$-i\Sigma^{+-} =$$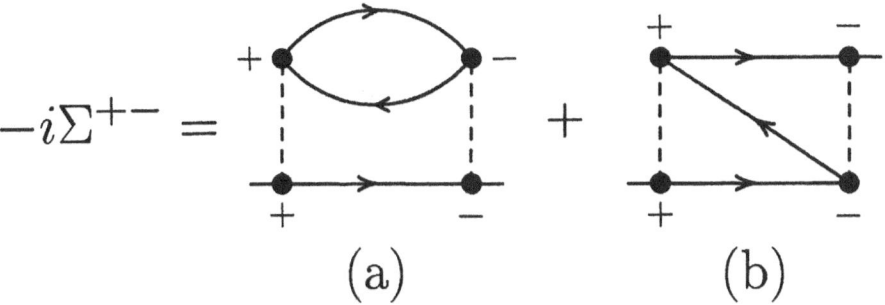

Fig. 1. — Contributions to the self-energy that are of higher order in $1/N_c$ than the Hartree term

it turns out, is essential for a complete and correct identification of Feynman amplitudes in casting the collision term into a Boltzmann-like form. The initial task is the evaluation of I_{coll} of Eq.(6), with scalar or pseudoscalar vertices. There are three possible combinations, Σ_σ, Σ_π and Σ_{coll}, which contain unity and $i\gamma_5$ throughout or unity or $i\gamma_5$ in a mixed fashion. For example, the scalar self-energy is determined to be

$$
\begin{aligned}
\Sigma_\sigma^{+-}(X,p) &= -4G^2\hbar^2 \int \frac{d^4p_1}{(2\pi\hbar)^4}\frac{d^4p_2}{(2\pi\hbar)^4}\frac{d^4p_3}{(2\pi\hbar)^4}(2\pi\hbar)^4\delta(p-p_1+p_2-p_3) \\
&\times \; [S^{+-}(X,p_1)\mathrm{tr}\left(S^{-+}(X,p_2)S^{+-}(X,p_3)\right) \\
&\quad -S^{+-}(X,p_1)S^{-+}(X,p_2)S^{+-}(X,p_3)],
\end{aligned}
\tag{15}
$$

while the other self-energies are similar [2].

As before, performing a spinor trace and integrating over a positive or negative energy range is required, so that the loss term, for example, is correspondingly given as

$$
J_{coll,\sigma}^{q,loss} = \frac{2\pi i}{2E_p}\mathrm{tr}\left[\Sigma_\sigma^{+-}(X,p_0=E_p,\vec{p})\sum_s u_s(p)\bar{u}_s(p)f_q(X,\vec{p})\right],
\tag{16}
$$

which, when evaluated using the quasiparticle approximations for the Green functions, leads to eight terms. After some algebra, one can show that only terms that represent elastic scattering survive energy integration. Other terms that contain vacuum fluctuations, as well as particle production and annihilation are present at this level, see Fig. 2, but are suppressed by the quasiparticle approximation. An off-shell ansatz is required for them not to vanish. It is also important to note that both diagrams of Fig. 1 are essential for the identification of the Feynman amplitudes for elastic scattering in the s, t and u channels, as appropriate. While graph (a) leads to the square of the individual amplitudes, graph (b) is necessary for the mixed terms that allow one to construct

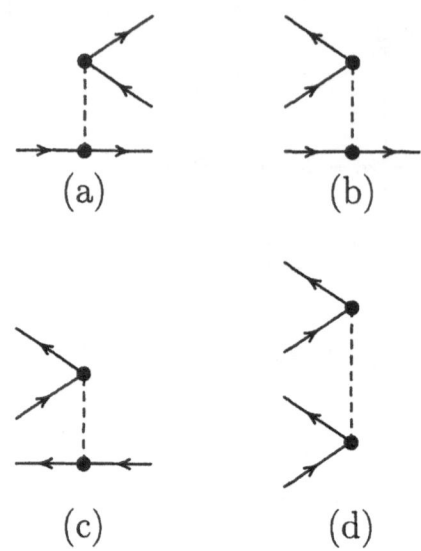

Fig. 2. — Processes of the form $q \to (q\bar{q})q$ and $(q\bar{q}) \to q$, as well as vacuum fluctuations that are suppressed in the quasiparticle approximation. These are heuristic graphs and are not Feynman graphs

a matrix element squared that involves several channels. The net result is

$$p^\mu \partial_\mu f_q(X, \vec{p}) + m(X) \partial_\mu m(X) \partial_p^\mu f_q(X, \vec{p}) =$$

$$N_c \int d\Omega \int \frac{d^3 p_2}{(2\pi\hbar)^3 2E_{p_2}} |\vec{v}_p - \vec{v}_2| 2E_p 2E_{p_2}$$

$$\times \quad \{\frac{1}{2} \frac{d\sigma}{d\Omega}|_{qq \to qq}(p2 \to 13)(f_q(p_1)\bar{f}_q(p_2)f_q(p_3)\bar{f}_q(p) - \bar{f}(p_1)f_q(p_2)\bar{f}_q(p_3)f_q(p))$$

$$+ \frac{d\sigma}{d\Omega}|_{q\bar{q} \to q\bar{q}}(p2 \to 13)(f_q(p_1)\bar{f}_{\bar{q}}(p_2)f_{\bar{q}}(p_3)\bar{f}_q(p) - \bar{f}_q(p_1)\bar{f}_{\bar{q}}(p_3)f_q(p_2)f_q(p))\},$$

$$(17)$$

where the scattering cross-sections contain s, t and u channel graphs as appropriate, in the Born approximation and which is the relativistic generalization of the result of Kadanoff and Baym. To lowest order in $\hbar$, arguments can be given to suggest that the constraint equation remains unaltered.

4.   DISCUSSION

The most important facets of this calculation are (i) in identifying precise Feynman amplitudes that contribute to the scattering processes in the collision term, and which represents the relativistic generalization of the Kadanoff-Baym work [7], and (ii) the recognition of additional processes that would be present should a quasiparticle assumption be relaxed, such as vacuum fluctuations and particle production and annihilation. The double expansion in $\hbar$ and $1/N_c$ lays the foundation for including additional higher order terms in the $1/N_c$ expansion, from which mesons are constructed, and which enables one to build a dynamical model for both mesons and quarks. For this, we refer the interested reader to [3, 8].

We have started numerical simulations of this chiral transport theory. A first study that solves the Vlasov equation by means of a finite difference technique and which inserts collision terms in the relaxation time approximation for three flavors is presented in Ref. [9]. In this, one finds that the collisions determine the high energy parts of the particle distribution, which is exponential. The mean field, on the other hand, produces an enhancement over the exponential fits at low energies. The behavior of the quark mass as a function of time is such that a dip occurs in the initial state, and this gradually flattens as the quarks flow out.

Acknowledgments

This work has been supported in part by the Deutsche Forschungsgemeinschaft DFG under the contract number Hu 233/4-4, and by the German Ministry for Education and Technology under contract number 06 HD 742.

References

[1] See for example, Laermann E., *Nucl. Phys.* **A610** (1996) 1.
[2] Klevansky S.P., Ogura A., and Hüfner J., Heidelberg Preprint HD-TVP-9702.
[3] Klevansky S.P., Rehberg P., and Hüfner J., in preparation.
[4] Schwinger J., *J. Math. Phys.* **2** (1961) 407. Keldysh L.V., *JETP* **20** (1965) 1018.
[5] Lifschitz E.M. and Pitajewski L.P., Physikalische Kinetik (Akademie Verlag, Berlin, 1986) p. 423.
[6] See for example, Mrówczyński S. and Heinz U., *Ann. Phys. (N.Y.)* **229** (1994) 1.
[7] Kadanoff L.P. and Baym G., Quantum Statistical Mechanics (Benjamin / Cummings, Massachusetts, 1962) p. 110.
[8] Klevansky S.P., Rehberg P., Ogura A., and Hüfner J., QCD Phase Transitions (GSI, Darmstadt, 1997) p. 397.
[9] Rehberg P. and Hüfner J., Heidelberg Preprint HD-TVP-97/03

On Deriving the Effective Interaction from the QCD Lagrangian

H. C. Pauli

Max-Planck-Institut für Kernphysik
D-69029 Heidelberg

The canonical front form Hamiltonian for non-Abelian SU(N) gauge theory in 3+1 dimensions is mapped non-perturbatively on an effective Hamiltonian which acts only in the Fock space of a quark and an antiquark. The approach is based on the novel method of iterated resolvents and on discretized light-cone quantization, driven to the continuum limit. It is free of the usual Tamm-Dancoff truncations of the Fock space, rather the perturbative series are consistently resumed to all orders in the coupling constant. Emphasis is put on dealing with the many-body aspects of gauge field theory. Important is that the higher Fock-space amplitudes can be retrieved self-consistently from these solutions.

1. THE STRUCTURE OF THE HAMILTONIAN

In canonical field theory the four components of the energy-momentum vector P^ν commute and are constants of the motion. In the front form of Hamiltonian dynamics [1] they are denoted [2] by $P^\nu = (P^+, \vec{P}_\perp, P^-)$. Its spatial components $\vec{P}_\perp$ and P^+ are independent of the interaction and diagonal in momentum representation. Their eigenvalues are the sums of the single particle momenta, $P^+ = \sum p^+$ and $\vec{P}_\perp = \sum \vec{p}_\perp$. Each single particle has four-momentum $p^\mu = (p^+, \vec{p}_\perp, p^-)$ and sits on its mass-shell $p^\mu p_\mu = m^2$. Each particle state "q" is then characterized by six quantum numbers $q = (p^+, \vec{p}_\perp, \lambda; c, f)$: the three spatial momenta, helicity, color and flavor. The temporal component

H.C. Pauli

n	Sector	1 q q̄	2 gg	3 q q̄ g	4 q q̄ q q̄	5 gg g	6 q q̄ gg	7 q q̄ q q̄ g	8 q q̄ q q̄ q q̄	9 gg gg	10 q q̄ gg g	11 q q̄ q q̄ gg	12 q q̄ q q̄ q q̄ g	13 q q̄ q q̄ q q̄ q q̄
1	q q̄	▨	·	◁	·	·	·	·	·	·	·	·	·	·
2	gg	·	▨	◁	·	◁	·	·	·	·	·	·	·	·
3	q q̄ g	◁	◁	▨	◁	·	◁	·	·	·	·	·	·	·
4	q q̄ q q̄	·	·	◁	▨	·	·	◁	·	·	·	·	·	·
5	gg g	·	◁	·	·	▨	·	·	·	◁	·	·	·	·
6	q q̄ gg	·	·	◁	·	◁	▨	◁	·	·	◁	·	·	·
7	q q̄ q q̄ g	·	·	·	◁	·	◁	▨	◁	·	·	◁	·	·
8	q q̄ q q̄ q q̄	·	·	·	·	·	·	◁	▨	·	·	·	◁	·
9	gg gg	·	·	·	·	◁	·	·	·	▨	◁	·	·	·
10	q q̄ gg g	·	·	·	·	·	◁	·	·	◁	▨	◁	·	·
11	q q̄ q q̄ gg	·	·	·	·	·	·	◁	·	·	◁	▨	◁	·
12	q q̄ q q̄ q q̄ g	·	·	·	·	·	·	·	◁	·	·	◁	▨	◁
13	q q̄ q q̄ q q̄ q q̄	·	·	·	·	·	·	·	·	·	·	·	◁	▨

Fig. 1. — The Hamiltonian matrix for a SU(N)-meson and a harmonic resolution $K = 4$. Only vertex diagrams are included. Zero matrices are marked by a dot ($\cdot$).

$P^- = 2P_+$ depends on the interaction and is a complicated non-diagonal operator [4, 6]. It propagates the system in the light-cone time $x^+ = x^0 + x^3$. The contraction of P^μ is the operator of invariant mass-squared,

$$P^\mu P_\mu = P^+ P^- - \vec{P}_\perp^2 \qquad \equiv H_{\rm LC} \equiv H \ . \tag{1}$$

It is Lorentz invariant and refered to somewhat improperly but conveniently as the 'light-cone Hamiltonian' $H_{\rm LC}$, or shortly H. One seeks a representation in which H is diagonal $H|\Psi\rangle = E|\Psi\rangle$). Introductory texts [4, 5, 6] are available.

The (light-cone) Hamiltonian is split for convenience into four terms: $H = T + V + F + S$. The kinetic energy T is diagonal in Fock-space representation, its eigenvalue is the free invariant mass squared of the particular Fock state. The vertex interaction V is the relativistic interaction *per se*. It has no diagonal matrix elements, is linear in g and changes the particle number by 1. The instantaneous interactions F and S are consequences of working in the light-cone gauge $A^+ = 0$. They are proportional to g^2. By definition, the seagull interaction S conserves the particle number and the fork interaction F changes it by 2. As a consequence of working in light-cone representation and in the light cone gauge, the vacuum state has no matrix elemens with any of the Fock states: The vacuum *does not fluctuate*.

The eigenvalue equation $H|\Psi\rangle = E|\Psi\rangle$ stands for an infinite set of coupled integral equations [6]. The eigenfunctions $|\Psi\rangle$ are superpositions of all Fock-space projections like $\langle q\bar{q}|\Psi\rangle$ or $\langle q\bar{q}\,g|\Psi\rangle$. However, if one works with discretized light-cone quantization (DLCQ [3]), one deals with a *finite set* of coupled matrix equations. As consequence of periodic boundary conditions the single particle momenta are discrete and the Fock states are denumerable and orthonormal. The Hamiltonian matrix is illustrated in Fig. 1 for the *harmonic resolution* $K \equiv P^+L/(2\pi) = 4$. Its rows and columns are the Fock states which are grouped into the Fock-space sectors $q\bar{q}$, $q\bar{q}\,g,\ldots$, denumerated for simplicity by $n = 1, 2, \ldots, 13$. Within each sector one has many Fock states, which differ by the helicities, colors, and momenta of the single particles, subject to fixed total momenta P^+ and $\vec{P}_\perp$. Since P^+ has only positive eigenvalues and since each particle has a lowest possible value of p^+, the number of particles in a Fock state is limited for any fixed value of K: The number of Fock space sectors is *finite*. The transversal momenta $\vec{p}_\perp$ can take either sign. Their number must be regulated by some convenient cutoff, (Fock-space regularization, *f.e.* see [2]). The most remarkable property of the Hamiltonian matrix is its block structure and its sparseness. Most of the blocks are zero matrices, marked by a dot ($\cdot$) in the figure. Only those blocks in Fig. 1 denoted by the graphical symbol of a vertex are potentially non-zero, with the actual matrix elements tabulated elsewhere [4, 6].

Here then is the problem, the bottle neck of any Hamiltonian approach in field theory: The dimension of the Hamiltonian matrix increases exponentially fast. Suppose, the regularization procedure allows for 10 discrete momentum states in each direction. A single particle has then about 10^3 degrees of freedom. A Fock-space sector with n particles has then roughly 10^{n-1} different Fock states. Sector 13 with its 8 particles has thus about 10^{21} and sector 1 ($q\bar{q}$) about 10^3 Fock states. What one wishes instead of is to derive an effective interaction which acts only in the comparatively small $q\bar{q}$-space, like a Coulomb interaction acts only between an electron and a positron. Loosely speaking, the aim of deriving an effective interaction can be understood as reducing the dimension in a matrix diagonalization problem from 10^{21} to 10^3!

The first attempt to formulate an effective interaction in field theory is the approach of Tamm and Dancoff (TDA [7, 8]). The present 'method of iterated resolvents' [9, 10] is closely related to TDA. Alternatively, one can apply a sequence of analytical and approximate unitary transformations in order to render the Hamiltonian matrix more and more 'band diagonal'. Two methods have been proposed recently, the 'similarity transform' of Glazek and Wilson, see [11], and the 'Hamiltonian flow equations' of Wegner [12]. All of these methods are under active research. Applications particularly by Trittmann *et al* [13, 14, 15], Brisudova *et al* [16, 17], Jones *et al* [18, 19], and Gubankova *et al* [20] are presented elsewhere in these proceedings. Common to most of them is that one truncates the Fock space before approximation methods are applied. No truncation of the Fock space is needed with the method of iterated resolvents, as to be shown below.

2.   THE METHOD OF ITERATED RESOLVENTS

The method of effective interactions is a well known tool in many-body physics which in field theory is known as the Tamm-Dancoff-approach [7, 8]. The method is ideally suited for a field theory, because Fock-space sectors $|i\rangle$ appear in the most natural way. The Hamiltonian matrix can be understood as a matrix of block matrices, whose rows and columns are enumerated by $i = 1, 2, \ldots N$ like in Fig. 1. The eigenvalue equation can be written as a coupled set of block matrix equations:

$$\sum_{j=1}^{N} \langle i|H|j\rangle\, \langle i|\Psi\rangle = E\, \langle n|\Psi\rangle \qquad \text{for all } i = 1, 2, \ldots, N \;. \tag{2}$$

In TDA, the rows and columns are split into a P-space, $P = |1\rangle\langle 1|$, and the rest, the Q-space. Explicitly written out, the eigenvalue equation (2) becomes a set of two coupled block matrix equations:

$$\langle P|H|P\rangle\, \langle P|\Psi\rangle + \langle P|H|Q\rangle\, \langle Q|\Psi\rangle \;=\; E\, \langle P|\Psi\rangle \;, \tag{3}$$
$$\text{and} \quad \langle Q|H|P\rangle\, \langle P|\Psi\rangle + \langle Q|H|Q\rangle\, \langle Q|\Psi\rangle \;=\; E\, \langle Q|\Psi\rangle \;. \tag{4}$$

Rewrite the second equation as $\langle Q|E - H|Q\rangle\, \langle Q|\Psi\rangle = \langle Q|H|P\rangle\, \langle P|\Psi\rangle$, and observe that the quadratic matrix $\langle Q|E - H|Q\rangle$ could be inverted to express the Q-space wave-function $\langle Q|\Psi\rangle$ in terms of $\langle P|\Psi\rangle$. But the eigenvalue E is unknown at this point. One introduces therefore a redundant parameter ω, and defines $G_Q(\omega) = [\langle Q|\omega - H|Q\rangle]^{-1}$. The so obtained effective interaction

$$\langle P|H_{\text{eff}}(\omega)|P\rangle = \langle P|H|P\rangle + \langle P|H|Q\rangle\, G_Q(\omega)\, \langle Q|H|P\rangle \;. \tag{5}$$

acts only in the much smaller P-space: $H_{\text{eff}}(\omega)|\Phi_k(\omega)\rangle = E_k(\omega)\,|\Phi_k(\omega)\rangle$. Varying ω one generates a set of energy functions $E_k(\omega)$. All solutions of the fixpoint equation $E_k(\omega) = \omega$ are eigenvalues of the full Hamiltonian H. In fact one can find all eigenvalues of H despite the fact that the dimension of H_{eff} is usually much smaller than that of the full H [9, 10]. The procedure is formal, since the inversion of the Q-space matrix is as complicated as its diagonalization. The adventage is that resolvents can be approximated systematically: The two resolvents

$$G_Q(\omega) = \frac{1}{\langle Q|\omega - T - U|Q\rangle} \quad \text{and} \quad G_Q^{(0)}(\omega) = \frac{1}{\langle Q|\omega - T|Q\rangle} \;, \tag{6}$$

defined once with and once without the non-diagonal interaction U, are identically related by $G_Q(\omega) = G_Q^{(0)}(\omega) + G_Q^{(0)}(\omega)\, U\, G_Q(\omega)$, or by the infinite series of perturbation theory. Albeit exact in principle, the Tamm-Dancoff-approach (TDA) suffers from a practical aspect: The approach is useful only if one truncates the perturbative series to its very first term. This destroys Lorentz and gauge invariance, and requires a sufficiently small coupling constant.

Truncation can be avoided by 'going backwards'. Reinterpreting the Q-space as the last sector N (sector 13 in Fig. 1) and the P-space with the rest, the above equations can be interpreted as to reduce the block matrix dimension from N to $N-1$, with an effective interaction acting now in the smaller space of $N-1$ blocks. This procedure can be iterated, from $N-1$ to $N-2$, and so on, until one arrives at block matrix dimension 1 where the procedure stops: The effective interaction in the Fock-space sector with only one quark and one antiquark is defined unambiguously. One has to deal then with 'resolvents of resolvents', or with iterated resolvents [10].

To be more explicit suppose one has arrived in the course of this reduction at block matrix dimension n, with $1 \le n \le N$. Denote the corresponding effective interaction $H_n(\omega)$. The eigenvalue problem reads then

$$\sum_{j=1}^{n} \langle i|H_n(\omega)|j\rangle \langle j|\Psi(\omega)\rangle = E(\omega)\, \langle i|\Psi(\omega)\rangle\,, \quad \text{for } i = 1, 2, \ldots, n\,. \quad (7)$$

Observe that i and j refer here to sector numbers, and that n refers to both, the last sector number, and the number of sectors. Now, like in the above, define the resolvent as the inverse of the Hamiltonian in the last sector

$$G_n(\omega) = \frac{1}{\langle n|\omega - H_n(\omega)|n\rangle} \quad (8)$$

$$\text{thus} \quad \langle n|\Psi(\omega)\rangle = G_n(\omega) \sum_{j=1}^{n-1} \langle n|H_n(\omega)|j\rangle\, \langle j|\Psi(\omega)\rangle\,. \quad (9)$$

The effective interaction in the $(n-1)$-space becomes then

$$H_{n-1}(\omega) = H_n(\omega) + H_n(\omega)G_n(\omega)H_n(\omega) \quad (10)$$

for every block matrix element $\langle i|H_{n-1}(\omega)|j\rangle$. Everything proceeds like in above, including the fixed point equation $E(\omega) = \omega$. But one has achieved much more: Eq.(10) is a *recursion relation* which holds for all $1 < n < N$! Since one has started from the bare Hamiltonian in the last sector, one has to convene that $H_N = H$. The rest is algebra and interpretation.

Applying the method to the block matrix structure of QCD, as displayed in Fig. 1, is particularly easy and transparent. By definition, the last sector contains only the diagonal kinetic energy, thus $H_{13} = T_{13}$. Its resolvent is calculated trivially. Then H_{12} can be constructed unambiguously, followed by H_{11}, and so on, until one arrives at sector 1. Grouping the so obtained results in a different order, one finds for the sectors with one $q\bar{q}$-pair:

$$H_{q\bar{q}} = \quad H_1 = T_1 + VG_3V + VG_3VG_2VG_3V\,, \quad (11)$$

$$H_{q\bar{q}\,g} = \quad H_3 = T_3 + VG_6V + VG_6VG_5VG_6V + VG_4V\,, \quad (12)$$

$$H_{q\bar{q}\,gg} = \quad H_6 = T_6 + VG_{10}V + VG_{10}VG_9VG_{10}V + VG_7V\,. \quad (13)$$

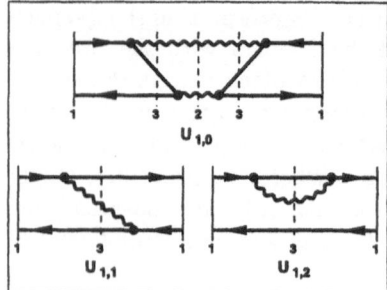

Fig. 2. — The three graphs of the effective interaction in the $q\bar{q}$-space. — The lower two graphs correspond to the chain $U = VG_3V$, the upper corresponds to $U_a = VG_3VG_2VG_3V$. Propagator boxes are represented by vertical dashed lines, with the subscript 'n' referring to the sector numbers.

The quark-gluon content of the sectors is added here for easier identification, in line with the notation in Fig. 1. Correspondingly, one obtains for the sectors with two $q\bar{q}$-pairs

$$H_{q\bar{q}q\bar{q}} = \quad H_4 \quad = \quad T_4 + VG_7V + VG_7VG_6VG_7V \ , \tag{14}$$

$$H_{q\bar{q}q\bar{q}\,g} = \quad H_7 \quad = \quad T_7 + VG_{11}V + VG_{11}VG_{10}VG_{11}V + VG_8V \ . \tag{15}$$

In the pure gluon sectors, the structure is even simpler:

$$H_{gg} = \quad H_2 \quad = \quad T_2 + VG_3V + VG_5V \ , \tag{16}$$

$$H_{gg\,g} = \quad H_5 \quad = \quad T_5 + VG_6V + VG_9V \ . \tag{17}$$

Note that these relations are all exact.

One is left with the eigenvalue problem in the $q\bar{q}$-space,

$$\sum_{q',\bar{q}'} \langle q; \bar{q}|H_{q\bar{q}}(\omega)|q'; \bar{q}'\rangle\psi_b(\omega)\rangle = M_b^2(\omega)\langle q; \bar{q}|\psi_b(\omega)\rangle \ . \tag{18}$$

By construction, the eigenvalues are identical with those of the full Hamiltonian and are enumerated by $b = 1, 2, \ldots$. They are the invariant mass^2 of a physical particle and its intrinsic excitations. The corresponding eigenfunctions $\langle q; \bar{q}|\psi_b(\omega)\rangle$ represent the normalized projections of $|\Psi\rangle$ onto the Fock states $|q; \bar{q}\rangle = b_q^\dagger d_{\bar{q}}^\dagger|vac\rangle$. The effective Hamiltonian as given by Eq.(11) has two types of interactions which are illustrated diagrammatically in Fig. 2. In the first of them, in VG_3V, the bare vertex interaction scatters the system virtually into the $q\bar{q}\,g$-space, where it propagates under impact of the full interaction until a second vertex interaction scatters the system into the $q\bar{q}$-space. The gluon can be absorbed on the same quark line as it was emitted, which contributes to the effective quark mass as illustrated by diagram $U_{1,2}$ of Fig. 2. The gluon absorbed by the other quark as in diagram $U_{1,1}$ of Fig. 2 providing a quark-antiquark interaction which cannot change the quark flavor. It includes all fine and hyperfine interactions, see for instance [14]. The second term in Eq.(11), the annihilation interaction $U_a = VG_3VG_2VG_3V$, potentially provides an interaction between different flavors and is illustrated in diagram $U_{1,0}$

of Fig. 2. As a net result the interaction scatters a quark with helicity λ_q and four-momentum p into a state with λ_q' and p'.

The knowledge of ψ_b is sufficient to retrieve all desired Fock-space components of the total wave-function. The key is the upwards recursion relation Eq.(9). Obviously, one can express the higher Fock-space components $\langle n|\Psi\rangle$ as functionals of $\psi_{q\bar{q}}$ by a finite series of quadratures, i.e. of matrix multiplications or of momentum-space integrations. One need not solve an other eigenvalue problem. This is quickly shown by way of example, by calculating the probability amplitude for a $|gg\rangle$- or a $|q\bar{q}\,g\rangle$-state in an eigenstate of the full Hamiltonian. The first two equations of the recursive set in Eq.(9) are

$$\langle 2|\Psi\rangle \;=\; G_2\langle 2|H_2|1\rangle\langle 1|\Psi\rangle\,, \tag{19}$$

$$\text{and}\quad \langle 3|\Psi\rangle \;=\; G_3\langle 3|H_3|1\rangle\langle 1|\Psi\rangle + G_3\langle 3|H_3|2\rangle\langle 2|\Psi\rangle\,. \tag{20}$$

The sector Hamiltonians H_n have to be substituted from Eqs.(12) and (16). In taking block matrix elements of them, the formal expressions are simplified considerably since most of the Hamiltonian blocks in Fig. 1 are zero matrices. One thus gets simply $\langle 2|H_2|1\rangle = \langle 2|VG_3V|1\rangle$ and therefore $\langle 2|\Psi\rangle = G_2VG_3V\langle 1|\Psi\rangle$. Substituting this into Eq.(19) gives $\langle 3|\Psi\rangle = G_3V\langle 1|\Psi\rangle + G_3VG_2VG_3V\langle 1|\Psi\rangle$. These findings can be summarized more succinctly as

$$|\psi_{gg}\rangle \;=\; G_{gg}VG_{q\bar{q}\,g}V\,|\psi_{q\bar{q}}\rangle\,, \tag{21}$$

$$\text{and}\quad |\psi_{q\bar{q}\,g}\rangle \;=\; G_{q\bar{q}\,g}V\,|\psi_{q\bar{q}}\rangle + G_{q\bar{q}\,g}VG_{gg}VG_{q\bar{q}\,g}V\,|\psi_{q\bar{q}}\rangle\,. \tag{22}$$

The finite number of terms is in strong contrast to the infinite number of terms in perturbative series. Iterated resolvents sum the perturbative series to all orders in closed form.

3. DISCUSSION AND PERSPECTIVES

All of the above relations are exact and hold for an arbitrarily large K [10]. They hold thus also in the continuum limit, where the resolvents are replaced by propagators and the eigenvalue problems become integral equations. The effective interaction is very simple in direct consequence of the structure of the gauge field Hamiltonian with its many zero matrices. No particle cut-off is required, and no assumption is made on the size of the coupling constant. The appoach is strictly non-perturbative. Instead of the inverting a huge matrix as in the Tamm-Dancoff approach, one has to invert only the comparatively small sector Hamiltonians H_n. The instantaneous interactions can be included easily ex post by the rule that every intrinsic line in graph in Fig. 2 has to be suplemented with the corresponding instantaneous line [10]. Once the general structure of the effective interaction has been formulated, one can proceed with simplifying assumptions. For example, one can replace the full by the free propagators to get a selected set of perturbative diagrams. The major effort of the ongoing work [21] is to find an approximation scheme which combines rigor with simplicity. Unfortunately the limited space prevents giving more details.

References

[1] Dirac P.A.M., Rev.Mod.Phys. **21**, 392 (1949).

[2] Lepage G.P. and Brodsky S.J., Phys.Rev. **D22**, 2157 (1980).

[3] Pauli H.C. and Brodsky S.J., Phys. Rev. **D32**, 1993 (1985).

[4] Brodsky S.J. and Pauli H.C., in *Recent Aspects of Quantum Fields*, Mitter and H. Gausterer Eds., Lecture Notes in Physics, Vol 396, (Springer, Heidelberg, 1991); and references therein.

[5] *Theory of Hadrons and Light-front QCD*, Glazek S.D. Ed., (World Scientific Publishing Co., Singapore, 1995).

[6] Brodsky S.J., Pauli H.C. and Pinsky S.S., "Quantum chromodynamics and other field theories on the light cone", Heidelberg preprint MPIH-V1-1997, Stanford preprint SLAC–PUB 7484, Apr. 1997, 203 pp. Submitted to Physics Reports.

[7] Tamm I., J.Phys. (USSR) **9**, 449 (1945).

[8] Dancoff S.M., Phys.Rev. **78**, 382 (1950).

[9] Pauli H.C., *The Challenge of Discretized Light-Cone Quantization*, in: "Quantum Field Theoretical Aspects of High Energy Physics", Geyer and E. M. Ilgenfritz Eds., NaturwissenschaftlichTheoretisches Zentrum der Universität Leipzig, 1993.

[10] Pauli H.C., 'Solving Gauge Field Theory by Discretized Light-Cone Quantization', Heidelberg Preprint MPIH-V25-1996, hep-th/9608035.

[11] Wilson K.G., Walhout T., Harindranath A., Zhang W.M., Perry R.J and Glazek S.D., Phys.Rev. **D49**, 6720 (1994); and references therein.

[12] Wegner F., Ann.Physik (Leipzig) **3**, 77 (1994).

[13] Trittmann U. and Pauli H.C., "Quantum electrodynamics at strong coupling", Heidelberg preprint MPI H-V4-1997, Jan. 1997, hep-th/9704215.

[14] Trittmann U. and Pauli H.C, "On rotational invariance in front form dynamics", Heidelberg preprint MPI H-V7-1997, Apr. 1997, hep-th/9705021.

[15] Trittmann U., "On the role of the annihilation channel in front front positronium", Heidelberg preprint MPI H-V17-1997, Apr. 1997, hep-th/9705072.

[16] Brisudova M. and Perry R.J, Phys.Rev. **D54**, 1831-1843 (1996).

[17] Brisudova M., Perry R.J. and Wilson K.G, Phys.Rev.Lett. **78**, 1227 (1997).

[18] Jones B. and Perry R., "The Lamb-shift in a light-front Hamiltonian approach", hep-th/9612163.

[19] Jones B. and Perry R., "Analytic treatment of positronium spin-splitting in light-front QED", hep-th/9605231.

[20] Gubankova E. and Wegner F., "Exact renormalization group analysis in Hamiltonian theory: I. QED Hamiltonian on the light front", hep-th/9702162.

[21] Pauli H.C., Proceedings on SCGT 96, Nagoya 1996, to appear 1997; and Heidelberg preprint 1997, in preparation.

Front form QED(3+1): The spin-multiplet structure of the positronium spectrum at strong coupling

U. Trittmann

Max–Planck-Institut für
Kernphysik
Postfach 10 39 80, D–69029 Heidelberg,
Germany

---

1. INTRODUCTION

The practitioner has to think about (at least) two obstacles, when he wants to solve a Hamiltonian field theory problem in *front form* dynamics. One is the problem that the theory is not manifestly rotational invariant. The other is the question, how one should construct effective theories at all in this framework. Of course, the first problem exists analogously in instant form dynamics, where the theory is not manifestly Lorentz invariant. Connected to this problem is the fact that any symmetry which is not manifest will break down immediately with the slightest approximation in a Hamiltonian field theory. There is broad agreement that the construction of effective theories is inevitable, if one wants to describe the low energy region, *i.e.* the bound states, of a strongly coupled theory. A variety of formalisms were suggested concerning this topic [1][2][3].

In this article I shall present quantitative investigations concerning the above two problems with a specified model, namely positronium as a QED(3+1) bound state. The model is represented by the energy diagrams of Fig. 1. To be more specific, an effective theory was constructed following the work of PAULI[3]. The effective matrix elements were calculated analytically, and the emerging integral equation was put on the computer to numerically solve for the eigenvalues and eigenfunctions of the Hamiltonian.

Proceeding in this manner, one can answer the following questions *quantitatively*: Are the results of front form theory rotationally invariant, *i.e.* do states of different J_z form multiplets? How good is the underlying effective theory

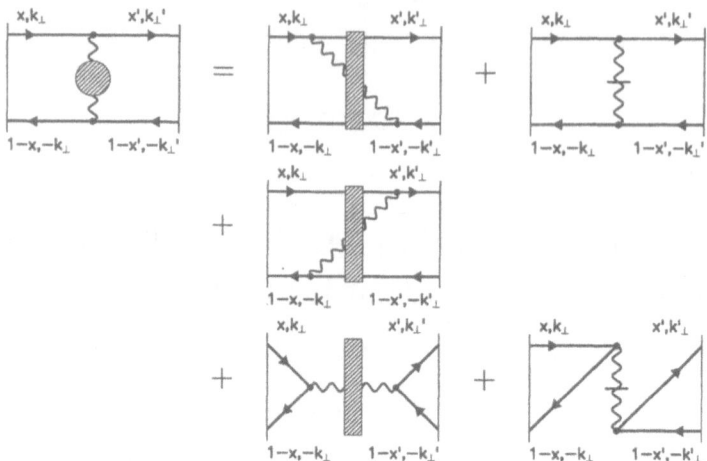

Fig. 1. — The graphs of the positronium model. Effective photon lines are labeled by hashed rectangles. The graphs of the annihilation interaction are at the bottom line.

concerning its cutoff dependence? Is the multiplet structure of the spectrum reproduced correctly? And: what is the numerical evidence for the degeneracy of corresponding states?

2. METHOD

A complete description of the applied method is surely beyond the scope of this article. I review shortly its major steps.

To get a finite dimensional Hamiltonian out of the infinite dimensional canonical Hamiltonian, we introduce a cutoff Λ on the kinetic energies. We then (formally) map an $N \times N$ onto a 2×2 block matrix by subsequent projections (Bloch-Feshbach formalism) and call these sectors P-space and Q-space. Note that the Q-space is already an effective sector by construction:

$$H_Q = H_Q^{(N)} + X_Q^{\text{eff}},$$

where $H_Q^{(N)}$ is a piece of the original Hamiltonian and X_Q^{eff} are the interactions generated by the projections. The Hamiltonian reads

$$H_{\text{LC}} = \begin{pmatrix} H_P & H_{PQ} \\ H_{QP} & H_Q \end{pmatrix}.$$

The final projection maps the Q-space onto the P-space. One solves for the state $\hat{Q}|\psi\rangle = |e\bar{e}\gamma\rangle$ with a resolvent involving the *redundant parameter* ω

$$G(\omega) := \langle Q|\omega - H_{\text{LC}}|Q\rangle^{-1},$$

and obtains the *nonlinear* equation

$$H_{\text{LC}}^{\text{eff}}(\omega)|\psi_n(\omega)\rangle = M_n^2(\omega)|\psi_n(\omega)\rangle,$$

with the effective Hamiltonian $(\hat{P}|\psi\rangle = |e\bar{e}\rangle)$

$$H_{\text{LC}}^{\text{eff}}(\omega) := \hat{P}H_{\text{LC}}\hat{P} + \hat{P}H_{\text{LC}}\hat{Q}(\omega - H_{\text{LC}})^{-1}\hat{Q}H_{\text{LC}}\hat{P}.$$

The fixing of the redundant parameter works as follows. We divide the sector Hamiltonian in Q-space into $H_Q = M_Q^2 + V_Q$, and the interaction into a diagonal part $\langle V_Q\rangle$ and a non-diagonal part δV_Q

$$V_Q = \langle V_Q\rangle + \delta V_Q.$$

With the definition $T^* := \omega - \langle V_Q\rangle$ we expand the resolvent around the diagonal interaction $\langle V_Q\rangle$

$$\frac{1}{\omega - H_Q} = \frac{1}{T^* - M_Q^2} + \frac{1}{T^* - M_Q^2}\delta V_Q \frac{1}{T^* - M_Q^2 - \delta V_Q}$$

$$\simeq \frac{1}{T^* - M_Q^2}. \tag{1}$$

The approximation in the last step is to consider the first term only. It has severe consequences: a collinear singularity occurs, proportional to

$$\frac{1}{\mathcal{D}}\frac{\Delta(x,k_\perp,x',k'_\perp;T^*)}{|x - x'|}, \qquad \text{with} \quad \mathcal{D}(x,x';T^*) := |x - x'|(T^* - M_Q^2).$$

We can calculate $\Delta(x,k_\perp,x',k'_\perp;T^*)$ with P-space graphs

$$\Delta = M_Q^2 - \omega - \frac{(k_\perp - k'_\perp)^2}{x - x'} - \frac{1}{2}\left(l_e^- - l_{\bar{e}}^-\right),$$

where $l_e^\mu := (k'_e - k_e)^\mu$ and $l_{\bar{e}}^\mu := (k_{\bar{e}} - k'_{\bar{e}})^\mu$ are the momentum transfers. One determines the parameter ω by demanding that the *collinear* singularity, induced by the approximation, Eq. (1), vanishes

$$\Delta(x,k_\perp,x',k'_\perp;T^*) = 0, \qquad \forall x,x',k_\perp,k'_\perp.$$

This allows for the calculation of an explicit expression for T^*

$$T^*(x,\vec{k}_\perp;x',\vec{k}'_\perp) = \frac{1}{2}\left(\frac{m_f^2 + \vec{k}_\perp}{x(1-x)} + \frac{m_f^2 + \vec{k}'_\perp}{x'(1-x')}\right).$$

The interpretation is that T^* is an approximation of the summed interactions of the higher Fock states.

Finally, the procedure yields an integral equation. The effective Hamiltonian operates only in *P*-space, and the continuum version of the eigenvalue problem is

$$0 = \left(\frac{m_f^2 + \vec{k}_\perp^2}{x(1-x)} - M_n^2\right)\psi_n(x, \vec{k}_\perp; \lambda_1, \lambda_2)$$

$$\frac{g^2}{16\pi^3}\sum_{\lambda_1', \lambda_2'}\int_D \frac{dx'\, d^2\vec{k}_\perp'}{\sqrt{xx'(1-x)(1-x')}}\,\frac{j^\mu(l_e, \lambda_e)j_\mu(l_{\bar{e}}, \lambda_{\bar{e}})}{l_e^\mu l_{e,\mu}}$$

$$\times\psi_n(x', \vec{k}_\perp'; \lambda_1', \lambda_2'),$$

with an integration domain D defined by a cutoff on the kinetic energy of the states. One observes that the effective interaction is gauge invariant and a Lorentz scalar

$$U_{\text{eff}} := \frac{j^\mu(l_e, \lambda_e)j_\mu(l_{\bar{e}}, \lambda_{\bar{e}})}{l_e^\mu l_{e,\mu}}.$$

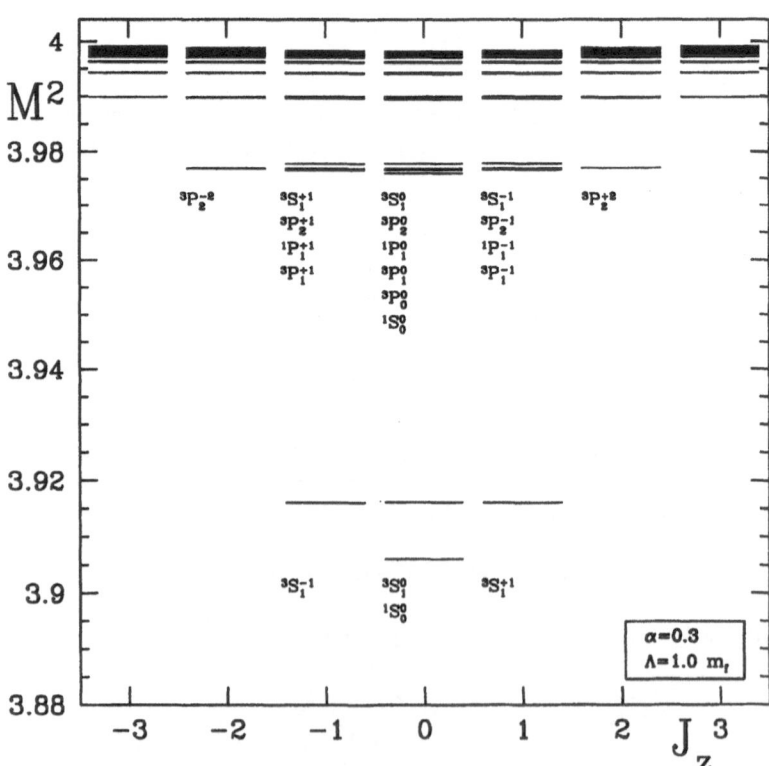

Fig. 2. — Positronium spectrum in different J_z sectors. $N_1 = N_2 = 21$.

3. RESULTS

As pointed out in the last section, the goal is to construct an effective Hamiltonian describing the positronium bound states. We showed how to find a reasonable *ansatz* for contributions of higher Fock states. To find out about the rotational symmetry of the theory, one has to observe that the rotation operator J_3 around the z axis is kinematic. Consequently, states can be classified according to the corresponding quantum number J_z. We have shown in Refs. [4][5] that closed expressions for the matrix elements of the *effective Hamiltonian* can be calculated, even for arbitrary J_z. This enables one to solve for the positronium eigenvalues in each sector of J_z separately and to compare the results to find out about possible degeneracies. A computer code for solving these eigenvalue problems, *i.e. integral equations*, has been generated with correct *Coulomb counterterms*. The latter is important to guarantee numerical stability and precision high enough to be able to give a quantitative statement concerning the (non)degeneracy of states. In fact, the convergence of the eigenvalues turns out to be exponential in the number of integration points. It is plotted in Fig. 4.

The results are compared to equal time perturbation theory (ETPT). The spectrum is shown in Fig. 2, which is the most important of this article. A one-to-one comparison between the multiplets for a Bohr quantum number $n = 2$ is depicted in Fig. 3. It is quite clear how to interpret these results. One reads off easily from the number of degenerate J_z levels the quantum number J of the complicated operator J^2, consisting of the angular momentum operators, *cf.* the theory of the Poincaré group in front form dynamics [4, Chapter 3.1].

It is worthwhile mentioning another non-trivial result of the formalism considered in this article. The annihilation channel can be included straightforwardly, which is an intrinsic check of the effective theory. Surprisingly the instantaneous and the dynamic graphs act in different J_z sectors. An interesting feature concerning the annihilation channel is the calculation of the hyperfine splitting, where this channel plays a major role and yields a well-known contribution. The coefficient of the hyperfine splitting is defined as

$$C_{hf} = \frac{M_t - M_s}{\alpha^4} = \frac{1}{2}\left[\frac{2}{3} + \left(\frac{1}{2}\right) - \frac{\alpha}{\pi}\left(\ln 2 + \frac{16}{9}\right) + \mathcal{O}(\alpha^2)\right].$$

The results for the two formalisms are

$$C_{hf} = \begin{cases} 0.56 & ;\text{this work} \\ \frac{7}{12} \simeq 0.58 & ;\text{ETPT}[\mathcal{O}(\alpha^4)], \end{cases}$$

which are in reasonable agreement.

Let us have a closer look at the results. One question listed above was, how good is the numerical evidence for degeneracies. One can plot the deviation of corresponding eigenvalues for $J_z=0$ and $J_z=1$ multiplets with growing number of integration points, as was done in Fig. 12 of Ref. [5]. A χ^2 fit of the difference function to

$$\Delta M^2(N) = a - b \exp\{(N - N_0)/c\}.$$

yields for the ground state triplet $1\,^3S_1$: $a = -(5.47 \pm 0.95) \times 10^{-5}$, $b = (1.88 \pm 0.03) \times 10^{-4}$, $c = 4.03 \pm 0.11$. It seems thus quite justified to consider these states degenerate.

The cutoff dependence of the eigenvalues was investigated, too. For the actual graphs, see Ref. [6]. The role of the annihilation channel seems to be to stabilize the eigenvalues. For the triplet ground state we can fit the dependence on $log\,\Lambda$ with

$$M_t^2(\Lambda) = \begin{cases} 3.90976 - 0.01858\log\Lambda + 0.00789\log^2\Lambda & ; \text{no annih.} \\ 3.91392 - 0.00029\log\Lambda + 0.00015\log^2\Lambda & ; \text{incl. annih.} \end{cases}$$

The decrease of the triplet with $log\,\Lambda$ is suppressed by the inclusion of the annihilation channel by a factor of 60! Care is, however, to be taken, since the annihilation channel is absent in all QED bound states not built out of a particle and its antiparticle [7].

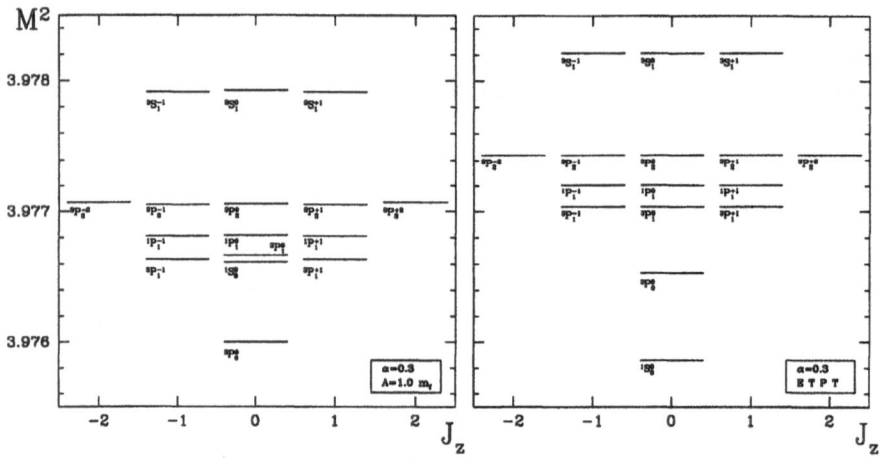

Fig. 3. — Comparison of multiplets for $n=2$: (a) results of the present work with $\alpha = 0.3$, $\Lambda = 1.0\,m_f$, $N_1 = N_2 = 21$; (b) equal-time perturbation theory (ETPT) up to order $\mathcal{O}(\alpha^4)$.

4.   CONCLUSIONS

To conclude, we review the results of the positronium theory in *front form* dynamics as presented in this article.

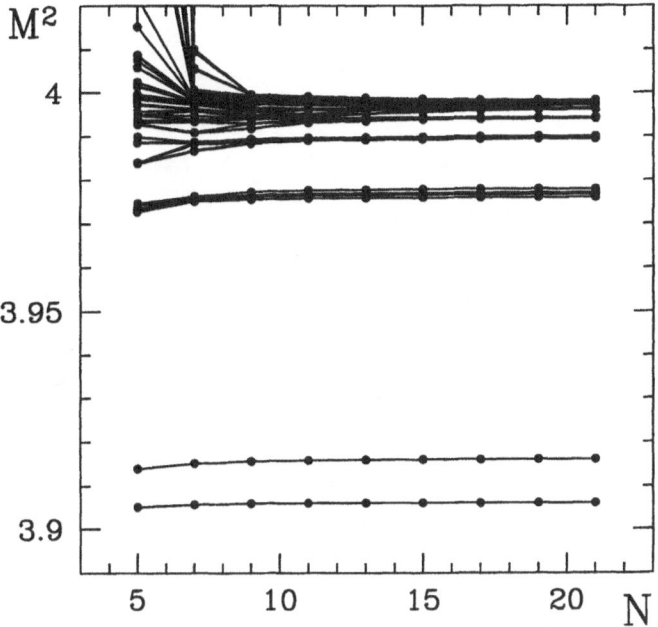

Fig. 4. — The positronium spectrum including the annihilation channel in the $J_z=0$ sector. Parameters of the calculation: $\alpha = 0.3, \Lambda = 1.0\,m_f$. The mass squared eigenvalues M_n^2 in units of the electron mass m_f^2 are shown as functions of the number of integration points $N \equiv N_1 = N_2$. The triplet states, especially 1^3S_1, are lifted up, the singlet mass eigenvalues are the same as without the annihilation channel.

The correct positronium spectrum is obtained in all J_z sectors. The J_z states form (degenerate) multiplets, which means that rotational invariance is restored in the solution of a Hamiltonian field theory in *front form* dynamics. Consequently, there is no need to diagonalize the complicated rotation operator $\mathcal{J}^2$, because its eigenvalues can be read off by counting the number of degenerate eigenstates within the J_z multiplets. The annihilation channel can be introduced into the theory straightforwardly. Furthermore, the annihilation channel seems important for the cutoff behavior. From these facts, we can deduce that the applied underlying effective theory is correct. The computer code which was generated to solve for the eigenstates is applicable also to other effective Hamiltonians. The application to WEGNER'S formalism [2] is work in progress. As an outlook one could think of plugging a running coupling constant into the code and to calculate the meson spectrum. Surely, this implies a deeper understanding of the numerical behavior of the occuring severe singularities.

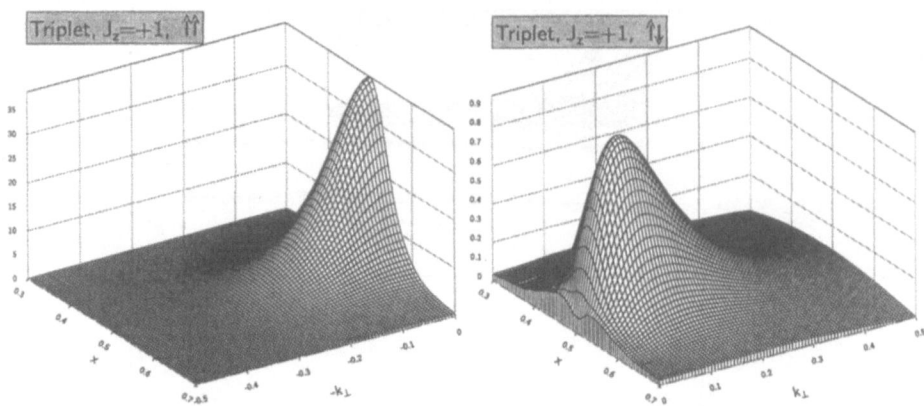

Fig. 5. — The triplet ground state wavefunction for $J_z = +1$ as a function of the longitudinal momentum fraction x and the transverse momentum $k_\perp$, omitting the dependence on the angle φ. The calculation was done with $\alpha = 0.3$, $\Lambda = 1.0\,m_f$, $N_1 = 41$, $N_2 = 11$. Shown are: (a) ($\uparrow\uparrow$)-component, (b) ($\uparrow\downarrow$)-component.

Acknowledgments

It is a pleasure to thank the organizers of the workshop for the invitation, hospitality and, last but not least, for the financial support.

References

[1] WILSON K.G., WALHOUT T.S., HARINDRANTH A., ZHANG W.-M. AND PERRY R.J., *"Nonperturbative QCD: A weak-coupling treatment on the light front"*, Phys. Rev. **D49** (1994) 6720.

[2] WEGNER F., *"Flow-equations for Hamiltonians"*, Ann. Phys. **3** (1994) 77–91.

[3] PAULI H.-C., *"Solving Gauge field Theory by Discretized Light-Cone Quantization"*, **hep-th/9608035**, submitted to Phys. Rev. **D**, 1996.

[4] TRITTMANN U. AND PAULI H.-C., *"Quantum electrodynamics at strong couplings"*, **hep-th/9704215**, 1997

[5] TRITTMANN U. AND PAULI H.-C. , *"On rotational invariance in front form dynamics,* **hep-th/9705021** submitted to Phys. Rev. **D**, 1997.

[6] TRITTMANN U., *"On the role of the annihilation channel in front form positronium"*,**hep-th/9705072**, submitted to Phys. Rev. **D**, 1997.

[7] BRODSKY S.J., remark made at the workshop.

Spectral Fluctuations of the QCD Dirac Operator

J.J.M. Verbaarschot

*Department of Physics,
SUNY at Stony Brook,
Stony Brook, NY 11794, USA*

1. INTRODUCTION

It is generally believed that QCD with massless quarks undergoes a chiral phase transition (see [1] for a review). This

leads to important observable signatures in the real world with two light quarks. The order parameter of the chiral phase transition is the chiral condensate $\langle \bar{\psi}\psi \rangle$, is directly related to the average spectral density of the Dirac operator [2]. However, the eigenvalues of the Dirac operator fluctuate about their average position. The question we wish to address in this lecture is to what extent such fluctuations are universal. If that is the case, they do not depend on the full QCD dynamics and can be obtained from a much simpler chiral Random Matrix Theory (chRMT) with the global symmetries of the QCD partition function.

This conjecture has its origin in the study of spectra of complex systems [3]. According to the Bohigas conjecture, spectral correlations of classically chaotic quantum systems are given by RMT. A first argument in favor of the universality in Dirac spectra came from the analysis of the finite volume QCD partition function. In particular, for box size L in the range $1/\Lambda \ll L \ll 1/m_\pi$, ($\Lambda$ is a typical hadronic scale and m_π is the pion mass) we expect that the global symmetries determine its mass dependence [4]. This implies that the fluctuations of Dirac spectra are constrained by Leutwyler-Smilga sum rules.

Recently, it has become possible to obtain *all* eigenvalues of the lattice QCD Dirac operator on reasonably large lattices [5]. This makes a direct verification of the above conjecture possible. This is the main objective of this lecture.

At nonzero chemical potential the QCD Dirac operator is nonhermitean with eigenvalues scattered in the complex plane. The possibility of a new type of universal behavior in this case will be discussed at the end of this lecture.

2. THE DIRAC SPECTRUM

The order parameter of the chiral phase transition, $\langle \bar{\psi}\psi \rangle$, is nonzero only below the critical temperature. As was shown in [2] $\langle \bar{\psi}\psi \rangle$ is directly related to the eigenvalue density of the QCD Dirac operator per unit four-volume

$$\Sigma \equiv |\langle \bar{\psi}\psi \rangle| = \frac{\pi \langle \rho(0) \rangle}{V}. \tag{1}$$

It is elementary to derive this relation. The Euclidean Dirac operator for gauge field configuration A_μ is given by $D = \gamma_\mu(\partial_\mu + iA_\mu)$. For Hermitean gamma matrices D is anti-hermitean with purely imaginary eigenvalues, $D\phi_k = i\lambda_k\phi_k$, and spectral density given by $\rho(\lambda) = \sum_k \delta(\lambda - \lambda_k)$. Because $\{\gamma_5, D\} = 0$, nonzero eigenvalues occur in pairs $\pm\lambda_k$. In terms of the eigenvalues of D the QCD partition function for N_f flavors of mass m can then be written as

$$Z(m) = \langle \prod_k (\lambda_k^2 + m^2)^{N_f} \exp(-S_{\mathrm{YM}}) \rangle, \tag{2}$$

where the average $\langle \cdot \rangle$ is over all gauge field configurations.

The chiral condensate follows immediately from the partition function (2),

$$\langle \bar{\psi}\psi \rangle = \frac{1}{VN_f} \partial_m \log Z(m) = \frac{1}{V} \langle \sum_k \frac{2m}{\lambda_k^2 + m^2} \rangle. \tag{3}$$

If we express the sum as an integral over the spectral density, and take the thermodynamic limit before the chiral limit so that we have many eigenvalues less than m we recover (1) (Notice the order of the limits.).

An important consequence of the Bank-Casher formula (1) is that the eigenvalues near zero virtuality are spaced as $\Delta\lambda = 1/\rho(0) = \pi/\Sigma V$. This should be contrasted with the eigenvalue spectrum of the non-interacting Dirac operator. Then $\rho^{\mathrm{free}}(\lambda) \sim V\lambda^3$ which leads to an eigenvalue spacing of $\Delta\lambda \sim 1/V^{1/4}$. Clearly, the presence of gauge fields leads to a strong modification of the spectrum near zero virtuality. Strong interactions result in the coupling of many degrees of freedom leading to extended states and correlated eigenvalues. On the other hand, for uncorrelated eigenvalues, the eigenvalue distribution factorizes and we have $\rho(\lambda) \sim \lambda^{2N_f+1}$, i.e. no breaking of chiral symmetry.

Because the QCD Dirac spectrum is symmetric about zero, we have two different types of eigenvalue correlations: correlations in the bulk of the spectrum and spectral correlations near zero virtuality. In the context of chiral symmetry we wish to study the spectral density near zero virtuality. Because the eigenvalues are spaced as $1/\Sigma V$ it is natural to introduce the microscopic spectral density [6]

$$\rho_S(u) = \lim_{V \to \infty} \frac{1}{V\Sigma} \rho \left(\frac{u}{V\Sigma} \right). \tag{4}$$

The dependence on the macroscopic variable Σ has been eliminated and therefore $\rho_S(u)$ is a perfect candidate for a universal function.

3. SPECTRAL UNIVERSALITY

Spectra for a wide range of complex quantum systems have been studied both experimentally and numerically (see [7] for a review). One basic observation has been that the scale of variations of the average spectral density and the scale of the spectral fluctuations separate. This allows us to unfold the spectrum, i.e. we rescale the spectrum in units of the local average level spacing. The fluctuations of the unfolded spectrum can be measured by suitable statistics. We will consider the nearest neighbor spacing distribution, $P(S)$, the number variance, $\Sigma_2(n)$, and the $\Delta_3(n)$ statistic. The number variance is defined as the variance of the number of levels in a stretch of the spectrum that contains n levels on average, and $\Delta_3(n)$ is obtained by a smoothening of $\Sigma_2(n)$.

These statistics can be obtained analytically for the invariant random matrix ensembles defined as ensembles of Hermitean matrices with independently distributed Gaussian matrix elements. Depending on the anti-unitary symmetry, the matrix elements are real, complex or quaternion real. The corresponding Dyson index is given by $\beta = 1$, 2, and 4, respectively. The nearest neighbor spacing distribution is given by $P(S) \sim S^{\beta} \exp(-a_{\beta}S^2)$. The asymptotic behavior of $\Sigma_2(n)$ and $\Delta_3(n)$ is given by $\Sigma_2(n) \sim (2/\pi^2\beta) \log(n)$ and $\Delta_3(n) \sim \beta\Sigma_2(n)/2$. For uncorrelated eigenvalues one finds that $P(S) = \exp(-S)$, $\Sigma_2(n) = n$ and $\Delta_3(n) = n/15$. Characteristic features of random matrix correlations are level repulsion at short distances and a strong suppression of fluctuations at large distances.

The main conclusion of numerous studies of eigenvalue spectra is that spectral correlations of a classically chaotic systems are given by RMT [7, 8].

4. CHIRAL RANDOM MATRIX THEORY

In this section we will introduce an instanton liquid inspired RMT for the QCD partition function. In the spirit of the invariant random matrix ensembles we construct a model for the Dirac operator with the global symmetries of the QCD partition function as input, but otherwise Gaussian random matrix elements.

The chRMT that obeys these conditions is defined by [6, 9, 10]

$$Z_\nu^\beta = \int DW \prod_{f=1}^{N_f} \det(\mathcal{D} + m_f) e^{-\frac{N\Sigma^2\beta}{4}\mathrm{Tr}W^\dagger W}, \quad \text{with} \quad \mathcal{D} = \begin{pmatrix} 0 & iW \\ iW^\dagger & 0 \end{pmatrix},$$

(5)

and W is a $n \times m$ matrix with $\nu = |n - m|$ and $N = n + m$. The matrix elements of W are either real ($\beta = 1$, chiral Gaussian Orthogonal Ensemble (chGOE)), complex ($\beta = 2$, chiral Gaussian Unitary Ensemble (chGUE)), or quaternion real ($\beta = 4$, chiral Gaussian Symplectic Ensemble (chGSE)). This model reproduces the following symmetries of the QCD partition function: *i)* The $U_A(1)$ symmetry. All nonzero eigenvalues of the random matrix Dirac operator occur in pairs $\pm\lambda$. *ii)* The topological structure of the QCD partition function. The Dirac matrix has exactly $|\nu| \equiv |n-m|$ zero eigenvalues. This identifies ν as the topological sector of the model. *iii)* The flavor symmetry is the same as in QCD. *iv)* The chiral symmetry is broken spontaneously with chiral condensate given by $\Sigma = \lim_{N\to\infty} \pi\rho(0)/N$. ($N$ is interpreted as the (dimensionless) volume of space time.) *v)* The anti-unitary symmetries. For fundamental fermions the matrix elements of the Dirac operator are complex for $N_c \geq 3$ ($\beta = 2$) but can be chosen real for $N_c = 2$ ($\beta = 1$). For adjoint fermions they can be arranged into real quaternions ($\beta = 4$).

Note that spectral correlations of chRMT in the bulk of the spectrum are given by the invariant random matrix ensemble with the same value of β. Both microscopic correlations near zero virtuality and in the bulk of the spectrum are stable against deformations of the ensemble. This has been shown by a variety of different arguments [11, 12, 13, 15, 14].

Below we will discuss the microscopic spectral density. For $N_c = 3$, N_f flavors and topological charge ν it is given by [9]

$$\rho_S(u) = \frac{u}{2} \left(J_a^2(u) - J_{a+1}(u)J_{a-1}(u) \right),$$

(6)

where $a = N_f + \nu$. The more complicated result for $N_c = 2$ is given in [16].

Together with the invariant random matrix ensembles, the chiral ensembles are part of a larger classification scheme. As pointed out in [17], there is a one to one correspondence between random matrix theories and symmetric spaces.

5. LATTICE QCD RESULTS

Recently, Kalkreuter [5] calculated *all* eigenvalues of the lattice Dirac operator both for Kogut-Susskind (KS) fermions and Wilson fermions for lattices as large as 12^4. In the the case of $SU(2)$ the anti-unitary symmetry of the KS and the Wilson Dirac operator is different [18, 19]. For KS fermions the Dirac matrix can be arranged into real quaternions, whereas the *Hermitean* Wilson Dirac matrix $\gamma_5 D^{\mathrm{Wilson}}$ can be chosen real. Therefore, we expect that the eigenvalue correlations are described by the GSE and the GOE, respectively [19].

In Fig. 1 we show results for $\Sigma_2(n)$, $\Delta_3(n)$ and $P(S)$. The results for KS fermions are for 4 dynamical flavors with $ma = 0.05$ on a 12^4 lattice. The results for Wilson fermion were obtained for two dynamical flavors on a $8^3 \times 12$ lattice. Other statistics are discussed in [20].

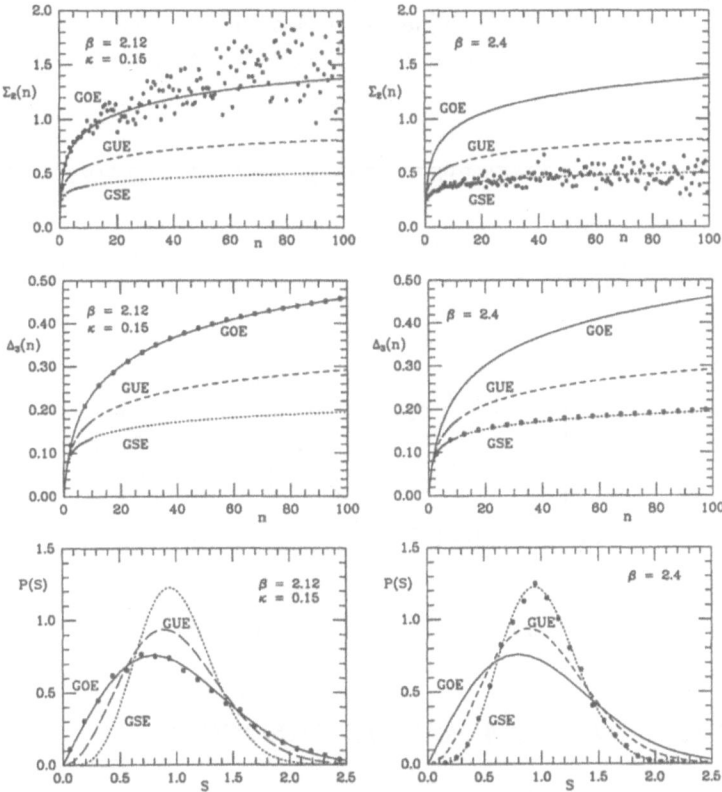

Fig. 1. — Spectral correlations of Dirac eigenvalues for Wilson fermions (upper) and KS-fermions (lower).

Recent lattice studies of the microscopic spectral density for quenched $SU(2)$ show perfect agreement between the microscopic spectral density and random matrix theory [21]. These calculations were performed at moderately strong coupling. Result for couplings in the scaling regime are in progress.

However, an alternative way to probe the Dirac spectrum is via the valence quark mass dependence of the condensate [22], i.e. $\Sigma(m) = \frac{1}{N} \int d\lambda \rho(\lambda) 2m/(\lambda^2 + m^2)$, for a fixed sea quark mass. In the mesoscopic range, $\Sigma(m)$ can be obtained analytically from the microscopic spectral density (6) [23],

$$\frac{\Sigma(x)}{\Sigma} = x(I_a(x)K_a(x) + I_{a+1}(x)K_{a-1}(x)),\qquad(7)$$

where $x = mV\Sigma$ is the rescaled mass and $a = N_f + \nu$. In Fig. 2 we plot this ratio as a function of x for lattice data of two dynamical flavors with mass $ma = 0.01$ and $N_c = 3$ on a $16^3 \times 4$ lattice. We observe that the lattice data for different values of β fall on a single curve. Moreover, in the mesoscopic range this curve coincides with the random matrix prediction for $N_f = \nu = 0$.

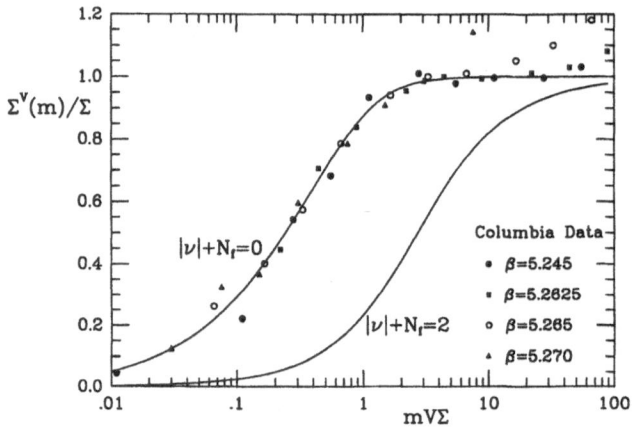

Fig. 2. — The valence quark mass dependence of the chiral condensate.

6.   CHIRAL RANDOM MATRIX THEORY AT $\mu \neq 0$

At nonzero temperature T and chemical potential μ a *schematic* random matrix matrix model of the Dirac operator in (5) is given by [24, 25, 27]

$$\mathcal{D} = \begin{pmatrix} 0 & iW + i\Omega_T + \mu \\ iW^\dagger + i\Omega_T + \mu & 0 \end{pmatrix},\qquad(8)$$

where $\Omega_T = T \otimes_n (2n+1)\pi\mathbf{1}$. Below, we will discuss a model with Ω_T absorbed in the random matrix and $\mu \neq 0$. For the three values of β the eigenvalues of $\mathcal{D}$ are scattered in the complex plane.

In the quenched approximation the eigenvalue distribution for $\beta = 2$ was obtained analytically [27] from the $N_f \to 0$ limit of a partition function with the determinant replaced by its absolute value. The same analysis can be carried out for $\beta = 1$ and $\beta = 4$. In the normalization (5), it turns out that the average spectral density does not depend on β [26]. However, the fluctuations of the eigenvalues are β dependent. In particular, one finds a very different behavior close to the imaginary axis and not too large values of μ. This regime of almost Hermitean random matrices was first identified by Fyodorov et al. [28]. For $\beta = 1$ we observe an accumulation of purely imaginary eigenvalues, whereas for $\beta = 4$ we find a depletion of eigenvalues in this domain. These results explain quenched instanton liquid calculations [29] and lattice QCD simulations with Kogut-Susskind fermions [30] at $\mu \neq 0$ (both for $SU(2)$), respectively.

7. CONCLUSIONS

We have shown that microscopic correlations of the QCD Dirac spectrum can be explained by RMT and have obtained an analytical understanding of the distribution of the eigenvalues near zero. An extension of this model to nonzero chemical potential explains previously obtained lattice QCD results.

Acknowledgments

This work was partially supported by the US DOE grant DE-FG-88ER40388. M. Halasz and M. Şener are thanked for a critical reading of the manuscript.

References

[1] DeTar C., *Quark-gluon plasma in numerical simulations of QCD*, in *Quark gluon plasma 2*, R. Hwa ed., World Scientific 1995.
[2] Banks T. and Casher A., Nucl. Phys. **B169** (1980) 103.
[3] Bohigas O. and Giannoni M., Lecture notes in Physics **209** (1984) 1.
[4] Leutwyler H. and Smilga A., Phys. Rev. **D46** (1992) 5607.
[5] Kalkreuter T., Comp. Phys. Comm. **95** (1996) 1.
[6] Shuryak E. and Verbaarschot J., Nucl. Phys. **A560** (1993) 306.
[7] Guhr T., Müller-Groeling A. and Weidenmüller H.A., Phys. Rep. (1997).
[8] Verbaarschot J., Proceedings Hirschegg 1997, hep-ph/9705355.
[9] Verbaarschot J., Phys. Rev. Lett. **72** (1994) 2531; Phys. Lett. **B329** (1994) 351; Nucl. Phys. **B427** (1994) 434.
[10] Verbaarschot J. and Zahed I., Phys. Rev. Lett. **70** (1993) 3852.
[11] Brézin E., Hikami S. and Zee A., Nucl. Phys. **B464** (1996) 411.
[12] Jackson A., Sener M. and Verbaarschot J., Nucl. Phys. **B479** (1996) 707.
[13] Nishigaki S., Phys. Lett. B (1996); Akemann G., Damgaard P., Magnea U. and Nishigaki S., hep-th/9609174.
[14] Jackson A., Sener M. and Verbaarschot J., hep-th/9704056.
[15] Guhr T. and Wettig T., hep-th/9704055.
[16] Verbaarschot J., Nucl. Phys. B426 (1994) 559.
[17] Zirnbauer M., J. Math. Phys. **37** (1996) 4986.
[18] Hands S. and Teper M., Nucl. Phys. **B347** (1990) 819.

[19] Halasz M. and Verbaarschot J., Phys. Rev. Lett. **74** (1995) 3920.
[20] Halasz M., Kalkreuter T. and Verbaarschot J., hep-lat/9607042.
[21] Berbenni-Bitsch M.E. et al., hep-lat/9704018.
[22] Chandrasekharan S., Lattice 1994, 475; Chandrasekharan S. and Christ N., Lattice 1995, 527; Christ N., Lattice 1996.
[23] Verbaarschot J., Phys. Lett. **B368** (1996) 137.
[24] Jackson A. and Verbaarschot J., Phys. Rev. **D53** (1996) 7223.
[25] Wettig T., Schäfer A. and Weidenmüller H., Phys. Lett. **B367** (1996) 28.
[26] Halasz M., Osborn J. and Verbaarschot J., hep-lat/9704007.
[27] Stephanov M., Phys. Rev. Lett. **76** (1996) 4472.
[28] Fyodorov Y., Khoruzhenko B. and Sommers H., Phys. Lett. **A 226**, 46 (1997); cond-mat/9703152.
[29] Shuryak E. and Schäfer Th., *private communication.*
[30] Baillie C. et al., Phys. Lett. **197B**, 195 (1987).

CHAPTER IV

New Developments

Collinear QCD Models

S. Dalley*

Theory Division, CERN, CH-1211 Geneva 23, Switzerland

In ancient times, 't Hooft studied the mesons in QCD_{1+1} [1] to illustrate the power of the large N limit [2] in the light-cone formalism. Recently some generalisations of the 't Hooft model have been studied, which retain a remnant of transverse degrees of freedom, based on a dimensional reduction of QCD to $1 + 1$ dimensions [3, 4, 5, 6, 7, 8, 9]. In this collinear approximation, quarks and gluons are artificially restricted to move in one space dimension, but retain their polarization degree of freedom. In this lecture, a problem which in principle involves a large number of partons will be addressed in the context of the collinear model at large N. The example treated here(^{1}) is the quark distribution function in a heavy meson, which is supposed to exhibit a version of Regge behaviour at small-x. The central idea involves high light-cone energy boundary conditions on wavefunctions — ladder relations — which typically connect Fock space sectors of differing numbers of partons. The same ideas carry over to $3 + 1$ dimensions [10].

We start from $SU(N)$ gauge theory in $3 + 1$-dimensions with one flavour of quarks. If we pick an arbitrary fixed space direction x^3 and restrict ourselves to zero momentum in the transverse directions for the gauge and quark fields, $\partial_{x_\perp} A_\mu = \partial_{x_\perp} \Psi = 0$, one finds an effectively two-dimensional gauge theory of adjoint scalars and fundamental Dirac spinors with action

$$
S = \int dx^0 dx^3 \quad -\frac{1}{4g^2} \text{Tr}\{F_{ab}F^{ab}\} + \text{Tr}\{-\frac{1}{2}\bar{D}_a\phi_\rho \bar{D}^a \phi^\rho
$$

(*) On leave from: Department of Applied Mathematics and Theoretical Physics, Cambridge University, Silver Street, Cambridge CB3 9EW, England.

(^{1}) Another example was treated in the lecture — asymptotic spectrum of glueballs — for which there was not enough room in these proceedings.

$$-\frac{tg^2}{4}[\phi_\rho,\phi_\sigma][\phi^\rho,\phi^\sigma] + \frac{1}{2}m_0^2\phi_\rho\phi^\rho\} + \frac{i}{\sqrt{2}}(\bar{u}\gamma^a D_a u + \bar{v}\gamma^a D_a v) \quad (1)$$

$$+\frac{m_F}{\sqrt{2}}(\bar{u}v + \bar{v}u) - \frac{sg}{\sqrt{2}}(\bar{u}(\phi_1 + i\gamma^5\phi_2)u - \bar{v}(\phi_1 - i\gamma^5\phi_2)v) \,,$$

where a and $b \in \{0,3\}$, $\rho \in \{1,2\}$, $\gamma^0 = \sigma^1$, $\gamma^3 = i\sigma^2$, $\gamma^5 = i\sigma^1\sigma^2$, $\phi_\rho = A_\rho/g$, $\bar{D}_a = \partial_a + i[A_a,.]$, $D_a = \partial_a + iA_a$, $\int dx^1 dx^2 = \mathcal{L}^2$, $g^2 = \tilde{g}^2/\mathcal{L}^2$, and $\tilde{g}$ is the four-dimensional coupling. The two-component spinors u and v are related to Ψ by

$$\Psi = \frac{1}{2^{1/4}\mathcal{L}}\begin{pmatrix} u_{R+} \\ u_{L+} \\ u_{L-} \\ u_{R-} \end{pmatrix} \,, \quad u = \begin{pmatrix} u_{L+} \\ u_{R+} \end{pmatrix} \,, \quad v = \begin{pmatrix} u_{L-} \\ u_{R-} \end{pmatrix} \,. \quad (2)$$

The suffices L (R) and $+$ ($-$) in (2) represent Left (Right) movers and $+ve$ ($-ve$) helicity, which is a conserved quantity. Thus u, v, ϕ_1, ϕ_2 represent the transverse polarisations of the $3+1$ dimensional quarks and gluons. Since the dimensional reduction procedure treats space asymmetrically, dimensionless parameters s and t, and a bare gluon mass m_0 can occur due to loss of transverse local gauge transformations. For the present application the precise choice of s and t will not be qualitatively important, so we set $s = 1$ and $t = 0$ for simplicity.

In the light-cone gauge $A_- = (A_0 - A_3)/\sqrt{2} = 0$, the fields A_+ and $u_{L\pm}$ are non-propagating in light-front time $x^+ = (x^0 + x^3)/\sqrt{2}$ so may be eliminated by their constraint equations

$$\partial_- u_{L\pm} = iF_\mp \,, \quad (\partial_-)^2 A_+ = g^2 J^+ \quad (3)$$

$$F_\pm = \frac{m_F}{\sqrt{2}}u_{R\pm} \pm gB_\mp u_{R\mp} \quad (4)$$

$$J^+ = i[B_-,\partial_- B_+] + i[B_+,\partial_- B_-] + u_{R+}u_{R+}^\dagger + u_{R-}u_{R-}^\dagger \quad (5)$$

and $B_\pm = (\phi_1 \pm i\phi_2)/\sqrt{2}$. The exchange of non-propagating particles associated with the constrained fields results non-local interactions in the light-cone hamiltonian

$$P^- = \int dx^- \ F_+^\dagger\frac{1}{i\partial_-}F_+ + F_-^\dagger\frac{1}{i\partial_-}F_- - \frac{g^2}{2}\text{Tr}\left\{J^+\frac{1}{(\partial_-)^2}J^+\right\} + \frac{m_0^2}{2}\text{Tr}\ \phi^2 \quad (6)$$

The zero momentum limit of the constraints (3) forces a condition on the quark-gluon combined system

$$\int_{-\infty}^{+\infty} dx^- F_\pm = 0 \,, \quad \int_{-\infty}^{+\infty} dx^- J^+ = 0 \,, \quad (7)$$

assuming u_L and $\partial_- A_+$ vanish at $x^- = \pm\infty$. This is required for finiteness of the non-local interactions in (6). The relation involving J^+, which amongst other things restricts one to gauge singlets, will not be discussed further here.

Introducing the harmonic oscillator modes of the physical fields[2]

$$u_{R\pm i} = \frac{1}{\sqrt{2\pi}} \int_0^\infty dk \left(b_{\pm i}(k) e^{-ikx^-} + d^\dagger_{\mp i}(k) e^{ikx^-} \right) \qquad (8)$$

$$B_{\pm ij} = \frac{1}{\sqrt{2\pi}} \int_0^\infty \frac{dk}{\sqrt{2k}} \left(a_{\mp ij}(k) e^{-ikx^-} + a^\dagger_{\pm ji}(k) e^{ikx^-} \right) \qquad (9)$$

in the quantum theory we can expand any hadron state $|\Psi(P^+)>$ of total momentum P^+ in terms of a Fock basis [11]. The operators $a^\dagger_\pm$ create gluons with helicity ± 1, while $b^\dagger_\pm$ and $d^\dagger_\pm$ correspond to quarks and antiquarks (respectively) with helicities $\pm\frac{1}{2}$. At large N a gauge-singlet meson is a superposition

$$|\Psi(P^+)> = \sum_{n=2}^\infty \int_0^{P^+} dk_i \delta\left(\sum_{i=1}^n k_i - P^+\right) \frac{f_{\alpha_1\ldots\alpha_n}(k_1, k_2, \ldots, k_n)}{\sqrt{N^{n-1}}}$$
$$\times d^\dagger_{\alpha_1 i}(k_1) a^\dagger_{\alpha_2 ij}(k_2) a^\dagger_{\alpha_3 jk}(k_3) \ldots a^\dagger_{\alpha_{n-1} lm}(k_{n-1}) b^\dagger_{\alpha_n m}(k_n)|0>$$

Introducing the fourier transform $\tilde{F}(w)$, one finds that in the quantum theory Eq.(7) can be meaningfully applied as an annihilator of physical states for the cases

$$\lim_{w\to 0^+} \tilde{F}_{\pm i}(w) \cdot |\Psi(P^+)> = 0 \ , \quad \lim_{w\to 0^-} \tilde{F}^\dagger_{\pm i}(w) \cdot |\Psi(P^+)> = 0 \ . \qquad (10)$$

The first relation yields a condition on the Fockspace wavefunctions f involving vanishing quark momentum $k = w > 0$, the second on vanishing anti-quark momentum $k = -w > 0$:

$$f_{\mp\pm\alpha_1\cdots\alpha_n}(k, k_1, \ldots, k_{n+1}) = \pm\lambda \left[\frac{f_{\pm\alpha_1\cdots\alpha_n}(k+k_1, k_2, \ldots, k_{n+1})}{\sqrt{k_1}} \right.$$
$$\left. + \int_0^\infty \frac{dpdq}{\sqrt{q}} \delta(p+q-k) f_{\pm\mp\pm\alpha_1\cdots\alpha_n}(p, q, k_1, \ldots, k_{n+1}) \right] \qquad (11)$$

$$f_{\mp\mp\alpha_1\cdots\alpha_n}(k, k_1, \ldots, k_{n+1})$$
$$= \pm\lambda \int_0^\infty \frac{dpdq}{\sqrt{q}} \delta(p+q-k) f_{\pm\mp\mp\alpha_1\cdots\alpha_n}(p, q, k_1, \ldots, k_{n+1}) \ , \qquad (12)$$

with $\lambda = \sqrt{g^2 N/2\pi m_F^2}$ and a similar set of relations for quarks; in (11)(12) the limit $k \to 0^+$ is understood. If we adopt the following momentum-space

---

[2] The superscript on k^+ has been dropped for clarity; $i, j \in \{1, \ldots, N\}$ are gauge indices and $\dagger$ is now understood as the quantum complex conjugate, so does not transpose them.

operator ordering in P^- (6)

$$\int_{-\infty}^{0} \frac{dw}{w} \tilde{F}_{\pm i}^{\dagger} \tilde{F}_{\pm i} - \int_{0}^{\infty} \frac{dw}{w} \tilde{F}_{\pm i} \tilde{F}_{\pm i}^{\dagger} , \qquad (13)$$

we then apparently have manifest finiteness as $w \to 0$ for physical states. Normal ordering the oscillator modes in P^- would spoil finiteness. Since we do not normal order the form (13), infinite quark self energies (self-inertias) are generated but no vacuum energies are generated.

However the above argument is flawed by the fact that infinities may also arise due to integration over the parton momenta in the wavefunction $|\Psi >$, since (11)(12) are to be interpreted at *fixed* k_i as $k \to 0^+$. This is evident from the light-cone Schrodinger equation, obtained by projecting $2P^+P^-|\Psi >= \mathcal{M}^2|\Psi >$ onto a specific n-parton Fock state

$$\left(\left[\frac{\mathcal{M}^2}{2P^+} \right] - \sum_{i=1}^{n} \frac{m_i^2}{2p_i} \right) f_{\alpha_1 \ldots \alpha_n}(p_1, \ldots, p_n) = \hat{V} \left[f_{\alpha_1 \ldots \alpha_n}(p_1, \ldots, p_n) \right] \qquad (14)$$

where $\hat{V}$ is the interaction kernel (including self-inertias), $\mathcal{M}$ the boundstate mass, and m_i is m_F (quark) or m_B (gluon). The ladder relations (11)(12) are necessary for finiteness of the internal integrations in $\hat{V}$ at fixed external momenta p_i. However further renormalisation of $\hat{V}$ is necessary since the ladder relations do not ensure finiteness when one or more external momenta p_i vanish in (14).([3]) In fact an explicit two-loop calculation of the fermion self-energy in light-cone Yukawa$_{1+1}$ [13] shows that divergences do not cancel for the same cutoff on all small momenta. In general the renormalisation that cures these divergences will depend on the precise cut-off(s) employed. It has been suggested to renormalise the fermion kinetic mass finitely to restore parity invariance in light-cone calculations [14], and this should coincide with ensuring finiteness of $\mathcal{M}$.([4])

Eqs.(11)(12) show that the meson wavefunction components do not vanish as the quark momentum vanishes. It will be demonstrated that this leads directly to a rising quark distribution function at small $x = k/P^+$. The probability to find an anti-quark — the answer is the same for a quark — with momentum fraction $x = k/P^+$ of the meson is

$$\begin{aligned} Q(x) &= \sum_{n=2}^{\infty} \sum_{\alpha_i} \int_{0}^{P^+} dk_i \, \delta(\sum_{i=1}^{n} k_i - P^+) \delta(k_1 - k) \, |f_{\alpha_1 \ldots \alpha_n}(k_1, k_2, \ldots, k_n)|^2 \\ &= \sum_{\alpha} < d_{\alpha i}^{\dagger} d_{\alpha i}(x) > \end{aligned} \qquad (15)$$

For the polarized version $\Delta Q(x)$ one inserts sgn(α). In order to make use of (11)(12) to evaluate $Q(x \to 0)$, it is helpful to eliminate the integral terms,

---

([3]) This point, and also the integral terms in (11)(12), were missed in ref.[7].
([4]) After this lecture was typed, a preprint appeared [15] which verifies this for certain examples with a Yukawa interaction only.

which generate renormalisation of the other (non-integral) terms. Although the integrals are over a set of measure zero, they are non-zero due to the singular behaviour of the integrand. This singular behaviour can be found from the (correctly renormalised) Schrodinger equation (14). Let us consider the helicity $+1$ meson with valence component f_{++}. The general idea is to use an expansion in λ and $\log 1/x$ to evaluate (15). The leading orders we shall calculate are independent of any additional fermion kinetic mass renormalisation in (14). The leading log approximation amounts to considering the integration region $k_{n-2} >> k_{n-2} >> \cdots >> k_2 >> k_1 = k$ in (15). In this region we may use the ladder relations iteratively to express every n-parton non-valence contribution in terms of f_{++}. For example, truncating to no more than one gluon we obtain $Q(x \to 0) = \lambda^2 < 1/y >_{++}$ from the f_{-++} component, where $< 1/y >_{++}$ is the average inverse momentum fraction in the f_{++} component.

Truncating to no more than two gluons, we can compute some of the subleading λ and $\log 1/x$ corrections. For this we need to use the $n = 4$ boundstate equation (14) (for which there is neither a fermion self-inertia nor a finite kinetic mass renormalisation) to evaluate the integral term in (11). The following results neglect the A_+ exchange process between quark and gluon in (14), whose effects cannot be explicitly resummed. From resumming instantaneous fermion (u_L) processes the correct ladder relations in this 2-gluon approximation become

$$f_{-++}(x \to 0, x_1, x_2) = \lambda^* \frac{f_{++}(x_1, x_2)}{\sqrt{x_1}} \tag{16}$$

$$f_{+-++}(x \to 0, x_1, x_2, x_3) = -\lambda \frac{f_{-++}(x_1, x_2, x_3)}{\sqrt{x_1}} \tag{17}$$

where

$$\lambda^* = \frac{\lambda}{1 + \lambda^2 \left[\frac{\log (m_F^2/m_B^2)}{1 - (m_B^2/m_F^2)} \right]} \tag{18}$$

Then to leading log the contributions from one and two gluon components of the wavefunction give

$$Q(x \to 0) \approx (\lambda^*)^2 (1 + \lambda^2 \log 1/x) < 1/y >_{++} . \tag{19}$$

An example of a next-to-leading log contribution comes from integrating the two-gluon contribution over the region $\int_0^k dk_2 \int_k dk_3$ using (14) at $n = 4$

$$\lambda^2 (\lambda^*)^2 \log \left(1 + \frac{m_F^2}{m_B^2} \right) < 1/y >_{++} \tag{20}$$

The analytic ladder results are compared with a non-perturbative DLCQ solution of (14) truncated to the same number of gluons in Figs. 1 and 2. The DLCQ calculations are formal at finite x since no additional fermion kinetic mass renormalisation has been carried out to ensure finite $\mathcal{M}$ as $K \to \infty$. In

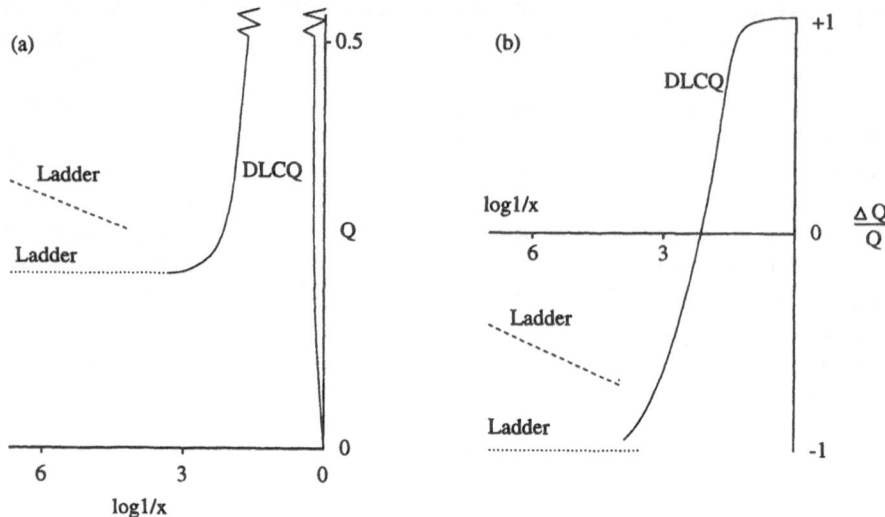

Fig. 1. — (a) Unpolarized distribution function for helicity +1 meson at $\lambda^2 = 0.1$: (Dotted) 1-gluon ladder prediction ($< 1/y >_{++}$ is indistinguishable from 2 for heavy quarks); (solid) DLCQ up to 1 gluon, K=24; (dashed) arbitrary-gluon tree-level ladder prediction.(b) helicity asymmetry $\Delta Q/Q$.

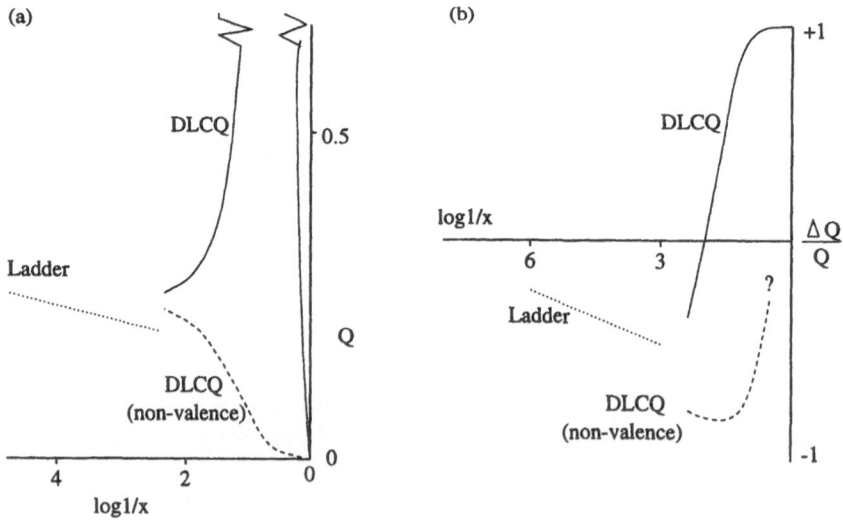

Fig. 2. — (a) Unpolarized distribution function for helicity +1 meson at $\lambda^2 = 0.1$, $m_B^2/m_F^2 = 0.1$: (Dotted) ladder prediction for up to 2 gluons and next-to-leading log; (solid) DLCQ up to 2 gluons, K=15; (dashed) non-valence part of the DLCQ calculation. (b) Helicity asymmetry $\Delta Q/Q$.

practice however, for heavy quarks the effects of this omission are very tiny at finite K, and the plots shown should be a good representation of the exact result at finite x (the same comment applies to plots in ref.[7]). At $x \sim 1/K$ the DLCQ results should in any case match onto the analytic predictions. There are many sources in the approximations which might account for the remaining discrepancy in the normalisation and slope seen in Fig.2 at small x, although the agreement is much better if the DLCQ calculation is repeated without the quark-gluon A_+ exchange.

More generally if we consider the ladder relations to leading order in λ, which means neglecting the integrals in (11)(12), an arbitrary number of gluons can be eliminated to yield an exponential sum of leading $\log 1/x$'s (Fig.1)

$$Q(x \to 0) \approx \lambda^2 x^{-\lambda^2} < 1/y >_{++} \tag{21}$$

The integral terms should only renormalise λ in the previous expression. All the results point to a rising small-x unpolarized distribution. At large N the polarization asymmetry changes sign and then vanishes at small x in the leading log approximation, $\Delta Q/Q \sim -x^{2\lambda^2}$.

Acknowledgments

I have had many useful discussions about collinear models with F. Antonuccio, S. Brodsky, M. Burkardt, H-C. Pauli and B. van de Sande. The DLCQ computer program is due to F. Antonuccio [16]. I thank P. Grangé for co-ordinating a stimulating workshop.

References

[1] 't Hooft G., *Nucl. Phys.* **B75** (1974) 461.

[2] 't Hooft G., *Nucl. Phys.* **B72** (1974) 461.

[3] Dalley S. and Klebanov I. R., *Phys. Rev.* **D47** (1993) 2517.

[4] Demeterfi K., Klebanov I. R. and Bhanot G., *Nucl. Phys.* **B418** (1994) 15.

[5] Pauli H. C., Kalloniatis A. C. and Pinsky S., *Phys. Rev.* **D52** (1995) 1176.

[6] Antonuccio F. and Dalley S., *Nucl. Phys.* **B461** (1996) 275.

[7] Antonuccio F. and Dalley S., *Phys. Lett.* **B376** (1996) 154.

[8] Pauli H. C. and Bayer R., *Phys. Rev.* **D53** (1996) 939.

[9] van de Sande B. and Burkardt M., *Phys. Rev.* **D53** (1996) 4628.

[10] Antonuccio F., Brodsky S. J. and Dalley S., preprint CERN-TH/97-88.

[11] Lepage G. P. and Brodsky S. J., *Phys. Rev.* **D22** (1980) 2157.

[12] Pauli H. C. and Brodsky S. J., *Phys. Rev.* **D32** (1985) 1993.

[13] Harindranath A. and Perry R., *Phys. Rev.* **D43** (1991) 4051.

[14] Burkardt M., Phys. Rev., D54, 1996, 2913.

[15] Burkardt M., preprint [hep-th/9704162].

[16] Antonuccio F., D.Phil Thesis, University of Oxford (1996).

LECTURE 12

A New Lattice Formulation of the Continuum

Yang Pang[1,2] and Haicang Ren[3]

([1]) *Department of Physics, Columbia University*
New York, NY, 10027, USA
([2]) *Department of Physics, Brookhaven National Laboratory*
Upton, NY 11973, USA
([3]) *Department of Physics, Rockefeller University*
New York, NY 10021, USA

1. INTRODUCTION

It is well known that the Dirac equation on a discrete lattice in D dimension has 2^D degenerate solutions. The usual method of removing these spurious solutions encounters difficulties with chiral symmetry when the lattice spacing $\ell \neq 0$, as demonstrated by the persistent problem of pion and kaon masses. On the other hand, we recall that in any crystal in nature, all the electrons do move in a lattice and satisfy the Dirac equation; yet there is not a single physical result that has ever been entangled with a spurious fermion solutions. Therefore it should not be difficult to eliminate these unphysical elements.

On a discrete lattice, particles hop from point to point, whereas in a real crystal the lattice structure is embedded in a continuum and electrons move continuously from lattice cell to lattice cell. In a discrete system, the lattice function are defined only on individual points (or links, as in the case of gauge fields). However, in a crystal the electron state vector is represented by the Bloch wave functions which are continuous functions in $\vec{r}$, and herein lies one of the essential differences.

In this new approach [1][2] we shall expand the field operator in terms of suitably chosen complete set of orthonormal Bloch functions

$$\{f_n(\vec{K} \mid \vec{r})\} \tag{1}$$

where $\vec{K}$ denotes the Bloch wavenumber restricted to the Brillouin zone, and n labels the different bands. Thus, $e^{-i\vec{K}\cdot\vec{r}} f_n(\vec{K} \mid \vec{r})$ has the periodicity of the lattice. The lattice approximation is then derived by either restricting it to only one band (say, $n = 0$), or to a few appropriately defined low-lying bands. Since the inclusion of all band is the original continuum problem, there is a natural connection between the lattice and the continuum in this method. By including the contributions due to more and more bands, one can systematically arrive at the exact continuum solution from the lattice approximation, as we shall see. There is a large degrees of freedom in choosing the Block functions (1), as the original continuum theory has no crystal structure. These extra degrees of freedom are analogous to gauge fixing; the final answer to the continuum problem is independent of the particular choice of Bloch functions.

2. SPURIOUS LATTICE FERMION SOLUTIONS

To see the origin of the spurious lattice fermion solutions, we may consider the replacement of the continuum equation $-i\partial\psi/\partial x = p\psi$ by its discrete form in one space dimension:

$$\frac{-i}{2\ell}(\psi_{j+1} - \psi_{j-1}) = p_L \psi_j \tag{2}$$

where ψ_j is the value of ψ at j^{th} site. The lattice-eigenvalue p_L is given by

$$p_L = \frac{1}{\ell} \sin K\ell \tag{3}$$

where

$$K\ell \equiv \theta \tag{4}$$

is between $-\pi$ and π. The spurious solution refers to the zero ($p_L(\theta) = 0$) at $\theta = \pi$ (which is the same as $\theta = -\pi$). This is a special case of the Nielsen-Ninomiya theorem [3].

2.1. Elimination of Spurious Lattice Fermion Solutions

We expand the continuum wave function $\psi(x)$ in terms of (1), in which the zeroth band ($n = 0$) is simply the linear interpolation of the discrete values $\{\psi_j\}$; i.e., in the zeroth-band approximation

$$\psi(x) = \sum_j \psi_j \Delta(x - j\ell) \tag{5}$$

where

$$\Delta(x) = \begin{cases} 1 - \frac{|x|}{\ell} & \text{for } |x| < \ell \\ 0 & \text{otherwise.} \end{cases} \tag{6}$$

Thus, at $x = j\ell$, $\psi(x) = \psi_j$. Substitute (5) into the continuum bilinear form

$$B(\psi(x)) \equiv -i \int \psi(x)^\dagger \frac{d\psi(x)}{dx} dx. \tag{7}$$

Setting $\partial B/\partial \psi_j = 0$ at a constant $\int \psi^\dagger \psi dx$, we find

$$\psi_j \propto e^{i\theta j} \tag{8}$$

with θ given by (4). Correspondingly, the zeroth-band Bloch function is

$$f_0(K \mid x) = \sqrt{\frac{3}{N\ell(2 + \cos\theta)}} \sum_j e^{i\theta j} \Delta(x - j\ell) \tag{9}$$

where N is the total number of lattice sites. It is not difficult to construct from $f_0(K \mid x)$ and the Fourier series a complete set of Bloch functions (1), which satisfy

$$\int_0^{N\ell} f_n(K \mid x)^* f_{n'}(K' \mid x) dx = \delta_{nn'} \delta_{KK'}; \tag{10}$$

e.g., these can be given by, for $n \neq 0$,

$$f_n(K \mid x) \equiv \frac{1}{\sqrt{N\ell}} e^{iKx + i2\pi nx/\ell} - \frac{a_n}{1 + a_0} \left[f_0(K \mid x) + \frac{1}{\sqrt{N\ell}} e^{iKx} \right] \tag{11}$$

where, for all n including 0,

$$a_n = \int_0^{N\ell} f_0(K \mid x)^* \frac{1}{\sqrt{N\ell}} e^{iKx + i2\pi nx/\ell}. \tag{12}$$

Let

$$\beta_n \equiv -i \int f_n(K \mid x)^* \frac{\partial}{\partial x} f_n(K \mid x) dx. \tag{13}$$

We find for $n = 0$

$$\beta_0 = \frac{3 \sin\theta}{2 + \cos\theta} \tag{14}$$

which, like (3), has a spurious zero solution at $\theta = \pi$.

If we substitute the full expansion

$$\psi(x) = \sum_n \sum_K q_n(K) f_n(K \mid x) \tag{15}$$

into (7), then $\delta B/\delta \psi(x) = 0$ at a constant $\int \psi^\dagger \psi dx$ gives $-i\partial \psi/\partial x = px$ where

$$p = K + \frac{2\pi m}{\ell} \tag{16}$$

with K given by (4) and $m = \cdots, -1, 0, 1, \cdots$.

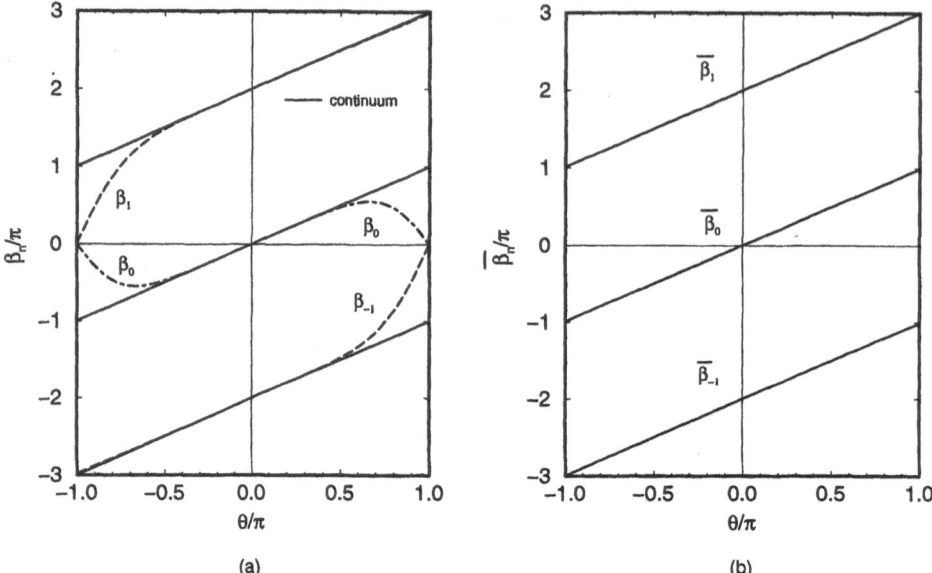

(a)   (b)

Fig. 1. — (a) The dash-dot line gives $\beta_0 = 3\sin\theta/(2+\cos\ell)$ vs. $\theta = K\ell$ and the dashed lines denote β_{-1} and β_1 defined by (13). The solid lines are the continuum free-particle spectrum $p\ell = K\ell + 2\pi m$ vs. θ, with $m = 0$ and ± 1. (b) Taking into account $n = 0$ and $n = \pm 1$, we have removed all spurious zero-mode solutions.

In Fig. 1, the abscissa is $\theta/\pi = K\ell/\pi$, the solid line gives the exact continuum value $p\ell/\pi$ and the different segments correspond to $m = -1, 0, 1$. The dashed line segments, in Fig. 1a are the corresponding β_n/π defined by (13). For $|n| > 1$, each β_n deviates from the exact continuum result (16) within $< 1\%$. For $|n| \leq 1$, we see that β_0 and β_{-1} are both 0 at $\theta = \pi$; likewise β_0 and β_1 are both 0 at $\theta = -\pi$. Thus, the spurious solutions also extend to $n = \pm 1$ bands. This additional unwanted degeneracy makes it easy to remove all spurious solutions, as we shall see.

Because $f_0(K \mid x)$ and $f_1(K \mid x)$ are not eigenfunctions of $-i\partial/\partial x$, the degeneracy between β_0 and β_{-1} at $\theta = \pi$ can be removed by considering the off-diagonal elements of $-i\partial/\partial x$. At $\theta = \pi - \epsilon$ where ϵ is a positive infinitesimal, we may consider only two bands, $n = 0$ and -1; in this subspace the operator $-i\partial/\partial x$ becomes the following 2×2 matrix:

$$\begin{pmatrix} 3\epsilon & \sqrt{10}/\ell \\ \sqrt{10}/\ell & -5\epsilon \end{pmatrix} \tag{17}$$

As $\epsilon \to 0$, its eigenvalues are

$$\pm\sqrt{10}/\ell = \pm 1.006 \times x/\ell. \tag{18}$$

Similar considerations apply to $\theta = -\pi$ by taking into account the coupling between $n = 0$ and $n = 1$ bands. Thus, by taking into account only $n = 0$ and ± 1, we have removed all spurious zero-mode solutions and, and in addition, the result differs from the exact continuum value by less than a few tenths of a percent for the entire range, see Fig. 1b.

3. LUMP FUNCTIONS

The function $\Delta(x)$, introduced in the zeroth-band approximation above is a special case of a general class of functions which we call lump functions. They are defined by

$$L_m(x) \equiv \int_{-\infty}^{\infty} \frac{dk}{2\pi} \frac{(2\sin\frac{k}{2})^m}{k^m} \times \begin{cases} e^{ikx/\ell} & \text{for } m \text{ even} \\ e^{ik(x-\ell/2)/\ell} & \text{for } m \text{ odd} \end{cases} \tag{19}$$

where m is a positive integer > 0 and x is the continuous coordinate variable. We see that for m even

$$\ell\frac{dL_m(x)}{dx} = -L_{m-1}(x) + L_{m-1}(x + \ell) \tag{20}$$

and for m odd

$$\ell\frac{dL_m(x)}{dx} = L_{m-1}(x) - L_{m-1}(x - \ell). \tag{21}$$

Another useful property is, within the range $0 < x < \ell$,

$$\sum_{j=-(m-2)/2}^{m/2} L_m(x - j\ell) = \text{constant} \quad \text{for } m \text{ even} \tag{22}$$

and

$$\sum_{j=-(m-1)/2}^{(m-1)/2} L_m(x - j\ell) = \text{constant} \quad \text{for } m \text{ odd} \tag{23}$$

which can be readily proven by noting that their derivatives are both zero, on account of (20) and (21). These type of functions are related to the mathematical problem called partition of one [4].

The lump functions of lower order, $m = 1, 2$ and 3, are particularly useful in our lattice formulations of quantum field theories. It is convenient to give these functions special names:

$$C(x) \equiv L_1(x) = \begin{cases} 1 & \text{for } 0 < x < \ell \\ 0 & \text{otherwise,} \end{cases} \tag{24}$$

$$\Delta(x) \equiv L_2(x) \tag{25}$$

and

$$S(x) \equiv L_3(x) = \begin{cases} \frac{1}{8}[(3 - 2\,|\,\frac{x}{\ell} - \frac{1}{2}\,|)^2 - 3(1 - 2\,|\,\frac{x}{\ell} - \frac{1}{2}\,|)^2] & \text{for } |\,x - \frac{\ell}{2}\,| < \frac{\ell}{2} \\ \frac{1}{8}[(3 - 2\,|\,\frac{x}{\ell} - \frac{1}{2}\,|)^2 & \text{for } \frac{\ell}{2}\,|\,x - \frac{\ell}{2}\,| < \frac{3\ell}{2} \\ 0 & \text{otherwise.} \end{cases} \tag{26}$$

4. NONCOMPACT LATTICE QCD

In order to extend the above consideration to QCD we have to construct an appropriate complete set of Bloch wavefunctions that is compatible to the gauge fixing condition. As we shall see, once that is done the restriction to the zeroth band $(n = 0)$ gives a noncompact formulation [2] of lattice QCD. The exact continuum theory can be reached through inclusion of all $n = 0$ and $n \neq 0$ bands, without requiring the lattice size $\ell \to 0$. This makes it possible, at a nonzero ℓ, for the lattice coupling g_ℓ to act as the renormalized continuum coupling. All physical results in the continuum are, of course, independent of ℓ.

To construct zeroth-band Bloch functions in the Coulomb gauge, we define

$$f(K \mid x) \equiv \sum_j e^{iKjl} \Delta(x - j\ell), \tag{27}$$

$$g(K \mid x) \equiv \sum_j e^{iK(j+\frac{1}{2})l} S(x - j\ell), \tag{28}$$

with $\mid K \mid \leq \pi/\ell$. Apart from a normalization constant, $f(K \mid x)$ is the same function given by (9). From (20), we see that

$$\ell \frac{\partial}{\partial x} g(K \mid x) = 2i \sin \frac{\theta}{2} f(K \mid x) \tag{29}$$

where $\theta \equiv K\ell$ and therefore $\mid \theta \mid \leq \pi$. Introduce

$$\begin{aligned}
F_x(\vec{K} \mid \vec{r}) &= c_1 g(K_1 \mid x) f(K_2 \mid y) f(K_3 \mid z) \\
F_y(\vec{K} \mid \vec{r}) &= c_2 f(K_1 \mid x) g(K_2 \mid y) f(K_3 \mid z) \\
F_z(\vec{K} \mid \vec{r}) &= c_3 f(K_1 \mid x) f(K_2 \mid y) g(K_3 \mid z)
\end{aligned} \tag{30}$$

where c_1, c_2 and c_3 are normalization constants. On account of (29), the derivatives

$$\frac{\partial F_x}{\partial x} \propto \frac{\partial F_y}{\partial y} \propto \frac{\partial F_z}{\partial z} \propto \prod_{a=1}^{3} f_a(K_a \mid x_a). \tag{31}$$

For each given $\vec{K}$, it is straightforward to introduce two unit vectors $\vec{\epsilon}_1(\vec{K})$ and $\vec{\epsilon}_2(\vec{K})$, orthogonal to each other, and the vector function $\vec{\mathcal{F}}_\tau(\vec{K} \mid \vec{r})$ whose spatial components are

$$\vec{\mathcal{F}}_\tau(\vec{K} \mid \vec{r})_i = \vec{\epsilon}_\tau(\vec{K})_i F_i(\vec{K} \mid \vec{r}); \tag{32}$$

so that

$$\vec{\nabla} \cdot \vec{\mathcal{F}}_\tau(\vec{K} \mid \vec{r}) = 0 \tag{33}$$

at all continuum $\vec{r}$ for the polarization index $\tau = 1$ or 2. The gauge field operator, in the zeroth-band approximation, is

$$\vec{A}^m(\vec{r}, t) = \sum_{\tau=1}^{2} \sum_{\vec{K}} q_\tau^m(\vec{K}, t) \vec{\mathcal{F}}_\tau(\vec{K} \mid \vec{r}). \tag{34}$$

The Coulomb gauge condition $\vec{\nabla} \cdot \vec{A}^m$ is satisfied everywhere. It is straightforward to generalize (11) for the construction of the higher-band Bloch functions by using Fourier series and the zeroth-band function given above.

In our formulation, the zeroth-band approximation is the noncompact lattice QCD, whose Hamiltonian H has exactly the same continuum form derived by Christ and Lee [5] in Coulomb gauge, except for the replacement of the gauge field operators, its conjugate momenta, and the bare coupling constant in the continuum with the corresponding lattice operators and lattice coupling constant.

Acknowledgments

The authors wish to thank R. Friedberg and T. D. Lee for collaborating on this project. Y. P. acknowledges the support of Alfred P. Sloan Foundation. This project is also supported in part by U.S. Department of Energy under grants DE-FG02-92 ER40699, DOE-91ER40651, Task B, and DE-AC02-76CH00016.

References

[1] Friedberg R., Lee T. D. and Pang Y., *J. Math. Phys.* **35** (1994) 5601.
[2] Friedberg R., Lee T. D., Pang Y. and Ren H. C., *Phys. Rev. D* **52** (1995) 4053.
[3] Nielsen H. B., *Ninomiya M.* **Nucl. Phys. B** (185) 1981,20.
[4] Grange P., private communication.
[5] Christ N. H. and Lee T. D., *Phys. Rev. D* **22** (1980) 939.

Colour-Dielectric Gauge Theory on a Transverse Lattice

B. van de Sande([1]), S. Dalley([2])

([1])*Institut Für Theoretische Physik III*
Staudstraße 7, D-91058 Erlangen, Germany
([2]) *Department of Applied Mathematics and Theoretical Physics*
Silver Street, Cambridge CB3 9EW, England

We investigate consequences of the effective colour-dielectric formulation of lattice gauge theory using the light-cone Hamiltonian formalism with a transverse lattice [1]. As a quantitative test of this approach, we have performed extensive analytic and numerical calculations for $2+1$-dimensional pure gauge theory in the large N limit. We study the structure of coupling constant space for our effective potential by comparing with results available from conventional Euclidean lattice Monte Carlo simulations of this system. In particular, we calculate and measure the scaling behaviour of the entire low-lying glueball spectrum, glueball wavefunctions, string tension, asymptotic density of states, and deconfining temperature.

The recent Euclidean Lattice Monte Carlo (ELMC) simulations of Teper [2] have shown that pure non-Abelian gauge theory behaves much the same way in three as in four dimensions: a discrete set of massive boundstates are generated by a linearly confining string-like force. Moreover, Teper has performed calculations for $N = 2$, 3, and 4, allowing an extrapolation to large N. The large-N limit is convenient, though not essential, for our Light-Cone Transverse Lattice (LCTL) formulation. Our formulation offers the rare possibility of describing the parton, constituent, and string behaviour of hadrons in one framework. The relationship between these pictures, each very different but equally successful, remains one of the outstanding enigmas of QCD.

We characterise a dielectric formulation as one in which gluon fields, or rather the $SU(N)$ group elements they generate, are replaced by collective variables which represent an average over the fluctuations on short distance scales, represented by complex $N \times N$ matrices M. These dielectric variables carry colour and form an effective gauge field theory with classical action minimised at zero field, meaning that colour flux is expelled from the vacuum at the classical level. The price one pays for starting with a simple vacuum structure is that the effective action will be largely unknown and must be investigated *per se*.

Starting with the Wilson lattice action [3], we take the continuum limit in the x^0 *and* x^2 directions, leaving the transverse direction x^1 discrete. Replacing the link variables $U \in SU(N)$ with $N \times N$ complex matrics M, one derives a transverse lattice action whose form was first suggested in Ref. [4]

$$A = \int dx^0 dx^2 \sum_{x^1} \left(\text{Tr}\left\{ D_\alpha M_{x^1} (D^\alpha M_{x^1})^\dagger \right\} - \frac{a}{4g^2} \text{Tr}\left\{ F_{\alpha\beta} F^{\alpha\beta} \right\} - V_{x^1}[M] \right)$$
(1)

where $\alpha, \beta \in \{0, 2\}$ and

$$D_\alpha M_{x^1} = \left(\partial_\alpha + i A_\alpha(x^1) \right) M_{x^1} - i M_{x^1} A_\alpha(x^1 + a) .$$
(2)

M_{x^1} lies on the link between x^1 and $x^1 + a$ while $A_\alpha(x^1)$ is associated with the site x^1. $V_{x^1}[M]$ is a purely transverse gauge invariant effective potential. Next we introduce light-cone co-ordinates $x^\pm = x_\mp = (x^0 \pm x^2)/\sqrt{2}$ and quantise by treating x^+ as canonical time. The theory has a conserved current

$$J_{x^1}^\alpha = i \left[M_{x^1} \overset{\leftrightarrow}{D^\alpha} M_{x^1}^\dagger + M_{x^1-a}^\dagger \overset{\leftrightarrow}{D^\alpha} M_{x^1-a} \right]$$
(3)

at each transverse lattice site x^1. If we pick the light-cone gauge $A_- = 0$ the non-propagating field A_+ satisfies a simple constraint equation at each transverse site $(\partial_-)^2 A_+(x^1) = g^2 J_{x^1}^+/a$. Solving this constraint leaves an action in terms of the dynamical fields M_{x^1}

$$A = \int dx^+ dx^- \sum_{x^1} \text{Tr}\left\{ \partial_\alpha M_{x^1} \partial^\alpha M_{x^1}^\dagger + \frac{g^2}{2a} J_{x^1}^+ \frac{1}{(\partial_-)^2} J_{x^1}^+ \right\} - V_{x^1}[M] .$$
(4)

At large N, Eguchi-Kawai reduction [5] introduces considerable simplification. For $P^1 = 0$ the theory is isomorphic to one compactified on a one-link transverse lattice, *id est* we can simply drop the argument x^1 or l from M in all of the previous expressions. Effectively one is now dealing with a $1 + 1$-dimensional gauge theory coupled to a complex scalar field in the adjoint representation (with self-interactions).

For the transverse effective potential $V[M]$, we will include all Wilson loops and products of Wilson loops up to fourth order in link fields M:

$$\begin{aligned} V[M] &= \mu^2 \text{Tr}\left\{ MM^\dagger \right\} + \frac{\lambda_1}{aN} \text{Tr}\left\{ MM^\dagger MM^\dagger \right\} \\ &\quad + \frac{\lambda_2}{aN} \text{Tr}\left\{ MM^\dagger M^\dagger M \right\} + \frac{\lambda_3}{aN^2} \text{Tr}\left\{ M^\dagger M \right\} \text{Tr}\left\{ M^\dagger M \right\} \end{aligned}$$
(5)

Note that the last term above, which might appear suppressed at large N, is in fact non-zero only for 2 particle Fock states.

Let us introduce creation/annihilation operators

$$M_{x^1}(x^-) = \frac{1}{\sqrt{4\pi}} \int_0^\infty \frac{dk}{\sqrt{k}} \left(a_{-1}(k, x^1) e^{-ikx^-} + \left(a_{+1}(k, x^1) \right)^\dagger e^{ikx^-} \right). \quad (6)$$

In the associated Fock space, we include only states annihilated by the charge $\int dx^- J^+$. This gives a Hilbert space formed from all possible closed Wilson loops of link modes $a_\pm$ on the transverse lattice. Thus, a typical p-link loop will be something like

$$\text{Tr}\left\{ a_{+1}^\dagger(k_1) a_{-1}^\dagger(k_2) a_{-1}^\dagger(k_3) \cdots a_{+1}^\dagger(k_p) \right\} |0\rangle \quad (7)$$

where the number of $+1$'s equals the number of -1's, $\sum_{m=1}^p k_m = P^+$, and $k_m \geq 0$. At large N we need only study the dynamics of single connected Wilson loops in the Hilbert space since the loop-loop coupling constant is of order $1/N$. These loops may be thought of as 'bare' glueballs, and the problem is to find the linear combinations that are on mass shell. Neglecting $k_m = 0$, which is consistent with expanding about the $M = 0$ solution of the dielectric regime, the Fock vacuum is an eigenstate of the full light-cone Hamiltonian $P^- |0\rangle = P^+ |0\rangle = 0$.

The theory possesses several discrete symmetries. Charge conjugation induces the symmetry $\mathcal{C} : a_{+1,ij}^\dagger \leftrightarrow a_{-1,ji}^\dagger$. There are two orthogonal reflection symmetries $\mathcal{P}_1$ and $\mathcal{P}_2$ either of which may be used as 'parity'. If $\mathcal{P}_1 : x^1 \rightarrow -x^1$, we have $\mathcal{P}_1 : a_{+1,ij}^\dagger \leftrightarrow a_{-1,ij}^\dagger$. If rotational symmetry has been restored in the theory, states of spin $\mathcal{J} \neq 0$ should form degenerate $\mathcal{P}_1$ doublets $| + \mathcal{J}\rangle \pm | - \mathcal{J}\rangle$ [2]. We use "spectroscopic notation" $|\mathcal{J}|^{\mathcal{P}_1 \mathcal{C}}$ to classify states.

For λ_1 and λ_2 small there is very little mixing between Fock states of different number of link modes p. In this case a mass eigenstate $|\Psi\rangle$ has predominantly a fixed p, the mass increasing with p. For a given p, the energy also tends to increase with the number of nodes in the wavefunction f due to the $J (\partial_-)^{-2} J$ term (1), which is in fact a positive contribution. Thus one expects the lowest two eigenstates to be approximately

$$\int_0^{P^+} dk\, f_{+1,-1}(k, P^+ - k)\, \text{Tr}\left\{ a_{+1}^\dagger(k) a_{-1}^\dagger(P^+ - k) \right\} |0\rangle \quad (8)$$

with the lowest state having a symmetric wavefunction $f_{+1,-1}(k, P^+ - k)$, corresponding to 0^{++}, and first excited state having $f_{+1,-1}$ antisymmetric with one node, corresponding to 0^{--}. The next highest states are either a 4-link state with positive symmetric wavefunctions $f_{+1,+1,-1,-1}$ and $f_{+1,-1,+1,-1}$ or a symmetric 2-link state with $f_{+1,-1}$ having two nodes. In the glueball spectrum we identify the latter states as 0_*^{++} and 2^{++}, respectively, although actual eigenstates are a mixture of these.

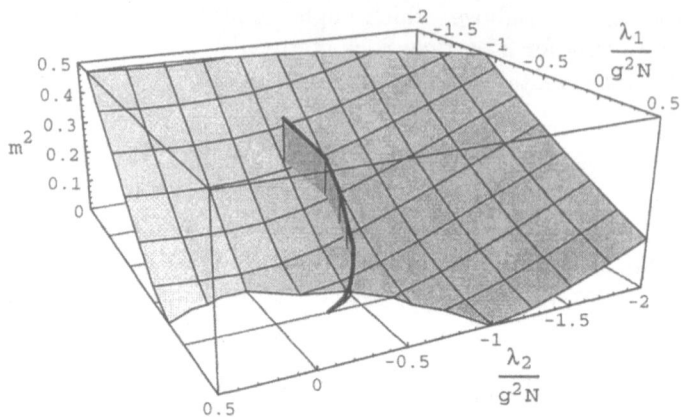

Fig. 1. — Mass $m^2 = \mu^2 a/(g^2 N)$ such that the lowest M^2 eigenvalue is zero vs $\lambda_1/(g^2 N)$ and $\lambda_2/(g^2 N)$ for $p \le 6$, $K = 14$, $\lambda_3 = 100 g^2 N$. Below this surface the spectrum is tachyonic and below $m^2 = 0$ our quantisation breaks down. The dark line is the scaling trajectory.

In order to fix the coupling constants in the effective potential, we perform a least χ^2 fit to Teper's ELMC large N extrapolated spectrum. As we shall later show, the mass in units of the coupling $m^2 = \mu^2 a/(g^2 N)$ is a measure of the lattice spacing while the other terms in our effective potential λ_1, λ_2, and λ_3 must be found from the fitting procedure. We will also determine $g^2 N/a$ based on a fit, which we check self-consistently with measurements of the string tension.

In our numerical solutions we restrict the number of link fields p in our basis states and discretise momenta by demanding antiperiodicity of the fields in $x^- \to x^- + L$. In order to minimise errors associated with these truncations, we take the spectra for various K and p truncations and extrapolate to the continuum, $K, p \to \infty$. Since we cannot measure $|\mathcal{J}|$ directly, we only classified states according to $\mathcal{P}_1$ and $\mathcal{C}$ during the fitting process. We estimate our one-sigma errors from finite K and p-truncation to be roughly $0.05 M^2$. As we fit the various couplings as a function of m^2, we find a narrow strip in parameter space — a scaling trajectory — where we obtain good agreement with the ELMC spectrum. As we shall see, moving along this strip corresponds to changing the lattice spacing a. The strip, where χ^2 has a local minimum, disappears when m^2 is sufficiently large, indicating that for large enough lattice spacing our truncation of the effective potential is no longer a good approximation.

The numerical bound from absence of tachyons is shown in Fig. 1 as a zero-mass surface. As the transverse lattice spacing vanishes the mass gap should vanish in lattice units. The fixed point for this, which we believe lies somewhere at negative m^2, should lie on the zero-mass surface, but is inaccessible to us

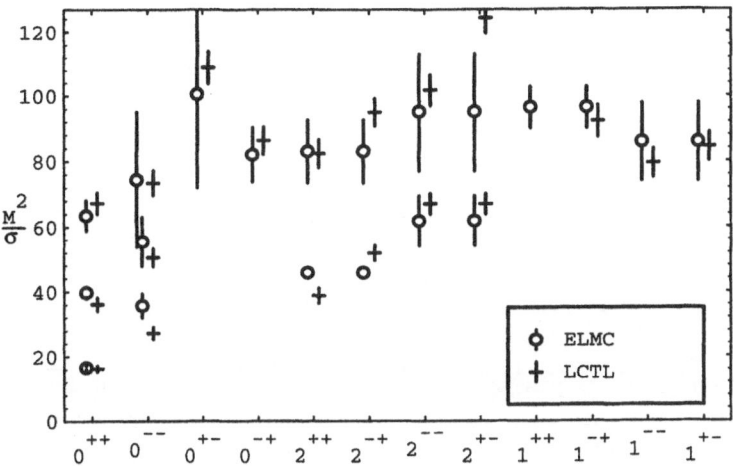

Fig. 2. — A fit of our LCTL extrapolated results against Teper's ELMC large N extrapolated spectrum. The M^2 eigenvalues are shown in units of (Teper's) string tension for various $|\mathcal{J}|^{P_1 C}$. The parameters are $m^2 = 0.065$, $\lambda_1 = -0.202g^2 N$, $\lambda_2 = -0.55g^2 N$, $\lambda_3 = 100g^2 N$, and $g^2 N = 3.47a\sigma$ with $\chi^2 = 23.5$. One finds similar spectra along the entire scaling trajectory in coupling constant space.

in the dielectric regime $m^2 > 0$. Nevertheless the scaling trajectory should gradually approach the zero-mass surface if it is to eventually encounter the fixed point.

In Fig. 2, we have plotted a typical spectrum along the scaling trajectory together with the ELMC results. For graphing purposes, we assigned $|\mathcal{J}|$ to our spectrum based on a best fit to Teper's results. Although the overall fit with the conventional lattice results is quite good, we see two deficiencies of our spectrum that cannot be attributed to K or p-truncation errors. First, we see that the lowest 0^{--} state is too low in energy. Second, we see that the lowest parity doublet $2^{\pm +}$ is not quite degenerate. We believe that these discrepancies must be due to our truncation of the effective potential. Finally, we have made no prediction for the lowest 1^{++} state since it lies too high in the $|\mathcal{J}|^{++}$ spectrum.

To measure the string tension in the x^1 direction (before Eguchi-Kawai reduction) consider a lattice with n transverse links and periodic boundary conditions. Constructing a basis of Polyakov loops, "winding modes," that wind once around this lattice, one may extract from the lowest eigenvalue M^2 vs n the lattice string tension $a\sigma = \Delta M_n / \Delta n$. String theory arguments indicate that oscillations of the winding mode transverse to itself yield a form $M^2/\sigma = \sigma a^2 n^2 - \pi/3$, the constant correction being due to the Casimir energy [6]. Fig. 3 shows a typical M^2 vs n plot for winding modes where we see a good fit to a quadratic. The constant term -3.5, however, does not agree well

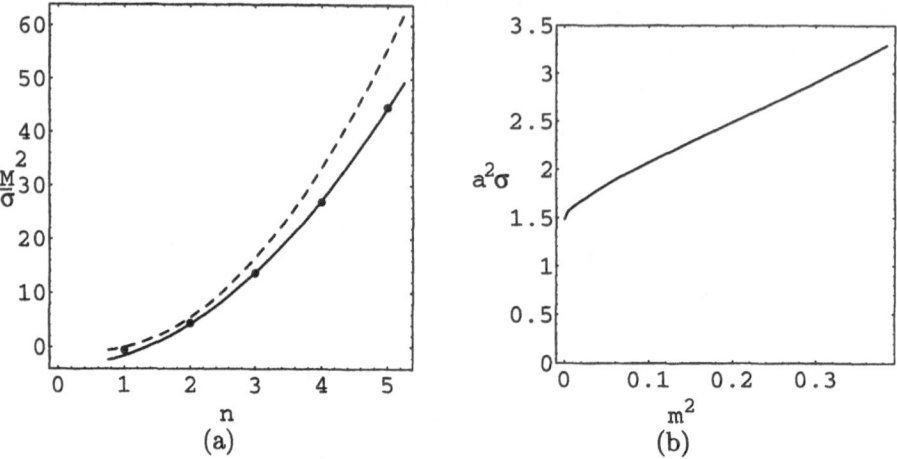

Fig. 3. — (a) Lowest M^2 eigenvalue vs n for winding modes. Here, $K = 10.5$ or 11, $p \leq n + 4$, and the couplings are taken from Fig. 2. Also shown is a numerical fit to $1.92n^2 - 3.5$. The dashed line is from an analytic estimate. (b) The lattice spacing in units of the string tension along the scaling trajectory. Here we plot $a^2\sigma$ vs m^2.

with the expected value of $-\pi/3$.

As a consistency check, we use σa^2 from the string tension measurement together with $g^2N/(a\sigma)$ from the spectrum to form the ratio σg^2 (as measured by us) to σg^2 (as measured by ELMC). The result is equal to one $\pm 5\%$ all along the scaling trajectory. If we then assume that our σ is equal to the ELMC value of σ, we can determine the lattice spacing a in units of the string tension. This demonstrates that the mass $m^2 = \mu^2 a/(g^2N)$ determines the lattice spacing and that the continuum limit occurs at $m^2 < 0$; see Fig. 3.

Another quantity of interest is the deconfinement temperature T_c which may be obtained from the Hagedorn behaviour [7] of the asymptotic density of mass eigenstates $\rho(M) \sim M^{-\alpha} \exp(M/T_c)$. The canonical partition function diverges for $T > T_c$. If $\alpha > (D+1)/2$ it is a phase transition, beyond which the canonical and microcanonical ensembles are inequivalent. If $\alpha < (D+1)/2$ the ensembles are equivalent and T_c represents a limiting temperature — the free energy diverges at T_c. It is essential to demonstrate that the spectrum is sufficiently converged in K, thus we only fit to the states between $0.5 < \log(t) < 5.5$ for the $K = 7$ data in Fig. 4. A numerical fit gives,

$$\log(t) = 3.99 + \frac{M}{0.78\sqrt{\sigma}} - 5.86 \log\left(\frac{M}{\sqrt{\sigma}}\right). \tag{9}$$

Since the density of states is $\rho(M) = dt/dM$, we find $T_c = 0.78\sqrt{\sigma}$ with an estimated error of at least 10%. Due to the large error, this result is compatible with the Euclidean lattice result [2, 8]. The fact that the power correction

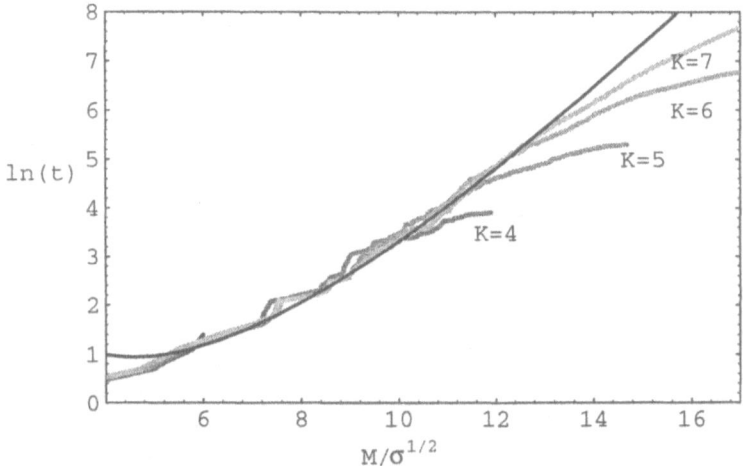

Fig. 4. — $\ln t$ vs the mass of the t-th eigenvalue M_t in units of the string tension. Here we have applied a cutoff in K only. Also shown is a least squares fit to the $K = 7$ data in the range $0.5 < \log(t) < 5.5$ and couplings from Fig. 2.

$\alpha \approx 4.86$ in the density $\rho(M)$ is much larger than $(D + 1)/2 = 2$ means we can safely say that the thermodynamic free energy remains finite at T_c, marking a true phase transition (presumably deconfinement) rather than a limiting temperature. In addition, we have an analytic estimate suggesting that T_c lies in the range $0.81\sqrt{\sigma} < T_c < 0.98\sqrt{\sigma}$, also in agreement with the ELMC result.

References

[1] Dalley S. and van de Sande B., Cambridge University report No. DAMTP-97-16, hep–ph/9704408.
[2] Teper M., *Phys. Lett.* **289B** (1992) 115; *Phys. Lett.* **311B** (1993) 223; *Nucl. Phys. (Proc. Suppl.)* **B53** (1997) 715; Oxford University Report OUTP-97-01P (unpublished) hep–lat/9701004.
[3] Wilson K.G., *Phys. Rev.* D **10** (1974) 2445.
[4] Bardeen W.A. and Pearson R.B., *Phys. Rev.* D **14** (1976) 547; Bardeen W.A., Pearson R.B. and Rabinovici E., *Phys. Rev.* D **21** (1980) 1037.
[5] Eguchi T. and Kawai H., *Phys. Rev. Lett.* **48** (1983) 1063.
[6] Lüscher M., *Nucl. Phys.* **B180** (1981) 317; de Forcrand P., Schneider H., Teper M., *Phys. Lett.* **B160** (1985) 137.
[7] Hagedorn R., *Nuovo Cimento Suppl.* **3** (1965) 147; Frautchi S., *Phys. Rev.* D **3** (1971) 2821; Carlitz R.D., *Phys. Rev.* D **5** (1972) 3231.
[8] Billo M., Caselle M., D'Adda A. and Panzeri S., *Int. J. Mod. Phys.* A **12** (1997) 1783.

CHAPTER V

Light Cone Quantization under Scrutiny

Light Front Treatment of Nuclei and Deep Inelastic Scattering

G. A. Miller([1,2,3])

([1]) *permanent address Department of Physics, Box 351560, University of Washington, Seattle, WA 98195-1560*
([2]) *Stanford Linear Accelerator Center, Stanford University, Stanford, California 94309*
([3]) *national Institute for Nuclear Theory, Box 35150, University of Washington, Seattle, WA 98195-1560*

1. INTRODUCTION

A light front treatment of the nuclear wave function is developed and applied, using the mean field approximation, to infinite nuclear matter. The nuclear mesons are shown to carry about a third of the nuclear plus momentum p^+; but their momentum distribution has support only at $p^+ = 0$, and the mesons do not contribute to nuclear deep inelastic scattering. This zero mode effect occurs because the meson fields are independent of space-time position.

2. DISCUSSION

The discovery that the deep inelastic scattering structure function of a bound nucleon differs from that of a free one (the EMC effect[1]) changed the way that physicists viewed the nucleus. With a principal effect that the plus momentum (energy plus third component of the momentum, $p^0 + p^3 \equiv p^+$) carried by the valence quarks is less for a bound nucleon than for a free one, quark and nuclear physics could not be viewed as being independent.

The interpretation of the experiments requires that the role of conventional effects, such as nuclear binding, be assessed and understood[2, 3]. Nuclear binding is supposed to be relevant because the plus momentum of a bound nucleon is reduced by the binding energy, and so is that of its confined quarks. Conservation of momentum implies that if nucleons lose momentum, other constituents such as nuclear pions[4], must gain momentum. This partitioning of the total plus momentum amongst the various constituents is the momentum sum rule. Pions are quark anti-quark pairs so that a specific enhancement of the nuclear antiquark momentum distribution is a testable [5] consequence of this idea. A nuclear Drell Yan experiment [6], was performed, but no influence of nuclear pion enhancement was seen. This led Ref. [7] to question the idea of the pion as a dominant carrier of the nuclear force.

3.   NUCLEAR CALCULATION

Here a closer look at the relevant nuclear theory is taken, and the momentum sum rule is studied. This talk is based on Ref.[8].

The structure function depends on the Bjorken variable x_{Bj} which in the parton model is the ratio of the quark plus momentum to that of the target. Thus $x_{Bj} = p^+/k^+$, where k^+ is the plus momentum of a nucleon bound in the nucleus, so a more direct relationship between the necessary nuclear theory and experiment occurs by using a theory in which k^+ is one of the canonical variables. Since k^+ is conjugate to a spatial variable $x^- \equiv t - z$, it is natural to quantize the dynamical variables at the equal light cone time variable of $x^+ \equiv t + z$. To use such a formalism is to use light front quantization, which requires a new derivation of the nuclear wave function, because previous work used the equal time formalism.

Such a derivation is provided here, using a simple model in which the nuclear constituents are nucleons ψ (or ψ'), scalar mesons ϕ[10] and vector mesons V^μ. The Lagrangian $\mathcal{L}$ is given by

$$\mathcal{L} = \frac{1}{2}(\partial_\mu \phi \partial^\mu \phi - m_s^2 \phi^2) - \frac{1}{4}V^{\mu\nu}V_{\mu\nu} + \frac{m_v^2}{2}V^\mu V_\mu$$
$$+\bar{\psi}'\left(\gamma^\mu(i\partial_\mu - g_v\,V_\mu) - M - g_s\phi\right)\psi' \qquad (1)$$

where the bare masses of the nucleon, scalar and vector mesons are given by $M, m_s,\ m_v$, and $V^{\mu\nu} = \partial^\mu V^\nu - \partial^\nu V^\mu$. This Lagrangian may be thought of as a low energy effective theory for nuclei under normal conditions.

This hadronic model, when evaluated in mean field approximation, gives[11] at least a qualitatively good description of many (but not all) nuclear properties and reactions. The aim here is to use a simple Lagrangian to study the effects that one might obtain by using a light front formulation. It is useful to simplify this first calculation by studying infinite nuclear matter.

The light front quantization procedure necessary to treat nucleon interactions with scalar and vector mesons was derived by Yan and collaborators[13,

14]. Glazek and Shakin[15] used a Lagrangian containing nucleons and scalar mesons to study infinite nuclear matter. Here both vector and scalar mesons are included, and the nuclear plus momentum distribution is obtained.

Next examine the field equations. The nucleons satisfy

$$\gamma \cdot (i\partial - g_v V)\psi' = (m + g_s\phi)\psi'. \tag{2}$$

The number of independent degrees of freedom for light front field theories is fewer than in the usual theory. One defines projection operators $\Lambda_\pm \equiv \gamma^0\gamma^\pm/2$ and the independent Fermion degree of freedom is $\psi'_+ = \Lambda_+\psi'$. The relation between ψ'_- and ψ'_+ is very complicated unless one may set the plus component of the vector field to zero. This is a matter of a choice of gauge for QED and QCD, but the non-zero mass of the vector meson prevents such a choice here. Instead, one simplifies the equation for ψ'_- by[14] transforming the Fermion field according to $\psi' = e^{-ig_v\Lambda(x)}\psi$ with $\partial^+\Lambda = V^+$ which yields

$$(i\partial^- - g_v\bar{V}^-)\psi_+ = (\vec{\alpha}_\perp \cdot (\vec{p}_\perp - g_v\vec{\bar{V}}_\perp) + \beta(M + g_s\phi))\psi_-$$
$$i\partial^+\psi_- = (\vec{\alpha}_\perp \cdot (\vec{p}_\perp - g_v\vec{\bar{V}}_\perp) + \beta(M + g_s\phi))\psi_+ \tag{3}$$

where

$$\partial^+\bar{V}^\mu = \partial^+V^\mu - \partial^\mu V^+ \tag{4}$$

The field equations for the mesons are

$$\partial_\mu V^{\mu\nu} + m_v^2 V^\nu = g_v\bar{\psi}\gamma^\nu\psi$$
$$\partial_\mu\partial^\mu\phi + m_s^2\phi = -g_s\bar{\psi}\psi. \tag{5}$$

We now introduce the mean field approximation[11]. The coupling constants are considered strong and the Fermion density large. Then the meson fields can be approximated as classical- the sources of the meson fields are replaced by their expectation values. In this case, the nucleon mode functions will be plane waves and the nuclear matter ground state can be assumed to be a normal Fermi gas, of Fermi momentum k_F, and of large volume Ω in its rest frame. We consider the case that there is an equal number of protons and neutrons. Then the meson fields are constants given by

$$\phi = -\frac{g_s}{m_s^2}\langle\bar{\psi}\psi\rangle$$
$$V^\mu = \frac{g_v}{m_v^2}\langle\bar{\psi}\gamma^\mu\psi\rangle = \delta^{0,\mu}\frac{g_v\rho_B}{m_v^2}, \tag{6}$$

where $\rho_B = 2k_F^3/3\pi^2$. This result that V^μ is a constant, along with Eqs. (4) and (6), tells us that the only non-vanishing component of $\bar{V}$ is $\bar{V}^- = V^0$. The expectation values refer to the nuclear matter ground state.

With this, the light front Schroedinger equation for the modes of the field operator $\sim e^{ik\cdot x}$ and can be obtained from Eq. (3) as[12]

$$(i\partial^- - g_v\bar{V}^-)\psi_+ = \frac{\vec{k}_\perp^2 + (M + g_s\phi)^2}{k^+}\psi_+. \tag{7}$$

The light front eigenenergy ($i\partial^- \equiv k^-$) is the sum of a kinetic energy term in which the mass is shifted by the presence of the scalar field, and an constant energy arising from the vector field. Thus the nucleons have a mass $M + g_s\phi$ and move in plane wave states. The nucleon field operator is constructed using the solutions of Eq. (7) as the plane wave basis states. This means that the nuclear matter ground state, defined by operators that create and destroy baryons in eigenstates of Eq. (7), is the correct wave function and that Equations (6) and (7) represent the solution of the approximate field equations, and the diagonalization of the Hamiltonian.

The computation of the energy and plus momentum distribution proceeds from taking the appropriate expectation values of $T^{\mu\nu}$:

$$P^\mu = \frac{1}{2} \int d^2 x_\perp dx^- \langle T^{+\mu} \rangle. \tag{8}$$

We are concerned with the light front energy P^- and momentum P^+. Within the mean field approximation one finds

$$T^{+-} = m_s^2 \phi^2 + 2\psi_+^\dagger (i\partial^- - g_v \bar{V}^-)\psi_+$$
$$T^{++} = m_v^2 V_0^2 + 2\psi_+^\dagger i\partial^+ \psi_+. \tag{9}$$

Taking the nuclear matter expectation value of T^{+-} and T^{++} and performing the spatial integral of Eq. (8) leads to the result

$$\frac{P^-}{\Omega} = m_s^2 \phi^2 + \frac{4}{(2\pi)^3} \int_F d^2 k_\perp dk^+ \frac{\vec{k}_\perp^2 + (M + g_s\phi)^2}{k^+} \tag{10}$$

$$\frac{P^+}{\Omega} = m_v^2 V_0^2 + \frac{4}{(2\pi)^3} \int_F d^2 k_\perp dk^+ k^+. \tag{11}$$

The subscript F denotes that $|\vec{k}| < k_F$ with k^3 defined by the relation

$$k^+ = \sqrt{(M + g_s\phi)^2 + \vec{k}^2} + k^3. \tag{12}$$

The expression for the energy of the system $E = \frac{1}{2}(P^+ + P^-)$[15], is the same as in the usual treatment[11]. This can be seen by summing equations (10) and (11) and changing integration variables using $\frac{dk^+}{k^+} = \frac{dk^3}{\sqrt{(M+g_s\phi)^2 + \vec{k}^2}}$. This equality of energies is a nice check on the present result because a manifestly covariant solution of the present problem, with the usual energy, has been obtained[16]. Moreover, setting $\frac{\partial E}{\partial \phi}$ to zero reproduces the field equation for ϕ, as is also usual. Rotational invariance, here the relation $P^+ = P^-$, follows as the result of minimizing the energy per particle at fixed volume with respect to k_F, or minimizing the energy with respect to the volume[15]. The parameters $g_v^2 M^2 / m_v^2 = 195.9$ and $g_s^2 M^2 / m_s^2 = 267.1$ have been chosen [17] so as to give the binding energy per particle of nuclear matter as 15.75 MeV with $k_F = 1.42$ Fm^{-1}. In this case, solving the equation for ϕ gives $M + g_s\phi = 0.56\,M$.

4. NUCLEAR PLUS MOMENTUM DISTRIBUTIONS

The use of Eq. (11) and these parameters leads immediately to the result that only 65% of the nuclear plus momentum is carried by the nucleons; the remainder is carried by the mesons. This is a much smaller fraction than is found in typical nuclear binding models[2, 3]. The nucleonic momentum distribution which is the input to calculations of the nuclear structure function of primary interest here. This function can be computed from the integrand of Eq.(11). The probability that a nucleon has plus momentum k^+ is determined from the condition that the plus momentum carried by nucleons, P_N^+, is given by $P_N^+/A = \int dk^+ \, k^+ f(k^+)$, where $A = \rho_B \Omega$. It is convenient to use the dimensionless variable $y \equiv \frac{k^+}{\bar{M}}$ with $\bar{M} = M - 15.75$ MeV. Then Eq.(11) and simple algebra leads to the equation

$$f(y) = \frac{3}{4} \frac{\bar{M}^3}{k_F^3} \theta(y^+ - y)\theta(y - y^-) \left[\frac{k_f^2}{\bar{M}^2} - (\frac{E_f}{\bar{M}} - y)^2 \right], \qquad (13)$$

where $y^\pm \equiv \frac{E_F \pm k_F}{\bar{M}}$ and $E_F \equiv \sqrt{k_F^2 + (M + g_s\phi)^2}$. Similarly the baryon number distribution $f_B(y)$ (number of baryons per y, normalized to unity) can be determined from the expectation value of $\psi^\dagger \psi$. The result is

$$f_B(y) = \frac{3}{8} \frac{\bar{M}^3}{k_F^3} \theta(y^+ - y)\theta(y - y^-)$$

$$\left[(1 + \frac{E_F^2}{\bar{M}^2 y^2})(\frac{k_f^2}{\bar{M}^2} - (\frac{E_F}{\bar{M}} - y)^2) - \frac{1}{2y^2}(\frac{k_F^4}{\bar{M}^4} - (\frac{E_F}{\bar{M}} - y)^4) \right], \qquad (14)$$

which is different than $f(y)$.

The nuclear deep inelastic structure function, F_{2A} can be obtained from the light front distribution function $f(y)$ and the nucleon structure function F_{2N} using the relation[3]

$$\frac{F_{2A}(x)}{A} = \int dy \, f(y) F_{2N}(x/y), \qquad (15)$$

where x is the Bjorken variable computed using the nuclear mass divided by A ($\bar{M}$): $x = Q^2/2\bar{M}\nu$. This formula is the expression of the convolution model in which one means to assess, via $f(y)$, only the influence of nuclear binding. Consider the present effect of having the average value of y equal to 0.65. Frankfurt and Strikman[3] use Eq. (15) to argue that an average of 0.95 is sufficient to explain the 15% depletion effect observed for the Fe nucleus. One may also compare the 0.65 fraction with the result 0.91 computed[18] for nuclear matter, including the effects of correlations, using equal time quantization. The present result then represents a very strong binding effect, even though this infinite nuclear matter result can not be compared directly with the experiments using Fe targets. One might think that the mesons, which cause this binding, would also have huge effects on deep inelastic scattering.

It is necessary to determine the momentum distributions of the mesons. The mesons contribute 0.35 of the total nuclear plus momentum, but we need to know how this is distributed over different individual values. The paramount feature is that ϕ and V^μ are the same constants for any and all values of the spatial coordinates $x^-, \vec{x}_\perp$. This means that the related momentum distribution can only be proportional to a delta function setting both the plus and $\perp$ components of the momentum to zero. This result is attributed to the mean field approximation, in which the meson fields are treated as classical quantitates. Thus the finite plus momentum can be thought of as coming from an infinite number of quanta, each carrying an infinitesimal amount of plus momentum. A plus momentum of 0 can only be accessed experimentally at $x_{Bj} = 0$, which requires an infinite amount of energy. Thus, in the mean field approximation, the scalar and vector mesons can not contribute to deep inelastic scattering. The usual term for a field that is constant over space is a zero mode, and the present Lagrangian provides a simple example. For finite nuclei, the mesons would carry a very small momentum of scale given by the inverse of the nuclear radius, under the mean field approximation. If fluctuations were to be included, the relevant momentum scale would be of the order of the inverse of the average distance between nucleons (about 2 Fm).

5. SUMMARY AND ASSESSMENT

The Lagrangian of Eq. (1) and its evaluation in mean field approximation for nuclear matter have been used to provide a simple but semi-realistic example. It would be premature to compare the present results with data. The specific numerical results of the present work are far less relevant than the emergent central feature that the mesons responsible for nuclear binding need not be accessible in deep inelastic scattering. Another interesting feature is that $f(y)$ and $f_B(y)$ are not the same functions.

More generally, we view the present model as being one of a class of models in which the mean field plays an important role. For such models nuclei would have constituents that contribute to the momentum sum rule but do not contribute to deep inelastic scattering. Thus the predictive and interpretive power of the momentum sum rule is vitiated.

Acknowledgments

This work is partially supported by the USDOE. I thank the SLAC theory group and the national INT for their hospitality. I thank S.J. Brodsky, L. Frankfurt, S. Glazek, C.M. Shakin and M. Strikman for useful discussions.

References

[1] Aubert J., *et al., Phys. Lett.* **B123** (1982) 275.

[2] Arneodo M., *Phys. Rep.* **240** (1994) 301.

[3] Frankfurt L.L. and Strikman M.I., *Phys. Rep.* **160** (1988) 235.

[4] Ericson M. and Thomas A.W., *Phys. Lett.* **B128** (1983) 112.

[5] Bickerstaff R.P., Birse M.C. and Miller G.A., *Phys. Rev. Lett.* **53** (1984) 2532.

[6] Alde D.M., et al. *Phys. Rev. Lett.* **64** (1990) 2479.

[7] Bertsch G.F., Frankfurt L. and Strikman M., *Science* **259** (1993) 773.

[8] Miller G.A., 1997 preprint, nucl-th/9702036, submitted to Phys. Rev. C.

[9] Our notation is that a four vector A^μ is defined by the plus, minus and perpendicular components as $(A^0 + A^3, A^0 - A^3, \vec{A}_\perp)$.

[10] The scalar mesons are meant to represent the two pion exchange potential which causes much of the medium range attraction between nucleons, as well as the effects of a fundamental scalar meson. Thus the pion is an important implicit part of the present Lagrangian.

[11] Serot B.D. and Walecka J.D., *Adv. Nucl. Phys.* **16** (1986) 1;IU-NTC-96-17, Jan. 1997 nucl-th/9701058

[12] The symbol for the nucleon field operator and the mode functions of that field is taken to be the same $-\psi$ to reduce the amount of notation.

[13] Chang S-J., Root R.G. and Yan T-M., *Phys. Rev.* **D7** (1973) 1133;*ibid* (1973) 1147.

[14] Yan T-M., *Phys. Rev.* **D7** (1974) 1760;*ibid* (1974) 1780.

[15] Glazek St. and Shakin C.M., *Phys. Rev.* **C44** (1991) 1012.

[16] Serot B.D. and Furnstahl R.J., *Phys. Rev.* **C43** (1991) 105.

[17] Chin S.A. and Walecka J.D., *Phys. Lett.* **B52** (1974) 24.

[18] Dieperink A.E. and Miller G.A., *Phys. Rev.* **C44** (1991) 866.

Light-cone strings

A. Neveu

*Laboratoire de Physique Mathématique, Université Montpellier II-CNRS,
F-34095 Montpellier, France*

1. INTRODUCTION

In this workshop, together with beautiful experimental work, we have heard reports of a lot of theoretical progress, some exciting, some puzzling, perhaps asking more questions than answering them. We have also seen that some problems which were around twenty-five years ago, like the problem of condensates and the vacuum in light-cone field theory, are still not completely solved, in spite of some recent progress. We are going to see that string theory presents the same pattern of unsolved problems.

2. LIGHT-CONE STRINGS

Indeed, the string seems to have been taylor made for the light-cone formalism. The classical string could have been solved ninety years ago, at the dawn of special relativity, but it had to wait for light-cone coordinates to be solved [1], classically in arbitrary space-time dimensions, and quantum mechanically in $D = 26$ dimensions.

The action for the motion of a point particle in space-time is the length of its space-time trajectory between some initial and final points. Similarly, the action for the motion of a string is the area of the surface it sweeps when moving in space-time. This surface can be parametrized by D functions $X^\mu(\sigma, \tau)$ of two variables σ and τ, $\mu = 0, 1, 2 \ldots D$. Using a dot for a τ derivative and a prime for a σ derivative, the area of this surface is:

$$S = -\frac{1}{2\pi\alpha'} \int d\sigma d\tau \sqrt{(\dot{X}^\mu X'_\mu)^2 - \dot{X}^2_\mu X'^{\,2}_\nu} \tag{1}$$

where α', the slope of the Regge trajectories, is a parameter with dimensions of an inverse mass squared. This area is a geometrical quantity, and thus invariant under reparametrization, i. e. arbitrary changes of variables $\tilde{\sigma} = \tilde{\sigma}(\sigma, \tau)$, $\tilde{\tau} = \tilde{\tau}(\sigma, \tau)$. This freedom of reparametrization is a gauge degree of freedom, and, correspondingly, the number of physical degrees of freedom is reduced.

This is a very non-linear action, which one could expand for example around the static solution of the string stretched between two points, but which can be made quadratic by an appropiate choice of parametrization. This special choice uses the fact that any surface can be parametrized with orthogonal coordinates:

$$(\dot{X}_\mu \pm X'_\mu)^2 = 0 \tag{2}$$

Since one has

$$4(\dot{X}^\mu X'_\mu)^2 - 4\dot{X}^2_\mu X'^{\,2}_\nu = (\dot{X}^2_\mu - X'^{\,2}_\mu)^2 - (\dot{X}_\mu + X'_\mu)^2(\dot{X}_\nu - X'_\nu)^2,$$

it is obvious that the action becomes effectively quadratic in these coordinates, and that the equation of motion is simply:

$$\ddot{X}_\mu - X''_\mu = 0 \tag{3}$$

which is indeed that of a vibrating string.

Orthogonal coordinates still leave freedom for further reparametrizations where the functions $\tilde{\sigma} = \tilde{\sigma}(\sigma, \tau)$, $\tilde{\tau} = \tilde{\tau}(\sigma, \tau)$ themselves satisfy the string equation (3). One can exploit this freedom to choose τ to be proportional to the light-cone time X^+ itself:

$$X^+ = c\tau$$

With this choice, the density of momentum in the + direction, $\mathcal{P}^+ \equiv \frac{\delta S}{\delta X^+}$ becomes a constant:

$$\mathcal{P}^+ = \frac{1}{2\pi\alpha'}\dot{X}^+ = \frac{c}{2\pi\alpha'}.$$

For a free string, the total momentum is conserved, so that the range of variation of the σ variable is a constant, proportional to the total momentum p^+ in the + direction. If we fix this range of variation of σ between 0 and π, we fix c:

$$c = 2\alpha' p^+$$

The next step is to use the parametrization conditions (2) to compute the momentum density and the position in the $-$ direction. These conditions can be rewritten as:

$$\mathcal{P}_\mu^2 + \frac{X_\mu'^2}{2\pi\alpha'} = 0 \tag{4}$$

$$X_\mu' \mathcal{P}_\mu = 0 \tag{5}$$

Eq. (4) gives:

$$\frac{2p^+}{\pi} \cdot \mathcal{P}^- + \mathcal{P}_i^2 + \frac{1}{(2\pi\alpha')^2} X_i'^2 = 0, \tag{6}$$

which determines $\mathcal{P}^-$ in terms of the transverse coordinates X_i and momentum densities $\mathcal{P}_i$.

Similarly, eq.(5) determines X^- in terms of X_i and $\mathcal{P}_i$, except for the center of mass position q^-, which shows up as the integration constant in this determination of X^-. Thus, one separates entirely the modes as transverse harmonic oscillator modes, plus the center of mass position and momentum.

The light-cone hamiltonian H is proportional to the p^- momentum. More precisely:

$$H = -cp^- = -2\alpha' p^- p^+ \tag{7}$$

Integrating eq. (6) over σ, one finds:

$$H = 2\pi\alpha' \int_0^\pi \frac{1}{2}[\mathcal{P}_i^2 + \frac{1}{(2\pi\alpha')^2}X_i'^2]d\sigma. \tag{8}$$

It is well known that this is the hamiltonian of an infinite set of transverse harmonic oscillators, together with the center of mass momentum in the transverse coordinates: Only $D - 2$ transverse vibration modes of the string are independent. This leads (after quantization) to a spectrum of states whose mass squared rises linearly with angular momentum.

Instead of using the eigenmodes of the string equation, one can also give a parton interpretation of the string, by approximating its transverse position by a polygon with M points $X_n^i, n = 1, 2 \ldots M$, each sharing a small (order $1/M$) fraction of the total p^+ momentum. The hamiltonian becomes

$$H = \alpha' M \sum_n [-\frac{\partial^2}{\partial X_n^i{}^2} + \frac{1}{(2\pi\alpha')^2}(X_{n+1}^i - X_n^i)^2]. \tag{9}$$

When M is very large, this is the hamiltonian of a very large number of very soft partons, arranged in a chain, each interacting with its nearest neighbors on the chain. This is rather far from an exact QCD picture, where presumably partons (gluons) should be allowed to carry a finite fraction of the total momentum, and furthermore are certainly not linked with harmonic forces.

After separation of the independent degrees of freedom in these light-cone coordinates, the string can be quantized by canonical quantization of the transverse modes and of the center of mass coordinates. It is well known that the result of this procedure is Lorentz invariant only in $D = 26$ space-time dimensions, and, furthermore, that the ground state is a tachyon, the first excited state being a transverse massless vector. In the case of the superstring, the dimension drops to $D = 10$, and the tachyon disappears.

After obtaining the light-cone dynamics of a free string, it took a couple of years to work out the interactions [2, 3]. In light-cone coordinates, the result turned out to be amazingly simple, at least conceptually: For open strings, there turned out to be one three-string vertex and one four-string interaction vertex.

The three-open string vertex is obtained by considering that two free strings come together at an end point of each at some value of the light-cone time, and that at later times, they form only one string. Together with this fusion vertex, there is a symmetric decay vertex, where one string, at some time, splits into two. In both cases, the interaction is just the condition that the initial string(s) coincide with the final string(s) at the fusion or splitting time, which is a product of an infinite number of delta functions of initial minus final string coordinates, one (or rather $D - 2$) for each value of σ.

The four-open string vertex is less intuitive at first, and corresponds to two strings coming together at some point other than their ends, at which point they exchange strands, forming two new strings. Here again, the interaction is just the condition that the initial strings coincide with the final strings at the exchange time.

These conceptually simple vertices reproduce the known string S-matrix. Now, one could ask the question: Since we have a rather explicit string field theory in light-cone coordinates, does this enable us, in the case of the bosonic string, to give some expectation value to the tachyon field (and perhaps to some of the other scalar fields) to go to some more stable vacuum, as happens in conventional field theory?

Covariant string field theory [4] was developed in the mid-eighties to overcome the potential problems with a non explictly Lorentz invariant formalism, partly to answer this question and find more interesting vacua. Although covariant string field theory in its present forms is satisfactory at the tree level, and has led to a number of interesting insights, it becomes unwieldy at the interacting level, and so far nobody knows the answer to this question of vacua. Attempts at a covariant field theory of superstrings have also been disappointing. In this case, too, the light-cone has decisive advantages for quantization, already at the level of free strings [5].

The indication seems to be that first treating the free theory, then the interactions, as has been done so far, is a dead end for the covariant approach within the present state of our technology: Indeed, by ananlogy, it is not the free piece of the standard elctroweak theory in the broken phase which tells us much about the symmetries of the theory, since they are broken. Rather, it is the interactions! The above approach to strings may be similar: Some beautiful dynamical symmetry is at present hidden from view. To unveil it, some totally new idea is needed, which may come once the fascinating aspects of string dualities are better understood. It may also be that a complete understanding of condensates in the light-cone field theory case would pave the way to a solution of the same problem for strings.

Another very suggestive link between the light-cone and strings comes from 't Hooft's analysis [6] of large-N_c (planar) field theory diagrams: using the same light-cone position $x^+ \sim \tau$ and momentum p^+ as occur so naturally in the string, he found that the propagators of a planar Feynman diagram, each one carrying momentum p^+ over a distance x^+, x^i, could be arrange as small rectangles partitioning a large one in the x^+, p^+ plane (we refer to the original paper and its figures for details). Now, such a propagator has the form

$$\frac{1}{2(2\pi)^3|p^+|}\theta(x^+p^+)\exp-\frac{p^+x_i^2}{2x^+}$$

Hence, the Feynman diagram is an integration over the $+$ and transverse positions x_n^+, x_n^i of the interaction vertices and the $+$ momenta p_{nm}^+ carried by the propagator between vertices n and m of an expression containing

$$\exp-\sum_{n,m}\frac{p_{nm}^+(x_n^i-x_m^i)^2}{2(x_n^+-x_m^+)}$$

Can one turn this into a theory which would be soluble in some sense, and which would have string features? For the moment, this works only in two dimensional QCD [7]. The introduction of transverse directions requires new ideas to be made effective. we have herd from S. Pinsky about the introduction of transverse polarizations, but this is only a small step. Perhaps somebody will be able to take off from two dimensions, perhaps going first to $2+\epsilon$ dimensions in some ϵ expansion which would give reliable numerical results up to four dimensions (for mass ratios for example), for the complicated string which is equivalent to four dimensional large-N_c gauge theory.

References

[1] Goddard P., Goldstone J., Rebbi C. and Thorn C.B., *Nucl. Phys.* **B56** (1973) 114.
[2] Cremmer E. and Gervais J.-L., *Nucl. Phys.* **B74** (1974) 209.
[3] Kaku M. and Kikkawa K. *Phys. Rev.* **D10** (1974) 1110.
[4] Nicolai H., Neveu A. and West P., *Phys. Lett.* **167B** (1986) 307.
[5] Green M.B. and Schwarz J.H., *Nucl. Phys.* **B243** (1984) 475.
[6] 't Hooft G., *Nucl. Phys.* **B72** (1974) 461.
[7] 't Hooft G., *Nucl. Phys.* **B75** (1975) 461.

Quantum Field Theory in Singular Limits *

Moshe Moshe [†]

Department of Physics
Technion - Israel Inst. of Technology
Haifa, 32000 ISRAEL

1. INTRODUCTION

It has been suggested [1] to approximate geometry by dense Feynman graphs of the same topology, taking the number of vertices to infinity. The dynamically triangulated random surfaces summed on different topologies are then viewed as the manifold for string propagation. Summing up Feynman graphs of the O(N) matrix model in d dimensions results in a genus expansion and it provides, in some sense, a nonperturbative treatment of string theory when the double scaling limit is enforced [2]. If these efforts could be extended to d > 1 dimensions, then a major progress would have been achieved in studying a long lasting problem in elementary particle theory. Namely, the relation between d dimensional quantum field theory and its possible formulation in terms of strings. The possibility of reaching a "stringy" representation of SU(N) and U(N) quantum field theory in a correlated singular limit was proposed some time ago in [3].

$O(N)$ symmetric vector models represent discretized filamentary surfaces - randomly branched polymers in the double scaling limit, in the same manner in which matrix models, in their double scaling limit, provide representations of dynamically triangulated random surfaces. Though matrix theories attract

(*) *A Short Summary - Les Houches February 1997*
(†) *This research was supported in part by a grant from the Israel Science Foundation*

most attention, a detailed understanding of these theories exists only for dimensions $d \leq 1$. On the other hand, in many cases, the double scaling limit of the $O(N)$ vector models[4, 5] can be successfully studied also in dimensions $d > 1$ thus, providing intuition for the search for a possible description of quantum field theory in terms of extended objects, string-like excitations, in four dimensions. This stimulated new studies [5]-[7] of the phase structure of O(N) vector quantum field theories that offered a new look into this long lasting problem. Most often, quantum field theories are described by their point-like quanta and a direct relation to a string-like structure is difficult to visualize; if there is a correspondence to an hadronic string, it must be of a less direct kind [3].

2. SINGULAR LIMITS OF O(N) SYMMETRIC MODELS

2.1. The Double Scaling Limit in O(N) Matrix Models

In O(N) matrix models the double scaling limit is enforced in the calculation of the partition function

$$Z_N(g) \;=\; \int D\hat{\Phi} exp\{-\beta \int d^d x Tr\{\hat{\Phi}(x)\hat{K}\hat{\Phi}(x) + V(\hat{\Phi}(x))\}\} \qquad [1]$$

where $\hat{\Phi}(x)$ is an NxN Hermitian matrix and V($\hat{\Phi}(x)$) is the potential depending on the coupling(s) constant(s) $\{g_i\}$.

In zero dimensions, after performing the integration on the angular variables, one is left with the integration on the eigenvalues λ_i

$$Z_N(g) \;=\; \Omega_N \int \prod_i^N d\lambda_i \; exp\{2\sum_{i,j}^N \ln|\lambda_i - \lambda_j| - \beta \sum_i U(\lambda_i)\}. \qquad [2]$$

In Eq. [2] one notes a Pauli repulsion between the eigenvalues, and a critical point $\{g_i\} = \{g_{iC}\}$ is found when the Fermi level reaches the extremum of the potential. The weak coupling limit, namely, $\frac{1}{N} \sim \frac{1}{\beta} \to 0$, gives a one-dimensional frozen Dyson gas and the planar graphs dominate. As $N \to \infty$ and $\frac{N}{\beta} \to 1$ (or $\{g_i\} \to \{g_{iC}\}$), the "melting" of the gas starts and the non-planar graphs become important. The genus (G) expansion of the free energy of the system is given by (S is the area which is proportional to the order of the Feynman graphs):

$$F = \ln Z_N \;=\; a + b\ln\beta + \sum_{G,S} N^{2(1-G)} (\frac{N}{\beta})^S \; F_S \;\sim \sum_G (\frac{1}{N})^{2G-2} \mathcal{A}_G\{g_i\} \; [3]$$

This topological series is not Borel summable due to the factorial growth of the positive $\mathcal{A}_G\{g_i\}$ with the genus G, and a nonperturbative approach is needed. As the set of coupling constants $\{g_i\} \longrightarrow \{g_{iC}\}$ approaches a set of critical values, the loop expansion at a given topology diverges. At a given topology, $\mathcal{A}_G\{g\}$ has a finite radius of convergence when expanded in powers of the coupling constant g. The limit $g \to g_c$ emphasizes the higher order terms in this expansion, namely, the denser Feynman graphs. Typically, for a potential $V(\hat{\vec{\Phi}})$ with a single coupling,

$$\mathcal{A}_G\{g\} \sim (g - g_C)^{-\kappa_G} \longrightarrow \infty \quad \{ \text{as } g \to g_c \} \tag{4}$$

In the suitable singular limit (the double scaling limit), all terms in the topological expansion in Eq. [3] are of equal importance. The nonperturbative framework needed here should be capable of reproducing the topological series as an asymptotic expansion but should also be a framework for nonperturbative calculations[2]. In the limit β=N $\to \infty$ and $g \to g_c$ in a correlated manner, the powers of $\frac{1}{N}$ in Eq. [3] are compensated by the growing $\mathcal{A}_G\{g\}$ in Eq.[4]. In this limit, Eq.[3] turns into an expansion in the string coupling constant since all genera are relevant now. The physical meaning of this formal limit will be soon clarified in the O(N) vector quantum field theory.

The model belongs to the same universality class of two dimensional conformally invariant matter field in gravitational background and thus, the critical exponents of this model are calculable from Eq. [3]. One may, indeed, find it quite remarkable that these relatively simple integrals possess the critical behavior of two dimensional gravity coupled to conformal matter.

2.2. The Double Scaling Limit in O(N) Vector Models

The double scaling limit of the O(N) vector models follows an analogous procedure,

$$Z_N(g) = \int \mathcal{D}\vec{\Phi} exp\{-\beta \int d^d x \{\vec{\Phi}(x)\hat{K}\vec{\Phi}(x) + V(\vec{\Phi}^2(x))\}\} \tag{5}$$

where $\vec{\Phi}$ is an O(N) vector. In zero dimensions, this turns into a simple N dimensional integral over the components of $\vec{\Phi}$. The expansion in terms of Feynman graphs in the large N limit resembles now an expansion of "randomly branched polymers"[4]. The graphical realization is obtained when the dual diagrams are defined from the Feynman graphs by interchanging the vertex and propagator in the Feynman graph by the bond and "molecule" in the dual graph, respectively, which describes now the branched polymer.

$$\ln Z_N = \sum_{h,b} N^{(1-h)} \, (\frac{N}{\beta})^b \, F_b \quad \sim \sum_h (\frac{1}{N})^{h-1} \mathcal{A}_h\{g_i\} \tag{6}$$

Following here the suitable scaling procedure, Eq. [5] turns into an expansion in a "polymer" coupling constant. For $V(\vec{\Phi}^2) \sim g\vec{\Phi}^2(x)^2$ in the correlated limit $N \to \infty$ and $g \to g_C$, the negative powers of N are compensated by $\mathcal{A}_h\{g\} \sim (g - g_C)^{-\gamma_h} \longrightarrow \infty \quad \{ \text{as } g \to g_c \}$. Criticality comes from the tuning of the potential in order to balance the centrifugal barrier. Whereas matrix models in dimensions $d > 1$ are very difficult to study, even in the leading large N limit or the double scaling limit, vector models are relatively easy to analyze in these limits even at $d \geq 2$.

There are several interesting physical questions that can be answered when the formal double scaling limit is enforced in $d \geq 2$ O(N) vector theories. A possible list of questions follows:

(1) What is the phase structure, spectrum and symmetries of the quantum field theory in this singular limit ?

(2) Does the flow $g \to g_C$ agree with the renormalization group flow ?

(3) How are the ultraviolet divergences taken into account at d>2 ? (Needless to say that these questions were not raised in matrix models in d≤1).

(4) Taking into account that we are discussing a renormalizable quantum field theory, what is the nature of the divergences needed for compensating the powers of N^{-h} in Eq. [6] so that all orders in N^{-1} are of equal importance?

These issues have been discussed in [5]-[7], while considering as an example the large N limit of the self interacting scalar O(N) symmetric vector model in $0 \leq d < 4$ Euclidian dimensions defined by the functional integral in Eq.[5] with $\hat{K} = \partial^2 D(\frac{-\partial^2}{\Lambda^2})$, (where D(z) is a positive non-vanishing polynomial with D(0)=1). The following results have been found [6, 7]:

(1) The physical meaning of the double scaling limit can be phrased as tuning the force between the O(N) quanta so that a singlet massless bound state is created in the spectrum. This is a singular limit in the sense that while $N \to \infty$ the coupling constant is often tuned to a negative value $g \to g_C$. The physical spectrum consists of the propagating O(N) quanta of small mass m in addition to the massless O(N) singlet.

(2) The flow of $g \to g_C$ is consistent with the renormalization group flow, provided the limits $N \to \infty$ and the cutoff $\Lambda \to \infty$ are appropriately correlated. This point was made also in Ref.[3].

(3) The ultraviolet divergences are dictating the effective field theory obtained in the singular limit of the double scaling limit. They enforce a detailed relation between Λ and N, mentioned in (2) above, $\Lambda = \Lambda(N)$.

(4) The divergences that compensate the decreasing powers of N^{-h} in Eq. [6] and make all orders in N^{-1} of equal importance, are infrared divergences. Thus, in the double scaling limit, the tuning of the forces that produce a massless singlet excitation, produce infrared divergences which are the essential ingredient of this limit.

(5) Following (4) above, in order that the compensating infrared singularities will show up, the effective field theory of the singlet bound state is super-renormalizable.

(6) In the critical dimensions (e.g. , Φ^6 in d=3) the massless O(N) singlet excitation is the Goldston boson of spontaneous breaking of scale invariance - the dilaton[8].

In the conventional treatment of the large N limit, Eq. [5] is expressed by:

$$Z_N = \int \mathcal{D}\rho \int \mathcal{D}\lambda \; exp\{-N \int d^d x \, [V(\rho) - \frac{1}{2}\lambda\rho] + \frac{1}{2}N \, tr \ln(-\partial^2 + \lambda). \quad [7]$$

For N large the integral is evaluated by the steepest descent. The saddle point value λ is the $\vec{\Phi}$-field mass squared $\lambda = m^2$ and $\rho = \rho_s$ at the saddle point. The matrix of the second partial derivatives of the effective action is:

$$N \begin{pmatrix} V''(\rho) & -\frac{1}{2} \\ -\frac{1}{2} & -\frac{1}{2}B_2(p; m^2) \end{pmatrix}, \quad [8]$$

where $B_2(p; m^2)$ is the appropriate "bubble graph". Since the integration contour for $\lambda = m^2$ should be parallel to the imaginary axis, a necessary condition for stability is that the determinant remains negative. For Pauli-Villars type regularization, the function $B_2(p; m^2)$ is decreasing so that this condition is implied by the condition at zero momentum

$$\det \mathbf{M} < 0 \Leftarrow 2V''(\rho)B_2(0; m^2) + 1 > 0 \quad [9].$$

For m small

$$B_2(0; m^2) = \frac{1}{2}(d-2)K(d)m^{d-4} - a(d)\Lambda^{d-4} + O\left(m^{d-2}\Lambda^{-2}, m^2\Lambda^{d-6}\right). \quad [10]$$

$V''(\rho)$ can be expanded now around the critical ρ.   From the saddle point condition:

$$\rho - \rho_c = -K(d)m^{d-2} + a(d)m^2\Lambda^{d-4} + O\left(m^d\Lambda^{-2}\right) + O\left(m^4\Lambda^{d-6}\right). \quad [11]$$

$$\rho_c = \frac{1}{(2\pi)^d} \int^\Lambda \frac{d^d k}{k^2}, \quad K(d) = -\frac{\Gamma(1-d/2)}{(4\pi)^{d/2}}, \quad a(d) = \frac{1}{(2\pi)^d} \int \frac{d^d k}{k^4}\left(1 - \frac{1}{D^2(k^2)}\right).$$

The constant $a(d)$ depends on the cut-off procedure, and one finds in Eq. [9] for a multicritical point:

$$(-1)^n \frac{d-2}{(n-2)!} K^{n-1}(d)m^{n(d-2)-d}V^{(n)}(\rho_c) + 1 > 0. \quad [12]$$

This condition is satisfied by a normal critical point since $V''(\rho_c) > 0$. For n even it is always satisfied while for n odd Eq.[12] is always satisfied above the critical dimension and never below. At the upper-critical dimension $2/(n-1) = d - 2$ we find a condition on the value of $V^{(n)}(\rho_c)$.

The mass-matrix has a zero eigenvalue which corresponds to the appearance of a new massless excitation other than the $\vec{\Phi}$ quanta (which has a mass m). Then

$$\det \mathbf{M} = 0 \Leftrightarrow 2V''(\rho)B_2(0; m^2) + 1 = 0. \quad [13]$$

In the two-space, the corresponding eigenvector has components $(\frac{1}{2}, V''(\rho))$. In the small m limit $V''(\rho)$ must be small and we are close to a multicritical point and

$$(-1)^{n-1} \frac{d-2}{(n-2)!} K^{n-1}(d) m^{n(d-2)-d} V^{(n)}(\rho_c) = 1. \qquad [14]$$

This equation has solutions only for $n(d-2) = d$, i.e. , at the critical dimension. The compatibility then fixes the value of $V^{(n)}(\rho_c) = \Omega_c$. If we take into account the leading correction to the small m behavior we find instead:

$$V^{(n)}(\rho_c)\Omega_c^{-1} - 1 \sim (2n-3) \frac{a(d)}{K(d)} \left(\frac{m}{\Lambda}\right)^{4-d}. \qquad [15]$$

This means that when $a(d) > 0$ there exists a small region $V^{(n)}(\rho_c) > \Omega_c$ where the vector field is massive with a small mass m and the $O(N)$ singlet bound-state is massless. The value Ω_c is a fixed point value. The analysis can be extended to a situation where the scalar field has a small but non-vanishing mass M and m is still small. In particular, the neighborhood of the special point $V^{(n)}(\rho_c) = \Omega_c$ can be explored. The vanishing of the determinant in Eq.[8] implies $1 + 2V''(\rho)B_2(iM; m^2) = 0$. Because M and m are small, this equation still implies that ρ is close to a point ρ_c where $V''(\rho)$ vanishes. Since reality imposes $M < 2m$, it is easy to verify that this equation has also solutions for only the critical dimension.

Of particular interest is the $\eta_0 (\vec{\Phi}^2)^3$ theory in three dimensions, discussed in the past in Ref.[8]. Taking the limit $N \to \infty, \Lambda \to \infty$ in a correlated manner with $\eta_0 \to \Omega_c$ one encounters a manifestation of dimensional transmutation at a nontrivial ultraviolet fixed point. In the massive phase described above, scale invariance is broken only spontaneously. Indeed, one finds that the trace of the energy momentum tensor stays zero at $\eta_0 = \Omega_c$:

$$< P'|\Theta_{\mu\nu}|P >= P'_\mu P_\nu + P'_\mu P_\nu - g_{\mu\nu} P' \cdot P + g_{\mu\nu} m^2 + (q_\mu q_\nu - q^2 g_{\mu\nu})(\frac{m^2}{q^2} + \frac{1}{4}) \quad [16]$$

where $q = P' - P$. A massless dilaton - the Goldston boson associated with this spontaneous breaking - appears in the ground state spectrum as a reflection of the Goldston realization of scale symmetry. The normal ordering of $\vec{\Phi}^6$ induces a $\vec{\Phi}^4$ interaction which guarantees the appearance of the dilaton pole in the physical amplitudes as $\eta_0 \to \Omega_c$.

The double scaling limit results here in a theory with an ultraviolet fixed point and the following properties are found: (a) There is a dynamical mass $m \neq 0$ for the scalar Φ particles. (b) The $\Phi - \Phi$ bound state pseudo-Goldston boson (dilaton) has a finite small mass and the interaction term of these scalars is calculable. This is an interesting mechanism to produce a low mass scalar in a spectrum with possible phenomenological implications. (c) The trace of the energy momentum tensor $\Theta_\mu{}^\mu$ is finite. (d) An induced Φ^4 coupling appears in the theory.

3. DISCUSSION

This is a short summary of the phase structure of O(N) symmetric quantum field theories in a singular limit, the double scaling limit. The main point emphasized here is that this formal singular limit, recently discussed mainly in $d = 0$ O(N) matrix models, has an intriguing physical meaning in $d \geq 2$ O(N) vector theories. In this limit all orders in $\frac{1}{N}$ are of equal importance since at each order infrared divergences compensate for the decrease in powers of $\frac{1}{N}$. The infrared divergences are due to the tuning of the strength of the force $(g \rightarrow g_C)$ between the O(N) quanta so that a massless O(N) singlet appears in the spectrum. At critical dimension an interesting phase structure is revealed, the massless excitation has the expected physical meaning: it is the Goldston boson of spontaneous breaking of scale invariance - the dilaton.

References

[1] David F., *Nucl. Phys.* **B257** **[FS14]** (1985) 45, 543.   Ambjørn J., Durhuus B. and Fröhlich J., *Nucl. Phys.* **B257** **[FS14]** (1985) 433.Kazakov V.A., *Phys. Lett.* **B150** (1985) 282.

[2] Douglas M. and Shenker S., *Nucl. Phys.* **B335** (1990) 635.Brézin E. and Kazakov V., *Phys. Lett.* **B236** (1990) 144. Gross D. and Migdal A.A., *Phys. Rev. Lett.* **64** (1990) 127.*Nucl. Phys.* **B340** (1990) 333.Douglas M. R., *Phys. Lett.* **B238** (1990) 176. F. David F., *Mod. Phys. Lett.* **A5** (1990) 1019.

[3] Neuberger H., *Nucl. Phys.* **B340** (1990) 703.David F. and Neuberger H., *Phys. Lett.* **B269** (1991) 134.

[4] Nishigaki S. and Yoneya T., *Nucl. Phys.* **B348** (1991) 787.Anderson A., Myers R.C. and Periwal V., *Phys. Lett.* **B254** (1991) 89.*Nucl. Phys.* **B360** (1991) 463.Di Vecchia P., Kato M. and Ohta N., *Nucl. Phys.* **B357** (1991) 495.

[5] Zinn-Justin J., *Phys. Lett.* **B257** (1991) 335.   Di Vecchia P., Kato M. and Ohta N., *Int. J. Mod. Phys.* **A7** (1992) 1391.   Yoneya T., *Prog. Theor. Phys. Suppl.* **107** (1992) 229.

[6] Di Vecchia P. and Moshe M., *Phys. Lett.* **B300** (1993) 49.

[7] Eyal G., Moshe M., Nishigaki S. and Zinn-Justin J., *Nucl. Phys.* **B470** (1996) 369.

[8] Bardeen W.A., Moshe M. and Bander M., *Phys. Rev. Lett.* **52** (1984) 1188. See also Bardeen W. A. and Moshe M., *Phys. Rev. D* **28** (1983) 1372.

CHAPTER VI

Gauge Theories and Topological Issues

On the Transition From Confinement to Screening in Large N Gauge Theory

J. Boorstein[1], D. Kutasov[1,2]

([1]) *Enrico Fermi Institute,*
University of Chicago,
5640 S. Ellis Ave.,
Chicago, IL 60637,
USA
([2])*Department of Physics of Elementary Particles,*
Weizmann Institute of Science, Rehovot,
Israel

The QCD string that describes confining large N $SU(N)$ gauge theory becomes tensionless at values of the couplings for which the gauge theory passes from confinement to screening. We study this phenomenon in two dimensional QCD coupled to adjoint matter. As the string tension goes to zero, the Hilbert space of "single string" states splits into disconnected Hilbert spaces describing massless and massive, single and multiparticle excitations. The enormous number of states that seem to come down to the same energy is replaced by a large number of degenerate decoupled sectors, labeled by discrete quantum numbers parametrizing a topological field theory. Our results point to subtleties in the usual treatment of "zero modes" in discretized light front quantization. (Talk presented at the conference "New Non-perturbative Methods and Quantization on the Light-Cone", Les Houches, February 24 – March 7, 1997.)

In this note we will discuss two dimensional Yang-Mills theory coupled to Majorana fermions in the adjoint representation of an $SU(N)$ gauge group. The model is described by the Lagrangian

$$\mathcal{L} = \frac{1}{g^2} F_{\mu\nu}^2 + \bar{\psi}\gamma^\mu D_\mu \psi + m\bar{\psi}\psi \tag{1}$$

where $F_{\mu\nu} = \partial_{[\mu}A_{\nu]} + i[A_\mu, A_\nu]$, $D_\mu\psi = \partial_\mu\psi + i[A_\mu, \psi]$, ψ_{ab} is a traceless hermitian anti-commuting matrix, m is the (bare) fermion mass and g the gauge coupling. In the past few years this model has received some attention (see $e.g.$ [1]-[25], and references therein), as a possibly tractable toy model of four dimensional confining gauge theories, particularly in the large N limit. Since this model has dynamical field theoretic degrees of freedom in the adjoint representation of the gauge group, it certainly shares with four dimensional gauge theory the non − triviality of the large N limit, in which the theory is supposed to be described by a "QCD string". The following properties of the theory (1) will be relevant for our subsequent discussion:

1. At large N the theory contains singlet bound states with a rich spectrum of masses M. The density of bound states ρ rapidly increases with their mass M; for large M it exhibits [2] the behavior:

$$\rho(M) \simeq M^\alpha e^{\beta_H M}. \tag{2}$$

2. The different bound states interact weakly at large N (the coupling between them is of order $1/N$).

3. Due to the rapid growth of $\rho(M)$ given in (2), the thermal canonical ensemble ceases naively to make sense at a critical temperature $T_H = \beta_H^{-1}$. Tr $\exp(-\beta H) \simeq \int dM \rho(M) \exp(-\beta M)$ diverges for $\beta < \beta_H$. Of course, T_H is not in fact a maximal temperature; rather, the system undergoes a Hagedorn transition, which is the large N manifestation of the deconfining transition. At high temperature one should not count the bound states (2) but rather the constituent adjoint quarks, of which there are $O(N^2)$.

4. For a critical value of the adjoint mass m, $m = g\sqrt{N}$, the theory becomes supersymmetric [2]. For other values of the mass it is *asymptotically supersymmetric* (*i.e.* the spectrum of highly excited bound states is almost supersymmetric, while for low energies SUSY is softly broken by the mass term; see [5] for further discussion).

5. When the constituent mass m vanishes, the model is equivalent to a theory of $N_f = N$ flavors of fermions transforming in the fundamental representation of the gauge group [10]. This has many consequences; one that we will use here is that states in the large N LCQ Hilbert space which describe single particles for nonzero m, turn into multiparticle states when $m = 0$. This phenomenon will be briefly described below.

6. A feature related to point 5. above is the peculiar behavior of the Wilson loop as a function of m [16]. For nonzero m, the Wilson loop exhibits the famous area law:

$$\langle W(C) \rangle \simeq e^{-\sigma A} \qquad (3)$$

where C is a large planar loop and A the area of the minimal surface with boundary C. As $m \to 0$, the string tension $\sigma \to 0$. For small $m (<< g\sqrt{N})$, one finds [16]:

$$\sigma \simeq 2\Sigma m \qquad (4)$$

where Σ is (minus) the fermion bilinear condensate, which is proportional to the gauge coupling, g.

7. Points 5. and 6. imply that as $m \to 0$ the system (1) passes from confining to screening [16]. At first sight it is surprising that heavy probe charges in the fundamental representation of $SU(N)$ can be screened by adjoint dynamical quarks. One way to understand that is by using the results of [10] which imply that the theory (1) contains solitons in the fundamental representation of $SU(N)$.

The properties 1. – 7. make this theory an interesting toy model for studying QCD strings, which should govern the physics of confining large N gauge theories. Adjoint QCD$_2$ is probably the simplest model exhibiting the behavior (2) and thus a Hagedorn transition. The asymptotic supersymmetry of the model is also intriguing, since in fundamental string theory it is a theorem that consistent theories with nonzero β_H are asymptotically supersymmetric [27]. It would be very interesting to understand whether an analog of this theorem holds for QCD strings.

In this note we will try to resolve certain confusions which arise when trying to reconcile the $m \to 0$ and $N \to \infty$ limits. In particular, we will attempt to answer the question, *how does the transition from confining to screening which occurs as $m \to 0$ manifest itself at large N?*

Since at large N the theory is an (almost) free field theory of bound states, the only effect of the $m \to 0$ limit should be to alter the density of bound states, *i.e.* the coefficients α, β_H in (2). A theory which screens rather than confines fundamental charge would naively be expected to have $\beta_H = 0$ (in the notation of eq. (2)), since such a theory is not expected to undergo a Hagedorn transition, which as we learned in point 3. is simply the large N manifestation of the deconfining transition.([1]) In terms of the QCD string, this could be achieved by taking the string tension to infinity, thereby pushing all the massive string states to large energies and in the limit decoupling most of the exponential density of states of the string.

However, the analysis of the Wilson loop [16] suggests that in the $m \to 0$ limit the string tension in fact goes to zero rather than to infinity (4). One

---

([1]) A theory that screens fundamental charge can have a continuous spectrum, but that should only affect α, not β_H.

would then expect infinite Hagedorn like towers of states to go to the same (ground state) energy and not disappear to infinite energies as $m \to 0$! Thus, two problems arise:

1. How can we reconcile the expected absence of Hagedorn transition when $m \to 0$ with the description of this limit in terms of an almost tensionless QCD string?

2. The theory (1) with $m = 0$ is clearly non-singular, however its description in terms of a tensionless QCD string seems sick. This raises the question: what is a tensionless QCD string?

In the rest of this note we will propose a physical picture that provides answers to the above questions.

The first clue towards the resolution of the puzzle comes from a result of [10] which we will briefly describe next. Recall that in light front quantization of (1) we are instructed to think of (say) x^- as space and x^+ as (light cone) time. Thus, the conjugate momentum to x^-, P^+, is diagonal on the Fock space, with all the dynamics residing in P^-. Normal ordering in the standard fashion, one finds that $P^\pm$ take the form:

$$
\begin{aligned}
P^+ &= \int_0^\infty dk k \psi_{ab}(-k)\psi_{ba}(k) \\
P^- &= m^2 \int_0^\infty \tfrac{dk}{k} \psi_{ab}(-k)\psi_{ba}(k) + g^2 \int_0^\infty \tfrac{dk}{k^2} J_{ab}^+(-k)J_{ba}^+(k)
\end{aligned}
\tag{5}
$$

where $\psi_{ab}(k)$ are Fourier modes satisfying canonical anti-commutation relations,

$$
\{\psi_{ab}(k), \psi_{cd}(k')\} = \delta(k+k')\delta_{a,d}\delta_{c,b}.
\tag{6}
$$

For simplicity of notation we have not substracted the trace of ψ, a decoupled fermion corresponding to $U(1) \in U(N)$; $a, b, c, d = 1, \cdots, N$. $\psi_{ab}(k)$ with $k \leq 0$ are creation operators, whereas the ones with $k > 0$ are annihilation operators. The light cone vacuum $|0\rangle$ is killed by annihilation operators:

$$
\psi_{ab}(k)|0\rangle = 0 \quad \forall k > 0.
\tag{7}
$$

The $SU(N)$ currents $J_{ab}^+(k)$ have the standard form:

$$
J_{ab}^+(k) = \int_{-\infty}^\infty dp \psi_{ac}(p)\psi_{cb}(k-p).
\tag{8}
$$

Standard large N theory asserts that states of the form

$$
|\text{phys}\rangle = \int dx_1 \cdots dx_k \Phi_l(x_1, \cdots, x_k) \text{Tr}\,(\psi(-x_1)\psi(-x_2)\cdots\psi(-x_k))\,|0\rangle
\tag{9}
$$

transform among themselves under the action of the light – cone Hamiltonian P^- (5). Mixing with states containing two or more traces occurs at higher order in $1/N$. Furthermore, states of the form (9) usually give rise (after solving the bound state equation for the light – cone wavefunctions Φ_l) to *single particle states.* Indeed, counting powers of N one finds that if a state of the form (9) has overlap with (say) both one and two particle states, then the coupling between the corresponding single particle states is of order one, in contradiction with the fact that interactions between the single particle bound states must go to zero as a power of N in the large N limit.

In [10] it was argued that at $m = 0$, the structure of the theory changes dramatically. In particular, it is no longer true that states of the form (9) have no overlap with multiparticle states. The main observation of [10] is that for $m = 0$ the light cone Hamiltonian P^- (5) is expressed solely in terms of $SU(N)$ currents $J_{ab}^+(k)$ (8), which form an affine $SU(N)$ Lie algebra at level N (see *e.g.* [26] for a review). The rearrangement of the Fock space into sectors corresponding to different representations ("blocks") of the current algebra corresponds to a change of basis on the Hilbert space. A generic state of the form (9) has overlap with many different current algebra blocks, and vice versa. The current algebra basis is particularly useful for $m = 0$ since then P^- (5) acts within the blocks, and thus the problem of diagonalizing it and finding the spectrum simplifies.

One of the results of the analysis of [10] is that out of the infinite number of current algebra blocks that can be realized by adjoint fermions in the large N limit only two give rise to single particle states. All the bosonic single particle states come from the current algebra block of the identity, and thus can be written in the form

$$|\text{phys}\rangle = \int dx_1 \cdots dx_k \, \Phi_l(x_1, \cdots, x_k) \text{Tr} \left(J(-x_1) J(-x_2) \cdots J(-x_k) \right) |0\rangle \quad (10)$$

All the fermionic bound states come from the adjoint current block, and have a form similar to (10) with an insertion of ψ inside the trace. Thus, there are much fewer single particle bound states in the theory with $m = 0$ than one would guess from a naive application of the large N limit.

What goes wrong with the standard argument mentioned two paragraphs above about the impossibility of states of the form (9) to couple to multiparticle states, is that the theory with $m = 0$ is not generic. The affine Lie algebra structure that underlies it forces all the dangerous couplings to vanish. A generic state of the form (9) has (for even l) non zero overlap both with the current algebra block of the identity, and with higher current algebra blocks. The former gives rise to single particle states; the latter, to multiparticle ones. However, the cross terms in the two point function $\langle \text{phys} | \text{phys} \rangle$ that would lead one to conclude that the coupling of three single particle states is of order one *vanish* due to the current algebra structure: the two point function of any two states that belong to different current algebra blocks is zero [26].

We see that the theory loses states as we send $m \to 0$ by turning single particle states into multiparticle states. For small mass it is useful to think of the single particle states of the $m \neq 0$ system as corresponding to bound states of the single particle states of the $m = 0$ one, with a binding energy which goes to zero as $m \to 0$. In the limit these bound states disassociate into their constituents (10).

How many of the "basic" bound states (10) are there? The answer to this question is not known, but the following argument suggests that the spectrum of (10) is similar to that of the ordinary 't Hooft model [28]. For $m = 0$ the problem of diagonalizing the light – cone Hamiltonian P^- (5) on states of the form (10) is a purely current algebraic problem that uses only the universal commutation relations of the currents (8) and therefore depends only on the integers N and k, the rank and level of the current algebra. We are interested in the spectrum for large N, k with $\alpha \equiv k/N = 1$.

For any finite k, and $N \to \infty$ it is known [28] that the spectrum of P^- on the current block of the identity (10) is the same, independent of k, and corresponds to a single "Regge trajectory" of states labeled by an integer, n, with $M^2 \simeq$ const $\times n$. Since all these cases correspond to $\alpha = 0$ (in the large N limit), it is possible that the spectrum has non-trivial α dependence and becomes much richer for finite α. However, it is physically unreasonable to expect a drastic change in the spectrum as we increase α from zero to one, and in particular one expects the density of single particle states of the form (10) to remain of order one (as in the 't Hooft model). Continuity in the dependence on k as well as the expected absence of a Hagedorn transition at $m = 0$ provide supporting circumstantial evidence for this scenario.

The fact that as $m \to 0$ many of the bound states (9) decompose into "more basic" bound states (10) means that the massless theory has much fewer states than the massive one. Could this be the solution to the problem posed above? In other words, could it be that the exponential density of single particle states (2) for finite m is to be reinterpreted as $m \to 0$ as the multiparticle density of states of a spectrum with far fewer states? The answer is *no*; it is a basic fact of life in field theory that the only way to get $\exp(\beta_H M)$ multiparticle states of energy $\leq M$ is to start with a single particle spectrum that has the density of states (2) as well; *e.g.* the density of multiparticle states for a finite number of free fields ϕ_i grows (in two dimensions) like $\rho(M) \propto \exp(c\sqrt{M})$, where c is a constant.

In fact, the solution appears to lie in a sector of the theory that we have so far ignored. Consider the massless theory, (1) with $m = 0$. In the extreme infrared the gauge coupling g diverges and one gets a "G/G model", a theory with gauged $G = SU(N)$ affine Lie algebra with no kinetic term for the gauge field. This is a CFT with central charge $c = (N^2 - 1)/2 - (N^2 - 1)/2 = 0$. It has no field theoretic degrees of freedom. However, it is not empty. In fact it contains a large number of discrete states with $P^\pm = 0$, one state for each representation of the $SU(N)$ affine Lie algebra that can be constructed out of the matter fields ψ. Such representations are of the form $R \times R^t$, where R is an arbitrary[2] irreducible representation of $SU(N)$ corresponding to a Young tableau R, and R^t corresponds to the transpose of the tableau R. A rough

---

[2] For finite N there is a restriction on R, due essentially to unitarity constraints of affine Lie algebra [26]. We will discuss the situation in the $N \to \infty$ limit.

way of thinking about these representations is as arising from states of the form $\psi_{a_1 b_1} \cdots \psi_{a_n b_n} |0\rangle$ where the $\{a_i\}$ can be symmetrized according to any Young tableau R with n boxes, and then the $\{b_i\}$ necessarily have the opposite symmetry, corresponding to R^t.

As shown in [10], for $m = 0$ the Hilbert space of the theory takes the form

$$\mathcal{H} = \mathcal{H}_{\text{massive}} \otimes \mathcal{H}_{\text{topological}} \tag{11}$$

where $\mathcal{H}_{\text{massive}}$ is the space of states (10) giving rise to massive bound states, while $\mathcal{H}_{\text{topological}}$ is the set of discrete states $|R\rangle$ corresponding to different Young tableaux R. The light – cone Hamiltonian P^- (5) acts only on $\mathcal{H}_{\text{massive}}$; $\mathcal{H}_{\text{topological}}$ provides a discrete label for all states, causing in effect an enormous degeneracy (recall that the number of Young tableaux with n boxes grows like the number of partitions of n). Massive states with energy E and momentum p satisfying $E^2 = p^2 + M_0^2$ have the form $|M_0\rangle \otimes |R\rangle$; their properties are independent of R. One can think of different R as corresponding to different vacua.[3]

When one turns on the adjoint mass m, the factorization (11) is spoiled. Different states $|M_0\rangle \otimes |R\rangle$ are mixed under the dynamics. From the point of view of the massless theory we have to solve an enormous exercise in degenerate perturbation theory for P^- (5). This has not been done but it is reasonable to expect the perturbation to split the degeneracy in R, and to give rise to a spectrum[4] such as:

$$M^2(R) \simeq M_0^2 + m^2 \frac{C_2(R)}{N} \tag{12}$$

Here $C_2(R)$ is the quadratic Casimir in the representation R and we ignored constants of order one. The interpretation is, as before, that single particle states in the massive theory can be thought of (for small m) as bound states of the basic massive bound states $|M_0\rangle$ of the $m = 0$ theory, and the discrete topological states $|R\rangle$. At $m = 0$ we think of $|M_0\rangle \otimes |R\rangle$ as a "two particle state"; for nonzero m it becomes a single particle state labeled by R, with mass (12). Since at large N, $C_2(R) \sim nN$ where n is the number of boxes in the Young tableau R, (12) describes a Hagedorn spectrum – the number of states with $M^2 \sim m^2 n$ grows like the number of partitions of n.

We end with a few comments:

1. It is now clear how the limit of vanishing string tension is reconciled with the disappearance of the deconfinement (Hagedorn) transition as $m \to 0$. For small non zero m the spectrum (12) is almost continuous, in agreement with what would be expected from a string with very small tension. As the towers of QCD string states come down to the ground state, the

---

[3] But note that the number of vacua is much larger than N, which is the value one would deduce from the Z_N symmetry of the theory (1).

[4] For small m, $m^2 \ll g^2 N$; as we shall mention later, corrections are expected at higher m.

spectrum splits (11); at $m = 0$ it describes relatively few massive particles living in one of an enormous number of decoupled vacua. This is also the way our tensionless QCD string manages to stay consistent.

2. An interesting feature of the model is that the string tension T goes like $T \sim m^2$, and not as $g^2 N$ as one would perhaps think.[5] An interesting scenario is that T increases until m^2 reaches $g^2 N$ and then saturates, such that for $m^2 > g^2 N$, $T \sim g^2 N$. The supersymmetric point $m^2 = g^2 N$ would then correspond to a phase transition, clarifying the origin of the enhanced symmetry that appears there.

3. In DLCQ one usually discretrizes P^+ and uses antiperiodic boundary conditions for the fermions around the x^- circle. For low m and moderate values of P^+ (for which it is feasible to perform the numerical analysis), this is problematic as it gives a rather poor approximation of the discrete states $|R\rangle$ corresponding to large tableaux R. In effect, the antiperiodic boundary conditions tend to push these states to higher P^-, and one needs to go to rather high P^+ to get a good estimate of their masses. In view of (12) we see that one can in general make large mistakes in estimating masses of rather low lying states. Adjoint QCD$_2$ is thus an example of a theory where it is very important to treat the zero modes correctly in order to get reliably even the low lying massive spectrum.

4. The whole construction should generalize straightforwardly to other $2d$ Yang-Mills theories with adjoint fermions (such as supersymmetric Yang-Mills models). It would be interesting to understand the analogs of the phenomena discussed here in $d > 2$.

Acknowledgments

We thank D. Gross, I. Klebanov, E. Rabinovici and A. Schwimmer for useful discussions. This work was partly supported by a DOE OJI grant.

References

[1] Dalley S. and Klebanov I. R., hep-th/9209049, *Phys. Rev.* **D47** (1993) 2517.
[2] Kutasov D., hep-th/9306013, *Nucl. Phys.* **B414** (1994) 33.
[3] Bhanot G., Demeterfi K. and Klebanov I. R., hep-th/9307111, *Phys. Rev.* **D48** (1993) 4980;hep-th/9311015, *Nucl. Phys.* **B418** (1994) 15.

[5] Note that there is no discrepancy with the result of [16] that $\sigma \propto m$ (4). σ and T are related but in general different quantities. The difference between the two is particularly clear when thinking about winding modes around Euclidean time in the context of finite temperature β^{-1}. The mass squared of the lightest winding mode $M^2(\beta)$ changes sign at $\beta = \beta_H$; β_H is determined by T. σ is related to the low temperature behavior of M: $M(\beta \to \infty) \sim \beta\sigma$. See [7] for further discussion.

[4] Kogan I. I., hep-th/9311164, *Phys. Rev.* **D49** (1994) 6799.

[5] Boorstein J. and Kutasov D., hep-th/9401044, *Nucl. Phys.* **B421** (1994) 263.

[6] Smilga A., hep-th/9402066, *Phys. Rev.* **D49** (1994) 6836;hep-th/9607007, *Phys. Rev.* **D54** (1996) 7757.

[7] Boorstein J. and Kutasov D., hep-th/9409128, *Phys. Rev.* **D51** (1995) 7111.

[8] Hansson T.H. and Tzani R., hep-th/9410235, *Nucl. Phys.* **B435** (1995) 241.

[9] Lenz F., Shifman M. and Thies M., hep-th/9412113, *Phys. Rev.* **D51** (1995) 7060.

[10] Kutasov D. and Schwimmer A. , hep-th/9501024, *Nucl. Phys.* **B442** (1995) 447.

[11] Antonuccio F. and Dalley S., hep-lat/9505009; hep-ph/9506456,*Nucl. Phys.* **B461** (1996) 275.

[12] Armoni A. and Sonnenschein J., hep-th/9508006, *Nucl. Phys.* **B457** (1995) 81.

[13] Pinsky S. S. and Kalloniatis A. C., hep-th/9509036, *Phys. Lett.* **B365** (1996) 225.

[14] Kogan I. and Zhitnitsky A., hep-ph/9509322, *Nucl. Phys.* **B465** (1996) 99.

[15] Zhitnitsky A., hep-ph/9510366, *Phys. Rev.* **D53** (1996) 5821.

[16] Gross D., Klebanov I. R., Matytsin A. V. and Smilga A., hep-th/9511104, *Nucl. Phys.* **B461** (1996) 109.

[17] Antonuccio F. and Dalley S., hep-th/9512106, *Phys. Lett.* **B376** (1996) 154.

[18] Pinsky S. and Robertson D., hep-th/9512135, *Phys. Lett.* **B379** (1996) 169.

[19] Paniak L.D., Semenoff G.W.and Zhitnitsky A.R., hep-th/9606194, *Nucl. Phys.* **B487** (1997) 191;hep-ph/9701270.

[20] Pinsky S. and Mohr R., hep-th/9610007, *Int. J. Mod. Phys.* **A12** (1997) 1063.

[21] Antonuccio F. and Pinsky S., hep-th/9612021, *Phys. Lett.* **B397** (1997) 42.

[22] McCartor G. and Robertson D. G., hep-th/9612083.

[23] Pinsky S., hep-th/9612073; hep-th/9703108.

[24] Frishman Y. and Sonnenschein J., hep-th/9701140.

[25] Fugleberg T., Halperin I. and Zhitnitsky A., hep-ph/9705281.

[26] Goddard P. and Olive D., *Int. Jour. Mod. Phys.* **A1** (1986) 303.

[27] Kutasov D. and Seiberg N.,*Nucl. Phys.* **B358** (1991) 600.

[28] 't Hooft G., *Nucl. Phys.* **B72** (1974) 461.

More on Screening and Confinements in 2D QCD

J. Sonnenschein

School of Physics and Astronomy,
Beverly and Raymond-Sackler Faculty of Exact Sciences,
Tel-Aviv University, Ramat-Aviv, Tel-Aviv 69978, Israel

1. Introduction

Gross et al. [1] have argued that massless dynamical fermions **screen** heavy external charges, even if they are in a representation which has zero "N_c-ality" ($Z_{N_c} \sim 0$), namely, even if the dynamical quarks cannot compose the external ones. Confinement is restored as soon as the dynamical fermions get some non-trivial mass (unless they can compose the representation of the external quarks). This conjecture was "proven" in [1] for the following systems (i) abelian case, (ii) $SU(N_c = 2)$ with the dynamical fermions in **3**, (iii) spinor **8** of $SO(8)$.

In the present talk I will describe two papers [2, 4] where we presented furhter evidence for screening of massless fundamental and adjoint fermions of any non-abelian group. In those works we have used the following methods: (i) Deducing the potential from solutions of the equations of motion of the bosonized action. (ii) Bosonizing also the heavy external charges. In this framework **Confinement** is manifested via the absence of soliton solutions of unbounded quarks, and **Screening** implies finite-energy quark static solutions (iii) Large N_f analysis of the system including analyzing also the next to the leading order behaviour.

1.1.   background remarks

(1) The screening mechanism in the massless Schwinger model could be attributed to the **exchange of the "massive photon"**. The non-abelian counterpart is clearly much more complicated. However, in the limit of large number of flavours N_f (with finite N_c), the non abelian theory resembes **a collection of $N_c^2 - 1$ abelian theories.** In that limit the spectrum includes Schwinger like $N_c^2 - 1$ massive modes that induce screening.

(2) The "massive gauge states" are an indication of the "non-confining" structure of the spectrum. Even though they are gauge invariant states, they are in the adjoint rep. of a "global color symmetry". These states had already been pointed out in an earlier work,[5] based on a BRST analysis and a special parametrization of the gauge configurations. However, in that paper we were not able to rigorusly show that indeed they were part of the BRST cohomology. Note, that even if they are not in space of physical states, **the massive states could nevertheless be responsible for the screening potential.**

(3) When passing from a screening picture at large N_f and finite N_c to the domain of a small N_f, one can anticipate two types of scenarios: (i) A **smooth transition**, namely, the screening behavior persists all the way down to $N_f = 1$. (ii) A **"phase transition"** at a certain value of N_f and confining nature below it.

(4) Are the massive modes of large N_f an artifact of the abelianization of the theory? To check this possibility we have searched for **non-abelian solutions** of the equations of motions. Indeed, we found new non-abelian gauge solutions that are also massive.[2] We believe that this presents certain evidence in favour of option (i).

(5) Low lying baryonic states, were determined in the semiclassical picture of the strong coupling limit.[7] Does it contradict the screening nature of the theory? It does not, since the baryons were discovered only for massive quarks and not for massless ones.

1.2.   The outline of the talk

(1) Review of Bosonization in QCD_2

  (2) Equations of motion of QCD_2 in the presence of external currents

  (3) Solutions of the equations without external quarks

  (4) Solutions of the equations with external current and the potential

  (5) Bosonized external currents

  (6) Large N_f expansion

  (7) $N = 1$ Super Yang-Mills

2. Review of Bosonization in QCD_2

2.1. Dirac fermions in the fundamental representation.

The action of massive QCD_2 [3]

$$
\begin{aligned}
S_{QCD_2} \quad = \quad & S_1(u) - \frac{1}{2\pi} \int d^2z Tr(iu^{-1}\partial u \bar{A} \\
+ \quad & iu\bar{\partial}u^{-1}A + \bar{A}u^{-1}Au - A\bar{A}) \\
+ \quad & \frac{m^2}{2\pi} \int d^2z : Tr_G[u + u^{-1}] : + \frac{1}{e_c^2} \int d^2z Tr_H[F^2]
\end{aligned}
$$

(1)

where $u \in U(N_f \times N_C)$; $S_k(u)$ is a level k WZW model; A and $\bar{A} \in$ algebra of $H \equiv SU(N_C)$; $F = \bar{\partial}A - \partial\bar{A} + i[A, \bar{A}]$; and where $m^2 = m_q\mu\frac{1}{2}e^\gamma$, where μ is the normal ordering mass.

Notice that the space-time has a Minkowski signature, and we use the following notations $B \equiv B_+$ and $\bar{B} \equiv B_-$.

The action for massless fermions can be simplified

$$
\begin{aligned}
S \quad = \quad & S_{N_f}(h) - \frac{N_f}{2\pi} \int d^2z Tr(ih^{-1}\partial h \bar{A} \\
+ \quad & ih\bar{\partial}h^{-1}A + \bar{A}h^{-1}Ah - A\bar{A}) \\
+ \quad & S_{N_C}(g) + \frac{1}{2\pi} \int d^2z[\partial\phi\bar{\partial}\phi] \frac{1}{e_c^2} \int d^2z Tr_H[F^2]
\end{aligned}
$$

(2)

using the following parametrization $u \equiv ghle^{i\sqrt{\frac{4\pi}{N_C N_f}}\phi}$ where $h \in SU(N_c)$, $g \in SU(N_f)$, $l \in \frac{U(N_c N_f)}{[SU(N_c)]_{N_f} \times [SU(N_f)]_{N_c} \times U_B(1)}$ and setting $l = 1$.

2.2. Majorana fermions in the adjoint representation.

An action of bosonized Majorana fermions in the adjoint representation[2]

$$
\begin{aligned}
S \quad = \quad & \frac{1}{2}S(h_{ad}) - \frac{1}{4\pi} \int d^2z Tr(ih_{ad}^{-1}\partial h_{ad}\bar{A} + iu\bar{\partial}h_{ad}^{-1}A \\
+ \quad & \bar{A}h_{ad}^{-1}Ah_{ad} - A\bar{A})\frac{m^2}{2\pi} \int d^2z \, Tr_G : [h_{ad} + h_{ad}^{-1}] : \frac{1}{e_c^2} \int d^2z Tr_H[F^2]
\end{aligned}
$$

(3)

where h_{ad} are $(N_c^2-1) \times (N_c^2-1)$ matrices; the Virasoro anomaly $c_{vir} = \frac{1}{2}(N_c^2 - 1)$, the affine Lie algebra anomaly $= N_c$ the conformal dimension $\Delta h_{ad} = \frac{1}{2}$ and the factor $\frac{1}{2}$ in $\frac{1}{2}S(h_{ad})$ comes from the reality of the Majorana fermions.

3. Equations of motion of QCD_2 in the presence of external currents

Studying the quantum system by analyzing the corresponding equations of motion is a justified approximation only provided that the **classical configurations dominate the functional integral**. Such a scenario can be achieved in the limit of a large number of flavours. where $\frac{1}{N_f}$ plays the role of $\hbar$.

The colored sectors of the action in ($A = 0$ gauge) takes the form

$$S = N_f \left\{ S_1(h) - \frac{1}{2\pi} \int d^2z Tr(ih^{-1}\partial h \bar{A}) + \frac{1}{\tilde{e}_c^2} \int d^2z Tr[\partial \bar{A}]^2 \right\}$$

where $\tilde{e}_c = e_c\sqrt{N_f}$.

The equations of motion which follow from the variation of the action (2) with respect to h are

$$\bar{\partial}(h^{-1}\partial h) + i\partial\bar{A} + i[h^{-1}\partial h, \bar{A}] = 0$$
$$\partial(h\bar{\partial}h^{-1}) - i\partial(h\bar{A}h^{-1}) = 0$$

$$(4)$$

A similar result holds for h_{ad}. External currents are coupled to the system by adding to the action (2) or (3)

$$\mathcal{L}_{ext} = \frac{1}{2\pi} \int d^2z Tr(J_{ext}\bar{A} + \bar{J}_{ext}A).$$

Now the equations of motion take the form

$$\partial^2\bar{A} + \alpha_c(iN_f h^{-1}\partial h + J_{ext}) = 0$$
$$\partial\bar{\partial}\bar{A} + [i\partial\bar{A}, \bar{A}] - \alpha_c[N_f(ih\bar{\partial}h^{-1} + h\bar{A}h^{-1} - \bar{A}) + \bar{J}_{ext}] = 0$$

$$(5)$$

where $\alpha_c = \frac{e_c^2}{4\pi}$. It follows from the equations of motion (4) and (5) that both the dynamical currents $j_{dy} = \frac{iN_f}{2\pi}h^{-1}\partial h$, $\bar{j}_{dy} = \frac{iN_f}{2\pi}[h\bar{\partial}h^{-1} - ih\bar{A}h^{-1} + i\bar{A}]$ as well as the external currents are covariantly conserved, which for $A = 0$ reads

$$\bar{D}j_{dy} + \partial\bar{j}_{dy} = 0 \qquad \bar{D}J_{ext} + \partial\bar{J}_{ext} = 0.$$

$$(6)$$

with $\bar{D} = \bar{\partial} - i[\bar{A},]$.

One can eliminate the dynamical current, and then one finds

$$\partial\bar{\partial}\bar{A} + [i\partial\bar{A}, \bar{A}] + \alpha_c(N_f\bar{A} - \bar{J}_{ext}) = 0$$

$$(7)$$

In fact the equation one gets $\partial[l.h.s](7) = 0$. The antiholomorphic function can be eliminated by fixing the residual gauge invariance $\bar{A} \to iu^{-1}\bar{\partial}u + u^{-1}\bar{A}u$, with $\partial u = 0$.

4.   Solutions of the equations without external quarks

We enlist several statements about the solutions:

(1) The equation of motion is not invariant but rather covariant with respect to the "global color" transformation $A \to u^{-1}Au$ with a constant u. So any solution is in the adjoint rep. of the "global color" group.

(2) **"abelian" massive mode**

Consider a configuration of the form $\bar{A} \equiv T^a \bar{A}^a(z, \bar{z}) = T^a \delta^{a,a_0} \bar{A}(z, \bar{z})$ then the commutator term vanishes and $\mathcal{A}$ has to solve

$\partial \bar{\partial} \bar{A} + \tilde{\alpha}_c \bar{A} = 0$ with $\tilde{\alpha}_c = N_f \alpha_c$.

(3) **No soliton solutions** Let us now check whether the equations admit soliton solutions. For static configurations

$$\partial_1^2 \bar{A} - \sqrt{2}[i\partial_1 \bar{A}, \bar{A}] - 2\tilde{\alpha}_c \bar{A} = 0$$

Multiplying the equation by $\bar{A}$, taking the trace of the result and integrating over dx one finds after a partial integration that $\int dx [Tr[(\partial \bar{A})^2 + 2\tilde{\alpha}_c A^2] = 0$ which can be satisfied only for a vanishing $\bar{A}$.

(4) **Abelian Bessel function** Consider an ansatz for the solution of the form $\bar{A} = zC(z\bar{z})$ and C is determined by the equation

$\rho C'' + 2C' + \tilde{\alpha}_c C = 0$, where $C' = \partial_{z\bar{z}} C$.

The solution for the gauge field

$$\bar{A} = \frac{\bar{A}_0 z}{\sqrt{\alpha_c z\bar{z}}} J_1(2\sqrt{\alpha_c z\bar{z}})$$

where $\bar{A}_0$ is an arbitrary constant matrix.

(3) **Non-abelian solution**

Consider in the special case of $SU(2)$ the configuarion $\bar{A} = e^{-i\theta\tau_0}\bar{A}_0 e^{i\theta\tau_0}$ with a constant matrix $\bar{A}_0 = e_0\tau_0 + \bar{e}\tau + e\bar{\tau}$. Plugging this ansatz into eqn. (7) with no external source one finds that there is a solution provided that $\theta = \theta_0 + k\bar{z} + \bar{k}z$

where k, $\bar{k}$ and θ_0 are constants. Indeed the following gauge field

$$\bar{A} = \frac{\tilde{\alpha}_c - k\bar{k}}{\bar{k}}\tau_0 + \sqrt{\frac{(k\bar{k} - \tilde{\alpha}_c)\tilde{\alpha}_c}{4\bar{k}^2}}[e^{-i\theta}\tau + e^{i\theta}\bar{\tau}] \tag{8}$$

is a " non-abelian solution". Setting $\theta_0 = 0$ requires that $(k\bar{k} - \tilde{\alpha}_c) > 0$. The solution is **truly non-abelian**. The corresponding F is $F = i\bar{k}e^{-i\theta\tau_0}(\bar{e}\tau - e\bar{\tau})e^{i\theta\tau_0}$. Performing a gauge transforamtion with $U = e^{-i\theta\tau_0} \to F_U = i\bar{k}(\bar{e}\tau - e\bar{\tau})$, $\bar{A}_U = \bar{A}_0 + k\tau_0$ $A_U = \bar{k}\tau_0$. $A_U, \bar{A}_U$ and F_U are constants and no two commuting. Furthermore an abelian gauge configuration of the form $\bar{A} = -i\bar{k}z(\bar{e}\tau - e\bar{\tau})$ and $A = 0$ that leads to the same F is not connected to $A_U, \bar{A}_U$ by a gauge transformation.

5. The energy-momentum tensor and the spectrum of non-abelian solution

First we have to compute the components of the energy momentum tensor $T \equiv T_{zz}, \bar{T} \equiv T_{\bar{z}\bar{z}}, T_{\bar{z}z}$ that corresponds to the action (2).

Only the colored part of the energy momentum tensor is relevant to our discussion.

$$T = \frac{\pi}{N_f + N_c} : Tr[j_{dy} j_{dy}] : \qquad T_{\bar{z}z} = \frac{1}{8\pi\alpha_c} Tr[(\partial \bar{A})^2]$$

where the dynamical currents which were defined below eqn.(5) are

$$\begin{aligned}
j_{dy} &= \frac{iN_f}{2\pi} h^{-1}\partial h = -\frac{1}{2\pi\alpha_c}\partial^2 \bar{A}; \\
\bar{j}_{dy} &= \frac{1}{2\pi\alpha_c}(\partial\bar{\partial}\bar{A} + i[\partial\bar{A}, \bar{A}]) = -\frac{iN_f}{2\pi} h^{-1}\bar{\partial} h = -\frac{N_f}{2\pi}\bar{A}
\end{aligned} \qquad (9)$$

It is natural in the light cone gauge to use a **light front quantization**. We take z to denote the space coordinate. $\bar{P}$ and P are integrals over T and $T_{\bar{z}z}$. The masses of the states = the eigenvalues of $M^2 = P\bar{P}$.

A proper normalization of the fields is introduced and discussed in [2]. Using this normalization we find the following expectation values of the energy momentum tensor

$$\begin{aligned}
<T> &= \frac{1}{8\pi}\frac{N_f^2}{N_f+N_c}(\bar{k})^2\frac{k\bar{k}}{\bar{\alpha}_c}(1 - \frac{\bar{\alpha}_c}{k\bar{k}}) \\
<T_{\bar{z}z}> &= \frac{N_f}{16\pi}(k\bar{k})(1 - \frac{\bar{\alpha}_c}{k\bar{k}})
\end{aligned} \qquad (10)$$

The corresponding momenta

$$\bar{P} = \frac{1}{8\pi}\frac{N_f^2}{N_f + N_c}(L\bar{k})\bar{k}\frac{k\bar{k}}{\bar{\alpha}_c}(1 - \frac{\bar{\alpha}_c}{k\bar{k}}) \qquad P = \frac{N_f}{16\pi}(L\bar{k})k(1 - \frac{\bar{\alpha}_c}{k\bar{k}})$$

The non-abelian state is characterized by masses

$$M^2 = \frac{n^2}{32}\frac{N_f^3}{N_f + N_c}(k\bar{k})\frac{k\bar{k}}{\bar{\alpha}_c}(1 - \frac{\bar{\alpha}_c}{k\bar{k}})^2$$

Note that
(1) $(k\bar{k} - \bar{\alpha}_c) > 0$ (the case of zero $k\bar{k} - \bar{\alpha}_c$ corresponds to vanishing $\bar{A}$).
(2) M starts from $M = 0$ and grows up linearly in $k\bar{k}$ for $k\bar{k} >> \bar{\alpha}_c$.
(3) the solution is singular for $\bar{\alpha}_c = 0$.

6. Solutions of the equations with external current

We turn now on a covariantly conserved (eqn. (6)) external current J_{ext}
(1) Abelian solutions of the equations of motion are easily constructed.

For instance for a pair of quark anti-quark as an external classical source
$\bar{J}^a ext = T^a \delta^{a,a_0} Q[\delta(x_1 - R) - \delta(x_1 + R)]$ the abelian solution is

$$\bar{A} = \frac{1}{2}\sqrt{2\alpha_c}T^a\delta^{a,a_0}Q[e^{-\sqrt{2\bar{\alpha}_c}|x_1 - R|} - e^{-\sqrt{2\bar{\alpha}_c}|x_1 + R|}].$$

Inserting this expression into $\frac{1}{2\pi}\int dx_1 Tr[A\bar{J}_{ext}]$
one finds the usual **screening potential**

$$V(r) = \frac{1}{2\pi}\sqrt{2\alpha_c}Q^2(1 - e^{-2\sqrt{2\alpha_c}|R|})Tr[(T^{a_0})^2]$$

(2) **"non-abelian" solutions**

The $SU(2)$ "non-abelian solution" is a solution also in case of a constant
external current $\bar{J} = T^a\delta^{a,0}J_0$ with the trivial modification that the coefficient
of $[e^{-i\theta}\tau + e^{i\theta}\bar{\tau}]$ is now $\sqrt{\frac{\bar{\alpha}_c[\frac{J_0 k}{N_f} + (k\bar{k} - \bar{\alpha}_c)]}{4\bar{k}^2}}$.

Consider now an external current of the form $\bar{J} = J_0(\bar{z})\tau_0$. A solution in
that case is
$$\bar{A} = (f_0 + J_0(\bar{z}))\tau_0 + [g(e^{[-i(k\bar{z}+\bar{k}z+I)]})\bar{\tau} + c.c]$$

where $\bar{\partial}I(\bar{z}) = \frac{1}{N_f}\bar{J}_0(\bar{z})$ with f_0 and g related to k and $\bar{k}$ as given in eqn. (8). In
the case of light-front "static" current $\bar{J}_{ext}(\bar{z})$ In particular a quark anti-quark
pair $\bar{J}^a_{ext} = \frac{1}{2}T^a\delta^{a,a_0}Q[\delta(\bar{z} - R) - \delta(\bar{z} + R)]$ The corresponding solution has
$\epsilon(\bar{z} - R)$ and $\epsilon(\bar{z} + R)$ factors in θ. The corresponding potential is a constant
thus non-cofining.

7.   Bosonized external currents

Another approach to the coupling of the dynamical fermions to external currents is to **bosonize the "external" currents.**

Let us briefly summarize first the abelian case. Consider external fermions
of mass M and charge qe described by the real scalar filed Φ and dynamical
fermions of unit charge e and mass m described by the scalar ϕ.

The Lagrangian of the combined system after integrating out the gauge fields
is given by

$$\begin{aligned}\mathcal{L} &= \frac{1}{2}(\partial_\mu\phi\partial^\mu\phi) + m\Sigma[cos(2\sqrt{\pi}\phi) - 1] + \frac{1}{2}(\partial_\mu\Phi\partial^\mu\Phi) \\ &+ M\Sigma[cos(2\sqrt{\pi}\Phi) - 1] - \frac{e^2}{2\pi}(\phi + q\Phi)^2\end{aligned}$$

(11)

Let us look for static solutions of the corresponding equations of motion
with finite energy. Take, without loss of generality, $\phi(-\infty) = \Phi(-\infty) = 0$.
From the M term we get $\Phi(\infty) = \sqrt{\pi}N$ with N integer. For $m \neq 0$ we also

get $\phi(\infty) = \sqrt{\pi}n$. Now from the e^2 term, $n + qN = 0$. Thus, for instance for $N = 1$ finite energy solutions occur only for $q = -n$. For $m = 0$ only $\Phi(\infty) = \sqrt{\pi}N$ and then a finite energy solution for $N = 1$ is if $\phi(\infty) = -\sqrt{\pi}q$. So, when $q = -n$, $\to$ the screening phase, $q \neq -n$ for $m \neq 0 \to$ confinement phase $m = 0 \to$ screening .

Proceeding now to the QCD case one can consider several different possibilities (J^F_{ext}, j^F_{dy}), (J^F_{ext}, j^{ad}_{dy}), (J^{ad}_{ext}, j^F_{dy}), $(J^{ad}_{ext}, j^{ad}_{dy})$ and with dynamical fermions that can be either massless or massive.

The system of dynamical adjoint fermions and external fundamental quarks can be described by an action which is the sum of (3) and (1).

Integrating over the gauge degrees of freedom one is left with the terms in (1) and (3) that do not include coupling to gauge field together with a current-current non-local interaction term.

For the interesting case of dynamical quarks in the adjoint and external in fundamental we get for the interaction term

$$\int d^2z \sum_a \left\{ \frac{1}{\partial} [Tr(T^a_F i u_{ext}^{-1} \partial u_{ext}) + \frac{i}{2} Tr(T^a_{ad} h^{-1} \partial h)] \right\}^2$$

where u is defined in (1), T^a_F are the $SU(N_c)$ generators expressed as $(N_c N_f) \times (N_c N_f)$ matrices in the fundamental rep. of $U(N_c \times N_f)$ T^a_{ad} the $(N_c^2 - 1) \times (N_c^2 - 1)$ matrices in the $SU(N_c)$ adjoint rep. For simplicity we discuss from here on the case of a single flavor.

Let us first consider the case of external adjoint quarks $u_{ext}(x) \in SU(N_c) \times U_B(1)$,

$$u_{ext} = \begin{pmatrix} e^{-i\Phi} & & & & \\ & e^{i\Phi} & & & \\ & & 1 & & \\ & & & \ddots & \\ & & & & 1 \end{pmatrix}$$

(Φ is not normalized canonically here).

The reason that we take a diagonal ansatz is that it corresponds, to a minimal energy configuration.

Ansatz (7) corresponds to $Q^1_{ext} \bar{Q}^2_{ext}$, namely, to an external adjoint state. We expect this state to be screened by the adjoint dynamical fermions. With this ansatz $\frac{1}{\partial}(i h_{ext}^{-1} \partial h_{ext})$ takes the form

$$\begin{pmatrix} \Phi & & & & \\ & -\Phi & & & \\ & & 0 & & \\ & & & \ddots & \\ & & & & 0 \end{pmatrix}$$

It is thus clear that only T_F^3 contributes to the trace in (7). To show the dynamical configuration that screens, take

$$
\log h_{ad} = \begin{pmatrix} 0 & -\phi & 0 & & \\ \phi & 0 & 0 & & \\ 0 & 0 & 0 & & \\ & & & \ddots & \\ & & & & 0 \end{pmatrix}
$$

with the matrix that contributes to the Tr in (7),

$$
T_{adj} = i \begin{pmatrix} 0 & 1 & 0 & & \\ -1 & 0 & 0 & & \\ 0 & 0 & 0 & & \\ & & & 0 & \\ & & & & \ddots \\ & & & & & 0 \end{pmatrix}
$$

which corresponds to the generator of rotation in direction 3 for the sub $O(3)$ of first three indices, thus obtaining the a term proportional to $(\Phi + \phi)^2$ emerging from (7). The mass terms for u_{ext} and h are now proportional to $(1 - cos\Phi)$ and $(1 - cos\phi)$ respectively. A boundary condition $\Phi(\infty) = 2\pi$ can be cancelled in the interaction term by the boundary condition $\phi(\infty) = -2\pi$.

Let us examine now the case of a single external quark

$$
u_{ext} = \begin{pmatrix} e^{-i\Phi} & & & \\ & 1 & & \\ & & \ddots & \\ & & & 1 \end{pmatrix}
$$

Its contribution to the interaction term is $\sum_{i=1}^{N_c - 1} (\eta_i \Phi + \text{``}dyn\text{''})^2$ where $\eta_i = \frac{1}{\sqrt{2i(i+1)}}$ and "dyn" is the part of the dynamical quarks.

If again we take a configuration of the dynamical quarks based on a single scalar like in (7) we get altogether an e^2 term of the form $(\frac{1}{2}\Phi + \phi)^2$.

Now if $\Phi(\infty) = 2\pi$ one cannot find a finite energy solution since from the mass term $\phi(\infty) = 2\pi n$, and thus there is no way to cancell the interaction term.

If, however, we consider $m = 0$ there is no constraint on $\phi(\infty)$ so it can be taken to be equal $-\pi$, and thus again a screening situation is achieved.

This argument should be supplemented by showing that one cannot find another configuration that may cancel the $\eta_i \Phi$ term in $(\eta_i \Phi + \text{``}dyn\text{''})^2$, for $SU(N_c)$ with $N_c \geq 3$.

8. Large N_f expansion

Consider a system of a quark in the fundamental representation placed at a distance of $2R$ from an anti-quark that transforms in the anti-fundamental representation. This can be expressed as the following classical c-number charge density

$$\rho^a = \delta^{a1}(\delta(x - R) - \delta(x + R)) \qquad (12)$$

Strictly speaking, one is allowed to introduce classical charges and neglect quantum fluctuations only if the external charges transform in a large color representation. (This is an analog of the statement that only for quantities of large angular momentum quantum fluctuations are suppressed.) Moreover, by choosing (12), there is an obvious "abelian" self-consistent solution of the equations for which all the dynamical quantities points in the '1' direction and thus all the commutators vanish.

One way to overcome these obstacles is to search for "truly" non-abelian solutions of the equations, as was discussed in section 4. Here we proceed by implementing Adler's[8] semi-classical approach for introducing static external quark charges. In this approach the quarks color charges satisfy non-abelian $SU(N_c)$ color algebra so that the external quark charge density takes the form

$$\rho^a = Q^a \delta(x - R) + \bar{Q}^a \delta(x + R) \qquad (13)$$

Q^a and $\bar{Q}^a$ are in $(\mathbf{N_C}, 1)$ and $(1, \bar{\mathbf{N}}_C)$ representations of $SU(N_C) \otimes SU(N_C)$ group respectively. The algebra of those operators was worked out in [8, 9]. is reviewed briefly in the appendix.

The expansion in $\frac{1}{N_f}$ is defined as usual by taking $N_f \to \infty$ while $\tilde{\alpha}_c = \frac{e^2 N_f}{4\pi}$ is kept fixed. One can solve the equations iteratively as follows. A solution for $\bar{A}$, J and J_{ext} of the equations expanded to a given order in e is inserted back to the equations as a source to determine the next order solution. A similar treatment in four dimensions is given in refs.[10, 11].

The formal expansion in e is as follows

$$\begin{aligned} \bar{A} &= eA^{(1)} + e^3 A^{(3)} + e^5 A^{(5)} + \dots \\ J &= J^{(0)} + e^2 J^{(2)} + e^4 J^{(4)} + \dots \\ J_{ext} &= e^2 j^{(2)} + e^4 j^{(4)} + e^6 j^{(6)} + \dots \end{aligned} \qquad (14)$$

After performing the interation one finds that the potential up to next to leading order takes the form

$$V(2R) = \mu \frac{\pi}{2N_f} \frac{N_c^2 - 1}{2N_c} (1 - e^{-\sqrt{\tilde{\alpha}}2R}) \qquad (15)$$

$$+ \mu (\frac{\pi}{2N_f})^3 \frac{N_c^2}{2} \frac{N_c^2 - 1}{2N_c} \left((1 - e^{-\sqrt{\tilde{\alpha}}2R})^2 - \sqrt{\tilde{\alpha}}2R \, e^{-\sqrt{\tilde{\alpha}}2R} (1 - e^{-\sqrt{\tilde{\alpha}}2R}) \right)$$

Thus, the potential that includes the first correction to the abelian one approaches a constant value at large distances where the force between the external quark and the anti-quark vanishes.

9.   Massive QCD_2

The bosonized action of massive QCD_2 with N_f fundamental representations was given in (1). We can now expand the mass term in a (non-local) power series in J as follows

$$h = 1 - \frac{2\pi}{iN_f}\frac{1}{\partial J} + \frac{2\pi}{iN_f}^2 \frac{1}{\partial J^2} - \frac{1}{\partial}\frac{1}{\partial J}J + \dots$$

Now the solution of the equation in the presence of (13) is

$$\bar{A} = -\frac{e}{\sqrt{2}}(1 + \frac{e^2N_f^2}{8\pi^2m^2})^{-1}(Q \mid x - R \mid + \bar{Q} \mid x + R \mid) \tag{16}$$

Substituting $\bar{A}$ in the potential yields,

$$V = -\frac{e^2}{2}(1 + \frac{e^2N_f^2}{8\pi^2m^2})^{-1}Q\bar{Q} \times 2R = \frac{\mu^2\pi}{2N_f}(1 + \frac{\sqrt{\bar{\alpha}}}{8\pi Cm_q})^{-1}C_2(\mathcal{R}) \times 2R \tag{17}$$

The same expression for the potential in the abelian case was obtained in [1]. Thus the dominant $\frac{1}{N_f}$ contribution exhibits a confinement behavior. It should be emphasized that in the above analysis it is assumed that the external charges cannot be composed by the dynamical ones.

10.   Supersymmetric Yang-Mills

The supersymmetric YM (SYM) action [12] is the following

$$S = \int d^2x \; tr\left(-\frac{1}{4}F_{\mu\nu}^2 + i\bar{\lambda}\,\slashed{D}\lambda + \frac{1}{2}(D_\mu\phi)^2 + 2ie\phi\bar{\lambda}\gamma_5\lambda\right), \tag{18}$$

The gluon equation of motion now reads

$$D_\mu D^\mu F = e\epsilon^{\mu\nu}D_\mu(J_\nu^\lambda + J_\nu^\phi), \tag{19}$$

where J_μ^λ denotes the gluino vector current and J_μ^ϕ denotes the scalar vector current. The equation for the divergence of the fermionic axial current.

$$\epsilon^{\mu\nu}D_\mu J_\nu^\lambda = -\frac{eN_c}{\pi}F + ie(\phi\bar{\lambda}\lambda + \bar{\lambda}\lambda\phi) \tag{20}$$

The factor N_c which appear in the anomaly term is due to the adjoint gluinos which run in the anomaly loop. A similar equation holds for the scalar current

$$\epsilon^{\mu\nu}D_\mu J_\nu^\phi = -\frac{eN_c}{2\pi}F - \epsilon^{\mu\nu}\partial_\mu(-2i[\phi, \partial_\nu\phi] + 2e[\phi, [A_\nu, \phi]]) \tag{21}$$

This equation was obtained by applying the point splitting technique to the scalar current.

Thus the quantum version of equation (19) is

$$(D_\mu D^\mu + \frac{3e^2 N_c}{2\pi})F = ie^2(\phi\bar{\lambda}\lambda + \bar{\lambda}\lambda\phi) - e\epsilon^{\mu\nu}\partial_\mu(-2i[\phi, \partial_\nu\phi] + 2e[\phi, [A_\nu, \phi]]) \quad (22)$$

which means that the gluon propagator has only a single pole which is massive.

The implication of the last equation on the potential between external charges is clear. The interaction mediated by the exchange of these massive modes is necessarily a screening one. The potential takes the form of eqn.(15) with a range that behaves like $\sim [e\sqrt{N_c}]^{-1}$.

References

[1] David J. Gross, Igor R. Klebanov,Andrei V. Matytsin and Andrei V. Smilga, *"Screening vs. Confinement in 1+1 Dimensions"*, Nucl. Phys. **B461** (1996) 109.

[2] Y. Frishman and J. Sonnenschein, *"QCD$_2$-Screening, Confinement and Novel non-abelian solutions"*, hep-th 9701140.

[3] Y. Frishman and J. Sonnenschein, *"Bosonization of Colored Flavored Fermions and QCD in Two-Dimensions"*, Nucl. Phys. **B294** (1987) 801 ; *"Bosonization and QCD in Two-Dimensions"*,Physics Reports **223** # 6 (1993) 309.

[4] A. Armoni, J. Sonnenschein, *" Screening and Confinement in the large N_f QCD$_2$ and in $N = 1$ SYM$_2$ hep-th 9703114.*

[5] Y. Frishman, A.Hanany and J. Sonnenschein, *"Subtleties in QCD in two dimensions "*, Nucl. Phys. **B429** (1994) 75

[6] G. 't Hooft, *"A Two-Dimensional Model for Mesons"*, Nucl. Phys. **B75** (1974) 461.

[7] G. Date, Y. Frishman and J. Sonnenschein, *Nucl. Phys.* **B283** (1987) 365

[8] S. L. Adler, *"Classical Algebraic Chromodynamics"*, Phys. Rev. **D17** (1978) 3212.

[9] R. Giles and L. McLerran, *"A Non-Perturbative Semi-Classical approach to the calculation of the Quark Force"*, Phys. Lett. **79B** (1978) 447.

[10] R. Jackiw, L. Jacobs and C. Rebbi, *"Static Yang-Mills fields with sources"*, Phys. Rev. **D20** (1979) 474.

[11] H. Arodź, *"On Classical Yang-Mills equations with weak external sources"*, Nucl. Phys. **B207** (1982) 288.

[12] S. Ferrara, *"Supersymmetric Gauge Theories in two Dimensions"*, Lett. Nuovo Cim. **13** (1975) 629.

Intermediate Volumes and the Role of Instantons

Pierre van Baal

Isaac Newton Institute for Mathematical Sciences,
20 Clarkson Road, Cambridge CB3 0EH, UK
and
*Instituut-Lorentz for Theoretical Physics,
University of Leiden, P.O.Box 9506,
NL-2300 RA Leiden, The Netherlands

1. Introduction

An outstanding problem is to understand the formation of a mass gap and the spectrum of excitations in a non-Abelian gauge theory. Non-perturbative aspects are believed to play a crucial role, but despite much progress a simple explanation is still lacking. Over the years we have been interested in addressing this problem in a finite volume, where its size can be used as a control parameter, which is conspicuously absent in infinite volumes, in particular for formulating the binding of gluons in glueballs. Much progress was made in intermediate volumes with a torodial geometry, where results can be directly compared to lattice Monte Carlo calculations in the same physical volume [1].

The essential features of this analysis are easily explained. At very small volumes the effective coupling constant is small, due to asymptotic freedom of non-Abelian gauge theories. In this domain ordinary perturbation theory can be used. For a torus, due to the presence of zero-momentum modes, for which the classical potential is quartic, this results in an expansion in powers of $g^{2/3}$ for the spectrum [2]. In a spherical geometry, where due to curvature of the manifold no zero-modes appear, perturbation theory is as usual [3].

(*) Permanent address.

2. The Role of Instantons

Irrespective of the geometry of the space on which the gauge theory is formulated there are low-energy modes in terms of which the wave functional at larger coupling (i.e. larger volume) will start to spread out over field space. Not only is the physical Yang-Mills field space a curved manifold [4], but also it has non-trivial topology [5]. In particular the latter is crucial for a better understanding of the non-perturbative dynamics. As an example, consider the instantons in the Hamiltonian formulation of the theory. They correspond to a path in field space associated with minimal action. The stability of the instanton is guaranteed because the path interpolates between vacua related by a topologically non-trivial gauge transformation. This guarantees that the path has non-trivial homotopy. Given the non-trivial action, there exists a non-zero potential barrier of minimal height which is called a sphaleron and exists because the size of the instantons is restricted by the size of the volume. It is the energy of this sphaleron that sets the scale beyond which the wave functional is no longer exponentially suppressed below the barriers separating different vacua. If this is the case, it is no longer possible to take instantons into account semiclassically. In essence, instanton solutions are used to find the relevant degrees of freedom in the Yang-Mills configuration space in whose directions the wave functional will first and foremost spread out.

3. Boundary Conditions in Field Space

One way of formulating the gauge field configuration space is to use a simple gauge condition as a parametrisation. Locally it is easily shown that this provides a unique description, but since the work of Gribov [6] one knows that such gauge conditions do not uniquely fix the gauge when moving away from the origin in field space. Using a background field gauge fixing, one can in principle cover field space by local neighbourhoods, with transition functions relating the different neighbourhoods [7]. These transition functions are gauge transformations relating gauges of overlapping patches [8]. Because field space is infinite dimensional, except for low-dimensional models, no satisfactory theory has been developed along these lines. Instead we introduce complete gauge fixing using a variational formulation of the Coulomb gauge [9], as in this gauge the Yang-Mills Hamiltonian has been studied extensively [10]. Minimising the L^2 norm of the vector potential, $A_i(x) = iA_i^a(x)\tau_a/2$, along the gauge orbit

$$\left\|{}^hA\right\|^2 = -\int_M d^3x \ \text{tr}\left(\left(h^{-1}A_ih + h^{-1}\partial_ih\right)^2\right),\qquad(1)$$

one *almost* uniquely fixes the gauge. Expanding around the minimum A using $h(x) = \exp(X(x))$, one finds:

$$\left\|{}^hA\right\|^2 = \|A\|^2 + 2\int_M \text{tr}\left(X\partial_iA_i\right) + \int_M \text{tr}\left(X^\dagger FP(A)X\right) + \mathcal{O}\left(X^3\right),\qquad(2)$$

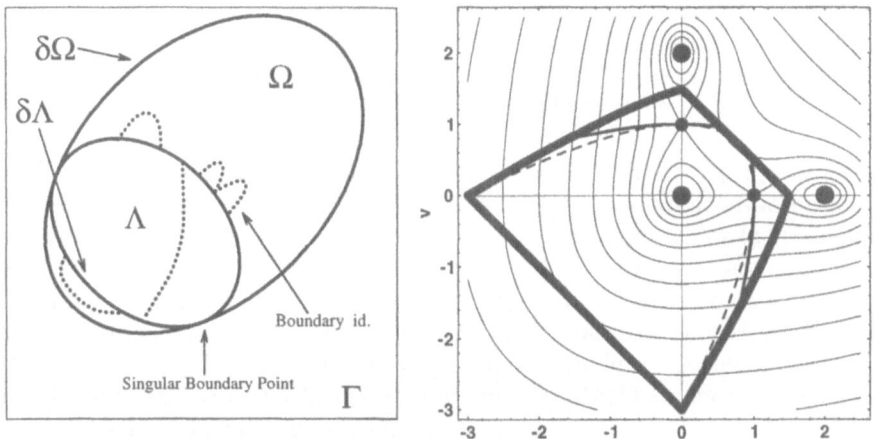

Fig. 1. — On the left a sketch of the fundamental and Gribov regions. The dotted lines indicate the boundary identifications. On the right a two dimensional cross section through the configuration space for S^3. Location of the classical vacua (large dots), sphalerons (smaller dots), the Gribov horizon (fat sections), the horizon for FP_f (dashed curves) which is contained in the boundary of the fundamental domain (full curves). Also indicated are the lines of equal potential in units of 2^n times the sphaleron energy.

where $FP(A) = -\partial_i D_i(A) = -\partial_i^2 - \partial_i \mathrm{ad}(A_i)$ is the Faddeev-Popov operator $(\mathrm{ad}(A)X \equiv [A, X])$. At any local minimum the vector potential is therefore transverse, $\partial_i A_i = 0$, and $FP(A)$ is a positive operator. The set of all these vector potentials is by definition the Gribov region Ω. Using the fact that $FP(A)$ is linear in A, Ω is seen to be a convex subspace of the set of transverse connections Γ. Its boundary $\partial\Omega$ is called the Gribov horizon. At the Gribov horizon, the *lowest* non-trivial eigenvalue of the Faddeev-Popov operator vanishes, and points on $\partial\Omega$ are associated with coordinate singularities, which can be shown to have a finite distance to the origin of field space [11].

The Gribov region, formed by the *local* minima, needs to be further restricted to the *absolute* minima to form a fundamental domain, denoted by Λ. One expects many relative minima, as local gauge functions $h(x) \in \mathcal{G}$ are like spin variables, noting similarity to the spin glass problem. We can write

$$\Lambda = \left\{ A \in \Gamma \Big| \min_{h \in \mathcal{G}} \left[\left\| {}^h A \right\|^2 - \|A\|^2 = \int \mathrm{tr}\left(h^\dagger FP_f(A)\, h \right) \right] = 0 \right\}, \quad (3)$$

where $FP_f(A) = -\partial_i^2 - iA_i^a \tau^a \partial_i$, is the SU(2) Faddeev-Popov operator, generalised to the fundamental representation. Since $FP_f(A)$ is linear in A, Λ is easily seen to be convex. Its interior is devoid of gauge copies, whereas its boundary $\partial\Lambda$ will in general contain gauge copies, associated to vector potentials where the absolute minimum of the norm functional are degenerate [12]. It can happen that for some points on the boundary the minimum is not quadratic, but of quartic (or higher) order. The Gribov horizon will touch the boundary of the fundamental domain at these so-called singular boundary points, see fig. 1.

It should be noted that the constant gauge degree of freedom is *not* fixed by the Coulomb gauge condition and therefore one still needs to divide by G to get the proper identification, $\Lambda/G = \mathcal{A}/\mathcal{G}$. Here Λ is considered to be the set of absolute minima modulo the boundary identifications, that remove the degenerate absolute minimum. It is these boundary identifications that restore the non-trivial topology of $\mathcal{A}/\mathcal{G}$. There is no problem in dividing out G by demanding wave functionals to be gauge singlets (colourless states) with respect to G. Because the boundary identifications are by gauge transformations, the wave functional will be identified up to a phase factor, possibly non-trivial when the associated gauge transformation is topologically non-trivial. The classical scale invariance of the theory guarantees that the fundamental domain and the Hamiltonian, when expressed in dimensionless fields LA, only depend on the shape but not on the size of the volume. The size dependence will appear solely due to the need of a short distance cut-off, giving rise to a scale dependent coupling constant. It is due to the increase of the effective coupling constant that wave functionals start to spread out over field space. The modes in which this spreading is largest are those associated to transitions over the sphaleron. Nonperturbative features become large when the wave functional bites its own tail through the boundary identifications. In the available examples this first happens at the sphalerons, which lie on the boundary of the fundamental domain and its boundary identifications are by gauge transformation with non-trivial topology, associated to instantons on whose tunnelling path the sphalerons lie.

4. Gauge Fields on the Three-Sphere

The conformal equivalence of $S^3 \times \mathbb{R}$ to $\mathbb{R}^4$ allows one to construct instantons explicitly [3]. This greatly simplifies the study of how to formulate θ dependence in terms of boundary conditions on the fundamental domain [13]. We embed S^3 in $\mathbb{R}^4$ by considering the unit sphere parametrised by a unit vector n_μ. Dependence on the radius R can be retrieved by rescaling the fields. We introduce $\sigma_\mu = (id, i\vec{\tau})$, which satisfy $\sigma_\mu \sigma_\nu^\dagger = \eta_{\mu\nu}^\alpha \sigma_\alpha$ and $\sigma_\mu^\dagger \sigma_\nu = \bar{\eta}_{\mu\nu}^\alpha \sigma_\alpha$, with η the 't Hooft symbols [14]. These can be used to define orthonormal framings on S^3, $e_\mu^a = \eta_{\mu\nu}^a n_\nu$ and $\bar{e}_\mu^a = \bar{\eta}_{\mu\nu}^a n_\nu$. Note that e and $\bar{e}$ have opposite orientations.

The (anti-)instantons in these framings are obtained from those for $\mathbb{R}^4$ by identifying the radius in $\mathbb{R}^4$ with the exponential of the time t in the space $S^3 \times \mathbb{R}$. The (anti-)instanton that tunnels through the (anti-)sphaleron, has for each time a constant energy density, and is particularly simple with respect to this framing. One finds $A_0 = 0$, $A_a = A^\mu e_\mu^a = -f(t)\sigma_a$ for the instanton (and $A_a = A^\mu \bar{e}_\mu^a = -f(t)\sigma_a$ for the anti-instanton) with $f(t) = 1/(1 + e^{-2t})$. The (anti-)sphaleron occurs in this parametrisation at $t = 0$. It is a saddle point of the energy functional with one unstable mode, corresponding to the direction of

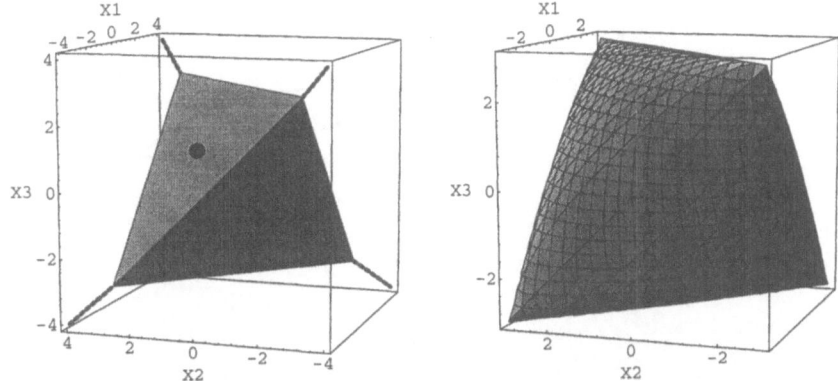

Fig. 2. — The fundamental domain (left) for constant gauge fields on S^3, with respect to the instanton framing e_μ^a, in the "diagonal" representation $A_a = x_a \sigma_a$ (no sum over a). By the dots on the faces we indicate the sphalerons, whereas the dashed lines represent the symmetry axes of the tetrahedron. To the right we display the Gribov horizon, which encloses the fundamental domain, coinciding with it at the singular boundary points along the edges of the tetrahedron.

tunnelling. At $t = \infty$, $A_a = -\sigma_a$ has zero energy and is a gauge copy of $A_a = 0$, by a gauge transformation $h = n \cdot \sigma^\dagger$ with winding number one. This gauge transformation also maps the anti-sphaleron to the sphaleron. The two dimensional space containing the tunnelling paths through the (anti-)sphalerons is parametrised by $A_\mu(u,v) = \frac{1}{2}(-ue_\mu^a - v\bar{e}_\mu^a)\sigma_a$. The gauge transformation $h = n \cdot \sigma$ with winding number -1 is easily seen to map $(u,v) = (w,0)$ into $(u,v) = (0, 2-w)$. The space of modes degenerate with these and of lowest energy is described by $A_\mu(c,d) = A_i(c,d)e_\mu^i = \frac{1}{2}(c_i^a e_\mu^i + d_j^a \bar{e}_\mu^j)\sigma_a$. The c and d modes are mutually orthogonal and satisfy the Coulomb gauge condition $\partial_i A_i(c,d) = 0$. The energy functional is given by [3]

$$
\begin{aligned}
\mathcal{V}(c,d) &\equiv -\int_{S^3} \frac{1}{2} \operatorname{tr}(F_{ij}^2) = \mathcal{V}(c) + \mathcal{V}(d) + \frac{2\pi^2}{3}\left\{(c_i^a)^2(d_j^b)^2 - (c_i^a d_j^a)^2\right\}, \\
\mathcal{V}(c) &= 2\pi^2 \left\{2(c_i^a)^2 + 6\det c + \frac{1}{4}[(c_i^a c_i^a)^2 - (c_i^a c_j^a)^2]\right\},
\end{aligned}
\tag{4}
$$

from which the degeneracy to second order in c and d can be verified. There are no modes with a lower zero-point frequency than these [13].

An effective Hamiltonian for the c and d modes is derived from the one-loop effective action [15]. To lowest order it is given by

$$
H = -\frac{g^2(R)}{2R}\left(\left(\frac{\partial}{\partial c_i^a}\right)^2 + \left(\frac{\partial}{\partial d_i^a}\right)^2\right) + \frac{1}{g^2(R)R}\mathcal{V}(c,d),
\tag{5}
$$

where $g(R)$ is the running coupling constant. It can be shown [13] that the boundary of the fundamental domain will touch the Gribov horizon $\partial\Omega$, such

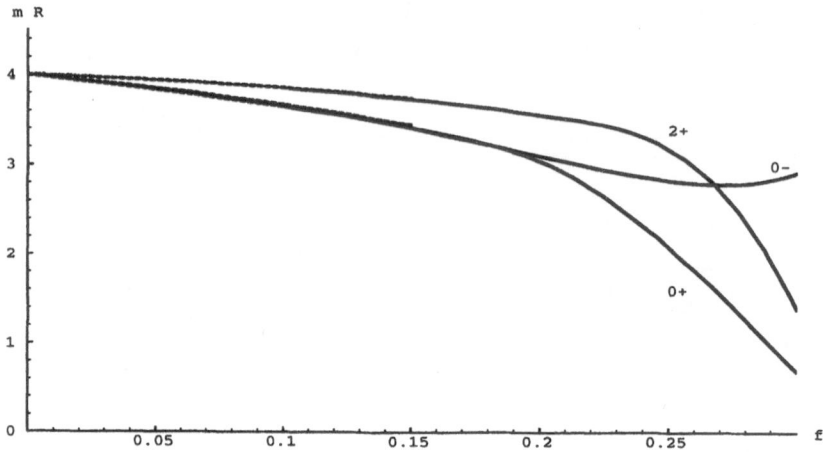

Fig. 3. — The full one-loop results for the masses of scalar, tensor and odd glueballs on S^3 as a function of $f = g^2(R)/2\pi^2$ for $\theta = 0$. The dashed lines correspond to the perturbative result.

that it contains singular points. This is illustrated in figure 2, which shows the fundamental and Gribov regions for $d = 0$, using the rotational and gauge invariance to rotate c to a "diagonal" form.

It is essential that the sphalerons do *not* lie on the Gribov horizon and that the potential energy near $\partial\Omega$ is relatively high as can be seen from figure 1. This is why we can take the boundary identifications near the sphalerons into account without having to worry about singular boundary points, as long as the energies of the low-lying states will be not much higher than the energy of the sphaleron. It allows one to study the glueball spectrum as a function of the CP violating angle θ, but more importantly it incorporates for $\theta = 0$ the noticeable influence of the barrier crossings, i.e. of the instantons.

The boundary conditions are chosen so as to coincide with the appropriate boundary conditions near the sphalerons, but such that the gauge and (left and right) rotational invariances are not destroyed. Projections on the irreducible representations of these symmetries turned out to be essential to reduce the size of the matrices to be diagonalised in a Rayleigh-Ritz analysis. Remarkably all this could be implemented in a tractable way [15]. Results are summarised in figure 3. One of the most important features is that the 0^- glueball is (slightly) lighter than the 0^+ in perturbation theory, but when including the effects of the boundary of the fundamental domain, setting in at $f \sim 0.2$, the $0^-/0^+$ mass ratio rapidly increases. Beyond $f \sim 0.28$ it can be shown that the wave functionals start to feel parts of the boundary of the fundamental domain which the present calculation is not representing properly [15]. This value of f corresponds to a circumference of roughly 1.3 fm, when setting the scale as for the torus, assuming the scalar glueball mass in both geometries at this intermediate volume to coincide.

5. Conclusion

Boundary identifications become relevant at large volumes, whereas at very small volumes the wave functional is localised around $A = 0$ and one need not worry about these non-perturbative effects. That these effects can be dramatic, even at relatively small volumes (above a tenth of a fermi across), was demonstrated for the case of the torus [1]. Here we have discussed the situation for S^3. Results for the spectrum are compatible with those of a torus in volumes around one fermi across [16], with $m(2^+)/m(0^+) \sim 1.5$ and $m(0^-)/m(0^+) \sim 1.7$. For more details and discussions see refs. [15, 17].

Acknowledgments

The author thanks the (session) organisers Yitzhak Frishman, Robert Perry, Simon Dalley and Pierre Grangé for their invitation. He also thanks the participants for many fruitful discussions on and off the slopes.

References

[1] van Baal P., Phys.Lett. **224B** (1989) 397; Nucl. Phys. **B351** (1991) 183.

[2] Lüscher M., Nucl.Phys. **B219** (1983) 233

[3] van Baal P. and Hari Dass N.D., Nucl.Phys. **B385** (1992) 185.

[4] Babelon O. and Viallet C., Comm.Math.Phys. **81** (1981) 515.

[5] Singer I., Comm.Math.Phys. **60** (1978) 7.

[6] Gribov V., Nucl.Phys. **B139**(1978) 1.

[7] Nahm W., in:IV Warsaw Sym.Elem.Part.Phys, 1981, ed. Z.Ajduk, p.275.

[8] van Baal P., in: Probabilistic Methods in Quantum Field Theory and Quantum Gravity, Damgaard P.H. et al, ed. (Plenum Press, New York, 1990) p31; Nucl.Phys. B(Proc.Suppl.)**20** (1991) 3.

[9] Semenov-Tyan-Shanskii M.A. and Franke V.A., Zapiski Nauchnykh Seminarov Leningradskogo Otdeleniya Matematicheskogo Instituta im. V.A. Steklov AN SSSR, **120** (1982) 159. Translation: (Plenum Press, New York, 1986) p.999; Zwanziger D., Nucl. Phys. **B209** (1982) 336.

[10] Christ N.M. and Lee T.D., Phys.Rev. **D22** (1980) 939.

[11] Dell'Antonio G. and Zwanziger D., Nucl.Phys. **B326** (1989) 333.

[12] van Baal P., Nucl.Phys. **B369** (1992) 259.

[13] van Baal P. and van den Heuvel B., Nucl.Phys. **B417** (1994) 215.

[14] 't Hooft G., Phys.Rev. **D14** (1976) 3432.

[15] van den Heuvel B.M, Phys.Lett. **B368** (1996) 124; **B386** (1996) 233; Nucl.Phys. **B488** (1997) 282.

[16] Michael C. and Teper M., Phys.Lett. **B199** (1987) 95.

[17] van Baal P., Global issues in gauge fixing, in: "Non-perturbative approaches to Quantum Chromodynamics", ed. D. Diakonov, (Gatchina, 1995) p.4.

Renormalization in the Coulomb Gauge

D. Zwanziger

Physics Department, New York University
New York, NY 10003, USA
E-mail address: daniel.zwanziger@nyu.edu

1. INTRODUCTION

The Coulomb gauge offers a simple confinement scenario. As originally proposed by Gribov [1], and supported recently by detailed arguments [2], $D_{0,0}$, the time-time component of the gluon propagator, provides a long-range confining force, while $D_{i,j}$, the 3-dimensionally transverse propagator of would-be physical gluons is suppressed at low momentum, reflecting the disappearance of gluons from the physical spectrum. This different behavior of spatial and temporal components of the gluon propagator in the Coulomb gauge obviously cannot hold in a covariant gauge. These properties of the Coulomb-gauge propagator are a consequence of the particular shape of the physical configuration space. For in a gauge theory, the physical configuration space must be restricted to the fundamental modular region (FMR), a region that is free of Gribov copies.

This intuitive and non-perturbative confinement scenario must be substantiated by detailed calculations. An essential step in such a program is to establish that the Coulomb gauge is renormalizable. The renormalization properties that hold in the Coulomb gauge have been studied recently [3]. The results are consistent the above picture of gluon confinement.

In a perturbative expansion of gauge-invariant quantities, the Coulomb gauge must agree order by order with the familiar Lorentz-covariant gauges, so it might seem that nothing much new is to be learned from perturbatively renormalizing in the Coulomb gauge. On the contrary, it is found that in the

Coulomb gauge a new type of Ward identity for the quantum effective action Γ holds at a fixed time t. It is an analog of Gauss's law in the BRST formalism, and we call it the "Gauss-BRST" identity. The familiar Zinn-Justin equation results when this identity is integrated over all t. A consequence of the Gauss-BRST identity is that $g^2 D_{0,0}$ is a renormalization-group invariant,

$$g^2 D_{0,0} = g_r^2 (D_{0,0})_r, \tag{1}$$

where g is the coupling constant, and the subscript r refers to renormalized quantities. Thus, in the Coulomb gauge, but not in covariant gauges, $g^2 D_{0,0}$ is independent of the cut-off and of the renormalization mass, and depends only on physical masses. It is conjectured that $g^2 D_{0,0}$ rises linearly at large separation when color is confined, and provides an order-parameter for color confinement [3].

The hamiltonian for non-Abelian gauge theory in the Coulomb gauge has been known for some time in its continuum version [4], and the lattice version has been found recently [2].

In section 2 we briefly review the relation of the fundamental modular region and the above-mentioned properties of the gluon propagator, and in section 3 we review recent results on the renormalization properties in the Coulomb gauge.

2. CONFINEMENT AND THE GRIBOV PROBLEM

An intuitive idea of why the restriction to the FMR leads to a singularity of $D_{0,0}(\vec{k})$ at $\vec{k} = 0$ is as follows. The argument should properly be made in the context of the lattice regularization, as is done in [2, 5], but we give here a continuum version. Gribov showed that at the nonperturbative level, the transversality condition $\nabla_i A_i = 0$ does not uniquely fix the gauge. A unique way to fix the gauge [6, 7] is to choose as representative on each gauge orbit that configuration $\vec{A}(x)$ which makes the Hilbert norm the *absolute* minimum with respect to all local gauge transformations $g(x)$, so

$$\|\vec{A}\|^2 \equiv \int d^4x \ \vec{A}^2 \le \|\vec{A}^g\|^2 \text{ for all g,} \tag{2}$$

where $\vec{A}^g$ is the gauge transform of $\vec{A}(x)$ by $g(x)$. (All vectors here and below are 3-dimensional.) The fundamental modular region Λ is the set of these absolute minima,

$$\Lambda \equiv \{\vec{A} : \|\vec{A}\|^2 \le \|\vec{A}^g\|^2 \text{ for all g}\}. \tag{3}$$

Under fairly general conditions, the minimum (2) exists [8], and is unique, apart from points on the boundary of Λ which must be identified topologically [9].

An absolute minimum is also a local minimum. To see what this implies, write $g(x) = \exp[\omega(x)]$ and expand in powers of ω. For every $\vec{A}$ in the fundamental modular region Λ, the inequality

$$\|\vec{A}^g\|^2 - \|\vec{A}\|^2 = -2(\nabla_i A_i, \omega) + (\omega, -D_i(\vec{A})\nabla_i \omega) + \dots \ge 0 \tag{4}$$

holds for all ω. This implies that $\vec{A}$ is transverse, $\nabla_i A_i = 0$, so we get back the Coulomb gauge condition in the new, well-defined gauge which we call the "minimal Coulomb gauge". *In addition* we find that the Faddeev-Popov operator is positive, $M(\vec{A}) = -D_i(\vec{A})\nabla_i \geq 0$, for all $\vec{A}$ in Λ. Thus all (non-trivial) eigenvalues of $M(\vec{A})$ are positive in the interior of Λ, so $M^{-1}(\vec{A})$ is well-defined there.

The fundamental modular region Λ is bounded in all directions [6, 7], and its boundary contains points where $M^{-1}(\vec{A})$ is singular [9]. When the Euclidean functional integral is restricted to Λ, then, in the limit of large lattice volume, it is found [2, 5] that entropy favors these points sufficiently that $\vec{k}^2 G(\vec{k})$ diverges at $\vec{k} = 0$, where the ghost propagator $G(\vec{k})$ is the fourier transform of $\langle [M^{-1}(\vec{A})]_{\vec{x},0} \rangle$. Indeed, the condition that $\vec{k}^2 G(\vec{k})$ diverge at $\vec{k} = 0$ is mathematically equivalent to the so-called "horizon condition" [2]. This condition assures that the functional integral be cut-off at the boundary of the fundamental modular region Λ. That the probability gets concentrated where $M^{-1}(\vec{A})$ is singular is also supported by numerical studies [10]. With the conclusion that $\vec{k}^2 G(\vec{k})$ is singular at $\vec{k} = 0$, the elements of a confining theory are in place. For the Gauss-BRST identity that holds in the Coulomb gauge implies that $D_{0,0}(\vec{k})$ contains $[\vec{k}^2 G(\vec{k})]^2$ as a factor. Confinement arises because this factor diverges at low $\vec{k}$ as a result of the restriction of the functional integral to the fundamental modular region.

It has been argued [11] and confirmed in models [12], that Gribov copies come in pairs that give equal and opposite contributions for gauge-invariant quantities, so one need not in fact cut off the functional integral at the boundary of Λ at all (at the cost however of having a Euclidean weight that is not positive). This does not alter the conclusion that $\vec{k}^2 G(\vec{k})$ diverges at $\vec{k} = 0$ in the minimal Coulomb gauge. The results of [3] hold to all orders of perturbation theory, and it is expected that they hold also non-perturbatively in the minimal Coulomb gauge.

3. THE PROBLEM OF COULOMB-GAUGE RENORMALIZA-TION AND ITS SOLUTION

In this section we first give a heuristic discussion of the problem based on a non-local action. We then exhibit the corresponding local, BRST-invariant phase-space action S, whose renormalization properties have been established in [3].

When applied to the Coulomb gauge, the standard Faddeev-Popov procedure gives the functional integral in configuration space,

$$Z = \int d^4 A \delta(\nabla_i A_i) \det(-\nabla_i D_i) \exp(-S_{cl}), \qquad (5)$$

where sources are suppressed, $\vec{D} \equiv \vec{D}(\vec{A})$ is the spatial gauge-covariant derivative, $[\vec{D}C]^a \equiv \vec{\nabla}C^a + gf^{abd}\vec{A}^b C^d$, $S_{cl} \equiv 2^{-1}\int d^4x(\vec{E}^2 + \vec{B}^2)$, and

$$\vec{E}^a \equiv \partial_0\vec{A}^a - \vec{\nabla}A_0^a + gf^{abc}A_0^b\vec{A}^c \tag{6}$$

$$B_i^a \equiv \nabla_j A_k^a - \nabla_k A_j^a + gf^{abc}A_j^b A_k^c, \tag{7}$$

for i, j, and k cyclic. This expression may be written in local form

$$Z = \int dAdCd\bar{C}d\lambda \exp(-S_{FP}), \tag{8}$$

where

$$S_{FP} = \int d^4x\{2^{-1}(\vec{E}^2 + \vec{B}^2) + \nabla_i\bar{C}\cdot D_iC - \nabla_i\lambda A_i\}, \tag{9}$$

and $\lambda = ib$, where b is a real field. Here and below, the center dot means contraction on color indices.

The problem with the configuration-space functional integral is that the closed ghost loops are not of renormalizable type. In lowest order, the ghost loop is given by

$$(2\pi)^{-4}\int dk_0 \int d^3k[(\vec{k}-\vec{p})^2\vec{k}^2]^{-1}, \tag{10}$$

where $\vec{p}$ and $\vec{k}$ are spatial vectors. The integrand is independent of k_0, which is characteristic of an instantaneous closed ghost loop. The divergent integral $\int dk_0$ multiplies a non-polynomial function of $\vec{p}$. This divergence occurs in any number of dimensions and the integral is not regularized by dimensional regularization.

One suspects that this divergence cancels against some other term. To organize the cancellation systematically, we introduce the Gaussian identity

$$1 = N\int d^3P \exp[-(1/2)(\vec{P} - i\vec{E})^2], \tag{11}$$

into the configuration-space functional integral, which converts it to the phase-space functional integral. To avoid a profusion of i's, we write $\vec{\Pi} \equiv i\vec{P}$, and obtain

$$Z = \int d^3\Pi dA_0 d^2 A^{tr}\ \det(-\nabla_i D_i)\ \exp(-S_1), \tag{12}$$

where

$$S_1 \equiv \int d^4x(\Pi_i\cdot E_i - 2^{-1}\vec{\Pi}^2 + 2^{-1}\vec{B}^2), \tag{13}$$

$\vec{B} = \vec{B}(\vec{A}^{tr})$, and $\vec{E} = \partial_0\vec{A}^{tr} - \vec{D}A_0$, and $\vec{D} = \vec{D}(\vec{A}^{tr})$. The superscript tr refers to 3-dimensionally transverse vectors $\nabla_i A_i^{tr} = 0$. The A_0 field appears linearly in the action, and not at all in the Faddeev-Popov determinant, and the dA_0 integration gives simply

$$Z = \int d^3\Pi d^2 A^{tr}\ \det(-\nabla_i D_i)\delta(D_i\Pi_i)\ \exp(-S_2), \tag{14}$$

where

$$S_2 \equiv \int d^4x(\Pi_i \cdot \partial A_i^{tr}/\partial t - 2^{-1}\vec{\Pi}^2 + 2^{-1}\vec{B}^2). \qquad (15)$$

The appearance of $\delta(D_i\Pi_i)$ in the functional integral means that Gauss's law, $D_i\Pi_i = 0$, is satisfied at every time. We decompose Π into its transverse and longitudinal parts

$$\vec{\Pi} = \vec{\Pi}^{tr} - \vec{\nabla}\Omega, \qquad (16)$$

so $d^3\Pi = d^2\Pi^{tr}d\Omega$. Gauss's law reads

$$-D_i\nabla_i\Omega = -gA_i^{tr} \times \Pi_i^{tr} \equiv \rho, \qquad (17)$$

where we have introduced the notation $(A \times B)^a \equiv f^{abc}A^bB^c$. Here ρ^a is the part of the color-charge density that comes from the dynamical fields $\vec{A}^{tr}$ and $\vec{\Pi}^{tr}$. (If quarks were present then there would be a quark contribution to ρ.) This fixes Ω,

$$\Omega = \Omega_{ph} \equiv [M(\vec{A}^{tr})]^{-1}\rho, \qquad (18)$$

and we call Ω_{ph} the color-Coulomb field. The Green's function M^{-1} is the inverse of the Faddeev-Popov operator $M(\vec{A}) \equiv -D_i(\vec{A})\nabla_i$, which is symmetric when $\vec{A}$ is transverse, $M(\vec{A}^{tr}) = -D_i(\vec{A}^{tr})\nabla_i = -\nabla_i D_i(\vec{A}^{tr}) = M^\dagger(\vec{A}^{tr})$.

The integral over $d\Omega$ eliminates the Faddeev-Popov determinant,

$$\int d\Omega \det(-\nabla_i D_i)\, \delta(-D_i\nabla_i\Omega + \rho)\, f(\Omega) = f(\Omega_{ph}). \qquad (19)$$

We obtain

$$Z = \int d^2\Pi^{tr}d^2 A^{tr} \exp(-S_{ph}), \qquad (20)$$

where

$$S_{ph} \equiv \int d^4x\{\Pi_i^{tr} \cdot \partial A_i^{tr}/\partial t - 2^{-1}[(\vec{\Pi}^{tr})^2 + (\vec{\nabla}\Omega_{ph})^2] + 2^{-1}\vec{B}^2\}. \qquad (21)$$

The functional integral is now expressed in terms of the dynamical fields $\vec{A}^{tr}$ and $\vec{\Pi}^{tr}$ for which the measure is Cartesian, as would be obtained from canonical quantization.

The last expression for Z makes manifest the physical content of the functional integral. The dynamical fields $\vec{A}^{tr}$ and $\vec{\Pi}^{tr}$ propagate dynamically in time, and interact instantaneously at a distance by M^{-1} which occurs in $\Omega_{ph} = M^{-1}\rho$. However, because $\det M$ does not appear in the partition function Z, *there are no closed instantaneous loops.*

Unfortunately, this expression is only heuristic because S_{ph} is a non-local action (because of the appearance of $\Omega_{ph} = M^{-1}\rho$), which we don't know how to renormalize. To obtain a renormalizable theory, we introduce the Gaussian identity (11) into the functional integral (8) with the local Faddeev-Popov action (9). This gives

$$Z = \int d^3 \Pi d^4 A dC d\bar{C} d\lambda \exp(-S), \tag{22}$$

where S is the local phase-space action

$$S \equiv d^4 x [\Pi_i \cdot E_i - 2^{-1} \vec{\Pi}^2 + 2^{-1} \vec{B}^2 + \nabla_i \bar{C} \cdot D_i(\vec{A}) C - \nabla_i \lambda \cdot A_i]. \tag{23}$$

The non-renormalizable instantaneous closed fermi-ghost loops now recur in the perturbative expansion but, according to the preceding discussion, they are canceled by corresponding instantaneous closed bose loops. This cancellation is verified explicitly in [3].

The equations of motion of A_0 and Π_i that one derives from S read

$$D_i \Pi_i = 0 \tag{24}$$

$$\Pi_i = E_i. \tag{25}$$

Thus despite the presence of ghost fields in the local Coulomb-gauge action (23), Gauss's law is satisfied at the level of equations of motion. This suggests that the Coulomb gauge is well-adapted for addressing the confinement problem, because one expects that Gauss's law is crucial to the long-range color-force.

The Faddeev-Popov action is invariant, $sS = 0$, under the BRST transformation

$$
\begin{aligned}
sA_\mu = D_\mu C \quad sC = -2^{-1} gC \times C \\
s\bar{C} = \lambda \quad s\lambda = 0 \quad s\Pi_i = g\Pi_i \times C
\end{aligned} \tag{26}
$$

which is nilpotent $s^2 = 0$. Because A_0 does not couple to ghosts, the BRST charge has the particularly simple form

$$Q = \int d^3 x (-C \cdot D_i \Pi_i). \tag{27}$$

This leads to the Gauss-BRST identity mentioned above [3].

Acknowledgments

This work was supported in part by the National Science Foundation under grant no. PHY93-18781. It is a pleasure to acknowledge many stimulating discussions with Martin Schaden on all aspects of this work. I recall with pleasure discussions with Richard Brandt, H. C. Ren, and Alberto Sirlin. The author wishes to thank the organizers Yitzhak Frishman and Pierre Grangé for their invitation.

References

[1] Gribov V. N., Nucl. Phys. B139 (1978) 1
[2] Zwanziger D., Nucl. Phys, B485 (1997) 185
[3] Zwanziger D. *Renormalization in the Coulomb Gauge and Order Parameter for Confinement*, Preprint no. NI97032-NQF, Newton Institute, Cambridge, UK
[4] Christ N. and Lee T.D., Phys. Rev. D, 22 (1980) 939
[5] Zwanziger D., Nucl. Phys. B 412 (1994) 657
[6] Zwanziger D., Nucl. Phys. B209 (1982) 236
[7] Semionov-Tian-Shanskii M.A. and Franke V.A., Zapiski Nauchnykh Seminarov Leningradskogo Otdeleniya Matematicheskogo Instituta im. V. A. Steklova AN SSSR 120 (1982) 159 [English translation: Journ. Sov. Math., 34 (1986) 1999]
[8] Dell'Antonio G. and Zwanziger D., Commun. Math. Phys. 138 (1991) 291
[9] van Baal P., Nucl. Phys. B369 (1992) 259
[10] Cucchieri A., Numerical Results in minimal Coulomb and Landau gauges: color-Coulomb potential and gluon and ghost propagators, Ph. D thesis, New York University (May, 1996); Infra-Red Behavior of the Gluon Propagator in Lattice Landau Gauge, Bielefeld Preprint, (to be published).
[11] Hirschfeld P., Nucl. Phys. B 157 (1979) 37
[12] Friedberg R., Lee T.D., Pang Y. and Ren H.C., Ann. Phys. (New York) 246 (1996) 381

LECTURE 21

Stable Knotlike Solitons

Antti J. Niemi

Department of Theoretical Physics, Uppsala University
P.O. Box 803, S-75108, Uppsala, Sweden

1. Introduction

The first to suggest that knotlike configurations might be of fundamental importance was Lord Kelvin. In 1867 [1] he proposed that atoms, at the time considered as elementary particles, could be understood as knotted vortex tubes in ether. Kelvin's theory led *e.g.* to an extensive classification of knots by Tait [2], which remains a classic contribution to the mathematical theory of knot.

Kelvin also conjectured [3], that (thin) vortex filaments that form torus knots [4] should be stable. This conjecture has attracted much interest but until recently [5] it has not even been properly addressed. This is due to the complex nature of dynamical models that could describe stable knots, we need a Lagrangian field theory where knots can appear as solitons.

There are numerous physical scenarios, where dynamical knots can be important. Indeed, in the spirit of Kelvin's original conjecture, it is now widely accepted that fundamental interactions are described by string theories [6], but with different elementary particles now corresponding to the vibrational excitations of a string. Even though some deep connections between modern string theory and knot theory [7] have been found, the possibility of a more intimate relationship *e.g.* at some nonperturbative level remains to be investigated. In a number of other physical scenarios the potential relevance of knotted structures has already been well established [4]: In early universe cosmology cosmic strings might form stable knotted configurations that could have important consequences. In QCD one could similarly expect that gluonic flux tubes that confine quarks might become tangled, and in pure Yang-Mills theory closed knotted flux tubes might appear as physical states such as a glueball. Knotted vortices could also be present in a variety of condensed matter physics scenarios, for example in type-II superconductors magnetic fields are confined within the cores of vortex-like structures and recent experiments with ^{3}He-A superfluids have also revealed interesting vortex structures that can be described by theoretical methods which are adopted from cosmic string models. The study of knotlike configurations is highly important in the context of chemical compounds such as polymenrs, and knot theory is rapidly becoming an important part of molecular biology, where entanglement of a DNA chain interferes with vital life processes of replication, transcription and recombination.

Until now, solitons [8] have been mainly investigated in 1+1 dimensions with the notable exceptions of the 2+1 dimensional vortex and nonlinear σ-model solitons, and skyrmeons and magnetic monopoles in 3+1 dimensions. These are all pointlike configurations, that can not be directly associated with knotted structures. But if a pointlike two dimensional soliton is embedded in three dimensions, it becomes a line vortex. Since the energy of a three dimensional vortex is proportional to its length, for a finite energy the length must be finite. This is possible if the core forms a closed, knotted structure. Indeed, in 1975 Faddeev [9] proposed that a torus-shaped closed vortex could be constructed in a definite dynamical model. The configuration suggested in [9] is constructed from a finite length line vortex, twisted once around its core before joining the ends. The twist ensures that the configuration is stable against shrinkage, and as a knot it corresponds to the *unknot* which is the topologically simplest knotlike configuration [4]. The existence of such a torus-shaped soliton in the model proposed in [9] has been recently verified [5]. Furthermore, there is also definite evidence for the existence of a soliton in the shape of a trefoil, which is the simplest possible torus knot. This is a strong indication, that the model proposed in [9] indeed realizes Kelvin's conjecture on the existence of stable torus knots.

2. A Hamiltonian for knots

In [5] we argue that knotted solitons can be described in terms of a three component vector field $\mathbf{n}(\mathbf{x})$ with unit length $\mathbf{n} \cdot \mathbf{n} = 1$ and static Hamiltonian [9]

$$H = E_2 + E_4 = \int d^3x \left(\frac{1}{2e^2}(\partial_i \mathbf{n})^2 + \frac{1}{4g^2}(\mathbf{n} \cdot \partial_i \mathbf{n} \times \partial_j \mathbf{n})^2 \right) \quad (1)$$

Here $\mathbf{n}$ is dimensionless so that e determines a length scale and g is a dimensionless coupling constant. This is the most general Hamiltonian for $\mathbf{n}$ that involves second and fourth order derivative terms and admits a relativistic interpretation. The first term E_2 coincides with the nonlinear O(3) σ-model action, that admits stable solitons in two dimensions. But with the fourth order derivative term E_4, stable finite energy solitons are also possible in three dimensions. This is suggested by the Derrick scaling argument: If we set $\mathbf{x} \to \rho \mathbf{x}$ we find

$$H = E_2 + E_4 \overset{\rho}{\to} \rho E_2 + \frac{1}{\rho} E_4 \qquad (2)$$

hence stable finite energy solitons may exist when E_4 is present. In particular, such solutions obey the virial theorem

$$E_2 = E_4 \qquad (3)$$

A knotlike soliton described by $\mathbf{n}(\mathbf{x})$ is a localized configuration in R^3, which at spatial infinity $|\mathbf{x}| \to \infty$ approaches a constant vector $\mathbf{n}_0$. Hence we compactify R^3 into a three dimensional sphere S^3, and view $\mathbf{n}(\mathbf{x})$ as a mapping from the compactified $R^3 \sim S^3 \to S^2$. Such mappings fall into nontrivial homotopy classes $\pi_3(S^2) \simeq Z$ that can be characterized by the Hopf invariant [4]. For this we introduce the closed two-form

$$F = (d\mathbf{n} \wedge d\mathbf{n}, \mathbf{n}) \qquad (4)$$

on the target S^2. Since $H_2(S^3) = 0$ its preimage $F_\star$ on the base S^3 is exact, $F_\star = dA_\star$ and the Hopf invariant Q_H coincides with the three dimensional Chern-Simons term,

$$Q_H = \frac{1}{4\pi^2} \int_{R^3} F \wedge A \qquad (5)$$

The existence of stable solitons in (1) with a nontrivial Hopf invariant is then strongly suggested by the lower bound estimate

$$H \geq c \cdot |Q_H|^{\frac{3}{4}} \qquad (6)$$

where c is a nonvanishing numerical constant [10], [11].

In soving the equations of motion, we find it convenient to introduce new variables, obtained by parametrizing our unit vector

$$\vec{n} = (\cos \varphi \sin \theta, \sin \varphi \sin \theta, \cos \theta) \qquad (7)$$

through stereographic coordinates

$$\varphi = -\arctan(\frac{V}{U}) \qquad (8)$$

$$\theta = 2 \arctan \sqrt{U^2 + V^2} \qquad (9)$$

After we properly scale our Hamiltonian we then find

$$H = G^2 \cdot E_2 + E_4$$

$$= \int d^3x \{ \frac{G^2}{(1 + U^2 + V^2)^2} (\partial_\mu U^2 + \partial_\mu V^2)$$

$$+ \frac{4}{(1 + U^2 + V^2)^4} (\partial_\mu U \partial_\nu V - \partial_\nu U \partial_\mu V)^2 \} \qquad (10)$$

3. Numerical simulations

The Euler-Lagrange equations for (10) are highly nonlinear and an analytic solution appears impossible. Indeed, it seems that the only tools available are numerical. However, even a numerical integration is quite nontrivial, there are several complications that need to be addressed.

For this we first investigate the problem at an abstract level, by considering a generic static energy functional $E(q)$ with some variables q_a. We introduce an auxiliary variable τ, and extend the (stationary) Euler-Lagrange equations of $E(q)$ to the following parabolic gradient flow equation

$$\frac{dq_a}{d\tau} = -\frac{\delta E}{\delta q_a} \tag{11}$$

Since

$$\frac{\partial E}{\partial \tau} = -\left(\frac{\delta E}{\delta q_a}\right)^2 \tag{12}$$

the energy decreases along the trajectories of (11). Furthermore, by squaring (11) and integrating from some initial value $\tau = T$ to $\tau \to \infty$ we get

$$\int_T^\infty d\tau \left(\frac{dq_a}{d\tau}\right)^2 = \int_T^\infty d\tau \left(\frac{\delta E}{\delta q_a}\right)^2 \tag{13}$$

Hence τ-bounded trajectories of (11) flow towards a stable critical point of $E(q)$. In particular, if we start at initial time $\tau = T$ from an initial configuration $q = q_0$ and follow a bounded trajectory of (11), in the $\tau \gg T$ limit we eventually flow to a stable critical point of $E(q)$

In (10) the fields $U(\mathbf{x})$, $V(\mathbf{x})$ correspond to the variables q_a, and by denoting $W_1 = U$, $W_2 = V$ the pertinent flow equation is

$$\frac{\partial W_a(\mathbf{x})}{\partial \tau} = -G^2 \cdot \frac{\delta E_2}{\delta W_a(\mathbf{x})} - \frac{\delta E_4}{\delta W_a(\mathbf{x})} \tag{14}$$

In a numerical simulation [5] we then see how the total energy indeed decreases towards a stable value when τ increases.

Notice that if we scale $\mathbf{x} \to \rho\mathbf{x}$ we find for the scaled Hamiltonian the flow equation

$$\frac{\partial W_a(\mathbf{x})}{\partial(\frac{1}{\rho}\tau)} = -(\rho^2 G) \cdot \frac{\delta E_2}{\delta W_a(\mathbf{x})} - \frac{\delta E_4}{\delta W_a(\mathbf{x})} \tag{15}$$

This implies that a flow towards a soliton with coupling constant G coincides with the flow towards a soliton with coupling constant $\rho^2 G$, provided we rescale the flow variable τ into $\frac{1}{\rho}\tau$. This means that the soliton is essentially unique; It is sufficient to consider the flow towards a soliton with a definite value for G, since solitons with other values of G are obtained from this configuration by a simple scale transformation.

Since the coupling constant G is the only dimensionfull quantity that appears in our equations, it determines the size of the soliton. In a numerical simulation we do not know *a priori* how the actual size depends on G, which poses a problem when we select the size of the lattice that we use in our numerical simulation. In order to resolve this, we introduce a simple renormalization procedure that allows us to select the initial configuration so that its overall size approximates properly an actual soliton. For this we first observe that the soliton obeys the virial theorem (3), in our present variables

$$G^2 E_2 = E_4$$

By demanding that this virial theorem is also obeyed during the flow (14), we promote the coupling constant G into a τ-dependent variable $G \to G(\tau)$,

$$G(\tau) = \sqrt{\frac{E_4(\tau)}{E_2(\tau)}}$$

When we approach a soliton as $\tau \to \infty$, the variable $G(\tau)$ must then flow towards an asymptotic value G^* which is the value of the coupling for the soliton,

$$G(\tau) \stackrel{\tau \to \infty}{\longrightarrow} G^* \qquad (16)$$

This renormalization procedure then allows us to choose the size of our lattice appropriately.

We have made extensive numerical simulations both in the case of an unknot and a trefoil configuration, and our simulations reveal very strong convergence both for the coupling constant $G(\tau)$ and for the total energy $E(\tau)$ towards fixed point values. Furthermore we find very strong pointwise convergence for the angular functions $\theta(\mathbf{x})$, $\varphi(\mathbf{x})$ the density of the Hopf invariant $Q_H(x, y, z)$ and for the energy density $\mathcal{E}(\mathbf{x})$ in (10), suggesting that the pertinent equations of motion indeed admit knotlike solitons.

Acknowledgments

We thank L. Faddeev for collaboration, and the Center for Scientific Computing in Espoo, Finland for providing us with an access to their high performance computers. Our work has been financed both by the NFR Grant F-AA/FU 06821-308 and by the Göran Gustafsson Foundation for Science and Medicine.

References

[1] Thomson W.H., *Trans. Royal Soc. Edinburgh* **25** (1869) 217
[2] Tait P.G. "On Knots I II, III" Scientific Papers, Cambridge University Press, (Cambridge) (1898)

[3] Thomson W.H., in "Mathematical and Physical Papers IV" (Cambridge) (1911)

[4] L.H., "Knots And Physics", World Scientific (Singapore) (1993)

[5] Faddeev L.D. and Niemi A.J., *Nature* **387** (1997) 58

[6] Green M.B., Schwarz J.H. and Witten E., "Superstring Theory I, II" Cambridge University Press (Cambridge) (1987)

[7] Witten E., *Commun. Math. Phys.* **121** (1989) 351

[8] Rebbi C. and Soliani G., "Solitons and Particles" World Scientific (Singapore) (1984)

[9] Faddeev L.D., *Quantisation of Solitons*, preprint IAS Print-75-QS70 (1975); and in *Einstein and Several Contemporary Tendencies in the Field Theory of Elementary Particles* in "Relativity, Quanta and Cosmology" vol. 1, Pantaleo M. and De Finis F. (eds.), Johnson Reprint (1979)

[10] Vakulenko A.F. and Kapitanski L.V., *Dokl. Akad. Nauk USSR* **248** (1979) 810

[11] Makhankov V.D., Rybakov Y.P. and Sanyuk V.I., "The Skyrme Model: Fundamentals, Methods, Applications" Springer-Verlag (Berlin) (1993)

CHAPTER VII

Structure Functions: Theory and Experiments

Theoretical Uncertainties in the Determination of α_s from Hadronic Event Observables: the Impact of Nonperturbative Effects

V.M. Braun

NORDITA, Blegdamsvej 17, DK–2100 Copenhagen Ø, Denmark

Precise determination of the strong coupling has become one of the most important tasks of the QCD theory and phenomenology [1]. From the theoretical point of view the cleanest measurement comes from the total e^+e^- annihilation cross section at high energies. The accuracy is dominated in this case by statistical errors; high statistics is needed because the effect proportional to α_s is a few percent fraction only of the total cross section. Thus, going over to various event shape observables which are proportional to α_s at the leading order presents a clear experimental advantage. The price to pay is that the QCD description becomes more complicated and making a defensible estimate of the theoretical accuracy presents a nontrivial task.

The uncertainty due to nonperturbative effects is a particularly delicate issue. From the experimentalist's point of view this is the uncertainty of 'hadronization corrections' which are applied to uncover the structure of the event at the *parton level* from the observed structure at the *hadron level*. The hadronization process is modelled in a certain way, and differences between models (say, Lund string fragmentation or parton showers) are taken to estimate the error. In statistically important regions the hadronization corrections are of order 10% (at the Z peak) while the claimed error is of order 2–3%. This is less than the uncertainties of existing perturbative calculations (estimated by the scale dependence) so that for an experimentalist the hadronization could be considered under control.

This procedure is very successful phenomenologically, but it is unsatisfactory from the theoretical point of view. In particular the separation of perturbative versus nonperturbative alias parton level versus hadronization effects is theoretically ill-defined, and the commonly accepted procedures might be plagued by double counting of infrared effects. The theoretical understanding is guided by the Wilson operator product expansion (OPE) which is applicable, most notably, to the deep inelastic scattering (DIS). As a representative example, consider the Gross–Llewellyn Smith sum rule (GLS):

$$\int_0^1 dx\, F_3^{\nu p + \bar{\nu} p}(x, Q^2) = 3\left[1 - \frac{\alpha_s}{\pi}(Q^2) + \cdots - \frac{8}{27}\frac{\langle\!\langle O \rangle\!\rangle}{Q^2}\right] + O(1/Q^4) \quad (1)$$

Here $\langle\!\langle O \rangle\!\rangle$ is the reduced matrix element of a certain quark-antiquark-gluon operator which quantifies the correlations between partons in the nucleon.

From the OPE one learns: (i) *Power counting* of nonperturbative effects; in this case that perturbation theory is valid to $O(1/Q^2)$ accuracy and that the description to order $O(1/Q^2)$ requires one dimensionful nonperturbative parameter; (ii) '*Universality*' of nonperturbative effects in the sense that the same quark-antiquark-gluon operator appears in different physical processes; the coefficient is calculable in perturbation theory, hence one can sacrifice one measurement to get the prediction for other ones.

Theoretical approaches to the nonperturbative corrections in hadron production would aim to get the similar structure. For a generic observable dominated by short distances one expects an expansion of the type

$$R = \sum_{n=0}^{\infty} r_n \alpha_s^n + \left(\frac{\Lambda_R}{Q}\right)^p (\ln Q/\Lambda)^\gamma \quad (2)$$

The goal is to understand the power counting, that is which powers p are present, calculate 'anomalous dimensions' γ and relate the nonperturbative parameters Λ_R in different processes (universality). Note that the question of actual magnitude of the nonperturbative parameters remains open.

For the precision α_s measurements the power counting is of primary importance. Indeed, assuming a $1/Q^2$ correction, for the effective hard scale $Q_{\text{eff}} \geq 10$ GeV and the intrinsic size of 'matrix elements' 1 GeV, one obtains a ball-park estimate for the nonperturbative effects $\sim 1\%$ which can be neglected. On the other hand, if the nonperturbative effects are $O(1/Q)$, they can be of order 10% and one has to estimate them quantitatively.

Nonperturbative power-suppressed corrections to physical observables are intimately connected with the same-sign factorial divergence $r_n \sim n!$ of the QCD perturbation theory in high orders. Divergence of perturbation theory implies that the sum of the series is only defined to power accuracy $\exp[-\text{const}/\alpha_s(Q)] \sim 1/Q^p$ and the ambiguity has to be remedied by adding nonperturbative corrections. In particular, the large-n behavior of the perturbative coefficients

$$r_n \sim \text{const} \cdot n^{\tilde{\gamma}} \cdot s^{-n} \cdot n!, \quad (3)$$

necessarily requires a nonperturbative correction in (2) with the powers p and γ related in a simple way to the coefficients s and $\tilde{\gamma}$, respectively.

This has two consequences. First, the structure of nonperturbative corrections to a large class of observables is related to the structure of divergences in the perturbative expansions, which attracted a lot of recent activity (see [2] for a recent review). Here is a short summary of results, related to α_s determinations:

- Most of the existing hadronic event observables are predicted to have nonperturbative corrections of order $1/Q$ [3, 4, 5].

- The thrust and heavy jet mass distributions have $1/Q$ corrections for the average values. However, outside the two-jet region nonperturbative effects have and extra suppression factor $\sim \alpha_s(Q)$ [6].

- The one-particle inclusive cross section in the e^+e^- annihilation (fragmentation function) has only $1/Q^2$ corrections for fixed energy fraction; however, the integrated longitudinal (and transverse) cross sections have $1/Q$ corrections [7, 8].

- Power counting of the nonperturbative effects in jet fractions depends on the jet finding algorithm; $1/Q$ corrections are intrinsic for JADE and are most likely absent for the Durham $k_\perp$-clustering method [9].

- There are no $1/Q$ corrections to the Drell-Yan (heavy quark production) cross section to leading order [10]; their existence to the $O(\alpha_s(Q)/Q)$ accuracy is still disputed [11].

The power counting of nonperturbative effects can be tested experimentally by the energy dependence of hadron event observables, subtracting the parton level prediction. This was done recently by DELPHI [12].

Perturbative calculations in certain regions of phase space may require resummations of large logarithms (threshold corrections). An important question is whether the power counting of nonperturbative effects is disturbed by the resummations. It was studied for the Drell-Yan production [13, 14, 10, 15]. Different resummation procedures which are equivalent in perturbation theory, can introduce different power-like corrections, and it was suggested that a criterium for a 'good' resummation technique is that it does not bring in nonperturbative effects which are absent in finite orders; this study emphasizes importance of large-angle soft gluon emission [10, 15]. It turns out that truncation of the resummations in moment and momentum spaces may introduce spurious $1/Q^p$ effects with a small power $p \sim \alpha_s$ [16].

The second consequence is that the separation of 'perturbative' and 'nonperturbative' effects is conceptually ambiguous. To make it meaningful one has to introduce an *IR matching scale* μ_{IR} [17], define 'nonperturbative' contributions as absorbing all effects from scales below μ_{IR} and subtract the small momentum contributions from perturbative series to avoid a double counting. Continuing the example with the GLS sum rule one obtains, schematically

$$\int_0^1 dx\, F_3^{\nu p + \bar{\nu}p}(x, Q^2) = 3\left[1 - \left(1 - \frac{\mu_{\text{IR}}^2}{Q^2}\right)\frac{\alpha_s}{\pi}(Q^2) + \ldots - \frac{8}{27}\frac{\langle\!\langle O^{(\mu_{\text{IR}})}\rangle\!\rangle}{Q^2}\right] + O(1/Q^4)$$

$$(4)$$

The premium for this refinement is that perturbation theory restricted to the contributions of scales above μ_{IR} is (almost) free from factorial divergences in high orders; the price to pay is that subtraction of small momenta is very awkward in practice and introduces an additional scale (and scheme) dependence.

The same applies to the 'hadronization corrections'. The lessons to be learnt from OPE are: (i) separation of the parton cascade and hadronization is ambiguous; (ii) high orders of the perturbative series can imitate a power correction; (iii) if extracted from comparison with the data, the power correction is expected to depend on the order of perturbation theory, factorization scheme and scale; (iv) numerical estimates suggest that 'true nonperturbative' contributions at small scales are of the same order as perturbative.

To see how it works, consider energy dependence of the mean value of thrust $\langle 1 - T \rangle$ in the e^+e^- annihilation. The experimental data are well described by the formula

$$\langle 1 - T \rangle(Q) = 0.335 \, \alpha_s(Q) + 1.02 \, \alpha_s^2(Q) + \frac{1\text{GeV}}{Q} \tag{5}$$

Here Q is the c.m. energy, the two first terms on the r.h.s. correspond to the perturbation theory and the last term is the nonperturbative correction. The common procedure is to fix the 1 GeV/Q correction (assume it comes from a certain model) and fit $\alpha_s(Q)$ to the data; this gives $\alpha_s(M_Z) = 0.120$. Then, *keeping the hadronization correction fixed*, one varies the scale $\alpha_s(Q) \to \alpha_s(\mu)$ to estimate the perturbative uncertainty coming from unknown higher orders.

The caveat is that if nonperturbative and higher-order perturbative effects are inseparable, the hadronization correction can be scale-dependent itself, an aspect which usually remains fogged. It should be fitted anew for each scale. Let us try $\mu = 0.13 \, Q$ which is of order of the gluon transverse momentum $k_\perp^2 \sim (1 - T)Q^2/4$. Keeping $\alpha_s(M_Z) = 0.120$ fixed, one obtains an equally good fit to the data by changing the $1/Q$ correction

$$\langle 1 - T \rangle(Q) = 0.335 \, \alpha_s(0.13Q) + 0.19 \, \alpha_s^2(0.13Q) + \frac{0.4\text{GeV}}{Q} \tag{6}$$

This illustrates that the reshuffling of higher-order perturbative corrections by changing the factorization scale can be compensated by the change of the hadronization correction. The only known way to separate the nonperturbative effects from the factorization scale dependence is to introduce an IR matching scale in the spirit of the OPE treatment as discussed above. This is attempted in the Dokshitser-Webber model [4], where contributions of small scales are subtracted from perturbation theory. First applications of this model to the extraction of α_s are encouraging [12]: The scale dependence is reduced by factor two compared to the traditional treatment, the IR matching scale dependence is small, and the value of the extracted nonperturbative parameter is stable.

Possible 'universality' of hadronization corrections needs further study and, ultimately, development of the OPE-like techniques. The present approaches to this problem rely on two assumptions: that nonperturbative effects are proportional to perturbative ambiguities and that present 'naive' estimates of these ambiguities are representative. These assumptions are not obvious and have to be tested. Of particular interest are predictions for the x-dependence of the $1/Q^2$ effects in DIS [9, 18, 19] and for the fragmentation functions [7, 8]. The Dokshitser-Webber model has a buit-in universality also for the event shapes.

In general, one should expect that the expansion is organized in powers of some physical scale (say the BLM scale) rather than the 'naive' c.m. energy, and differences in the effective scales for various processes (together with the power counting) can give a rough idea of the difference in nonperturbative corrections. For example, the twist-4 effects in DIS at $x \to 1$ are proportional to $\Lambda^2/[(1-x)Q^2]$ while they are of order $\Lambda^2/[(1-x)^2 Q^2]$ for the Drell-Yan cross section. The difference reflects different hard scales for the gluon emission.

Viewed this way, the problem of power corrections is inseparable from the familiar problem of the scale dependence and scale fixing. The specifics of hadron event observables is that the physical scale (estimated e.g. by any of existing scale fixing prescriptions) appears to be very low – of order $1/10$ of the c.m. energy. It was argued [20] that the data do not show any preference for a low scale since the spread of the results for α_s between different observables is not reduced. From my point of view the observed spread rather indicates an intrinsic accuracy of the present treatment, with the separation of perturbative and hadronization effects as independent entities. (^{1})

To summarize, what is done and what is necessary to do to improve the theoretical accuracy of the determinations of α_s from hadronic event shapes?

- Theoretical tools are developed to determine the power counting of nonperturbative effects in hadronic event observables. Some of the event shapes are predicted to have smaller corrections than the others, and it is advisable to concentrate on them rather than make 'global fits'.

- One should try to design new event shapes. An 'event shape of the year' should be measurable, calculable to $O(\alpha_s^3)$ accuracy and have small nonperturbative effects. Some work in this direction is reported in [8].

- In addition to the traditional procedures, one has to apply alternative methods for the data analysis with the hadronization corrections fitted to the data rather than taken from models. One should always bear in mind that hadronization corrections do not have objective meaning unless a clear scale separation is made.

(^{1}) Using low scales is more consistent with hadronization corrections taken from parton shower models which imply $\alpha_s(k_\perp)$ for each gluon emission. Note, however, that these corrections by construction parametrize contributions of small gluon virtualities, which is in spirit of the OPE. Combining them with the fixed-order perturbative calculations one has to subtract contributions of the low scales from the latter to avoid the double counting, at least in principle. In practice this is difficult to do because the parton showers do not include NLO radiative corrections consistently.

- The existing procedures should be checked for double counting of perturbative contributions at low scales. The complete next-to-leading order parton shower models would help a lot.

- A convention for the scheme and scale setting is badly needed.

My personal feeling is that present theoretical accuracy of $\alpha_s(M_Z)$ can be improved up to factor two if the above questions are clarified.

Acknowledgments

I am grateful to M. Beneke, K. Hamacher and L. Lönnblad for the discussions. Special thanks are due to P. Chiappetta for the invitation to this school.

References

[1] Bethke S., *Nucl. Phys. Proc. Suppl.* **54A** (1997) 314;
    Schmelling M., MPI-H-V39-1996 (hep-ex/9701002).
[2] Beneke M., SLAC-PUB-7277 (hep-ph/9609215).
[3] Webber B., *Phys. Lett.* **B339** (1994) 148.
[4] Dokshitser Yu. and Webber B., *Phys. Lett.* **B352** (1995) 451.
[5] Akhoury R. and Zakharov V.I., *Phys. Lett.* **B357** (1995) 646; *Nucl. Phys.* **B465** (1996) 295.
[6] Nason P.and Seimour M.H., *Nucl. Phys.* **B454** (1995) 291.
[7] Dasgupta M. and Webber B., *Nucl. Phys.* **B484** (1997) 247.
[8] Beneke M., Braun V.M. and Magnea L., NORDITA-96/79 P (hep-ph/9701309).
[9] Dokshitser Yu., Marchesini G. and Webber B., *Nucl. Phys.* **B469** (1996) 93.
[10] Beneke M. and Braun V.M., *Nucl. Phys.* **B454** (1995) 253.
[11] Korchemsky G.P., LPTHE-ORSAY-96-78 (hep-ph/9610207).
[12] DELPHI collab., DELPHI 96-107 CONF 34, paper pa02003 submitted to ICHEP'96, Warsaw.
[13] Contopanagos H. and Sterman G., *Nucl. Phys.* **B419** (1994) 77.
[14] Korchemsky G.P. and Sterman G., *Nucl. Phys.* **B437** (1995) 415.
[15] Beneke M. and Braun V.M., *Nucl. Phys. Proc. Suppl.* **51C** (1996) 217.
[16] Catani S., *et al.*, *Nucl. Phys.* **B478** (1996) 273.
[17] Novikov V., *et al.*, *Nucl. Phys.* **B249** (1985) 445.
[18] Stein E., *et al*, *Phys. Lett.* **B376** (1996) 177;
     Meyer-Hermann M., *et al.*, *Phys. Lett.* **B383** (1996) 463.
[19] Dasgupta M. and Webber B., *Phys. Lett.* **B382** (1996) 273.
[20] Burrows P.N., *et al.*, *Phys. Lett.* **B382** (1996) 157.

Phenomenology of Renormalons in Inclusive Processes

L. Mankiewicz

Institute for Theoretical Physics TU-München, D-85747 Garching, Germany

Measurements of QCD observables have reached such high precision that power corrections to the structure functions can often be extracted with a reasonable accuracy from the existing data. The situation on the theoretical side is much less clear. In the best understood case of deep inelastic scattering, the relevant contributions can be attributed in the framework of operator product expansion (OPE) to matrix elements of higher-twist operators [1], but their determination in QCD is ambiguous due to occurrence of power divergences [2]. From the phenomenological point of view, however, attempts to compute these matrix elements using e.g., QCD sum rules have provided results which seem to have at least the right order of magnitude [3] as compared with available experimental estimates.

Recently there has been some interest in another phenomenological approach to power-suppressed corrections in QCD [4], based on the fact that the only possibility to interpret the higher-order radiative corrections in a consistent manner (i.e. as asymptotic series) requires the existence of power-suppressed terms [5]. The IR-renormalon contributions occur because certain classes of higher-order radiative corrections to twist-2 are sensitive to large distances, contrary to the spirit of OPE. Although these divergences have to cancel with UV power divergences in matrix elements of twist-4 operators i.e., they are totally spurious, there are arguments which suggest that they reflect the correct shape, if not the magnitude, of the power-suppressed terms [6, 7]. In this sense we define the prediction of the renormalon model of higher-twist corrections to a DIS structure function as the power-suppressed uncertainty which occurs in the perturbative expansion of the Wilson coefficient to the twist-2 contribution.

Commonly the divergent series of radiative corrections is regarded as an asymptotic series and defined by its Borel integral. The actual calculations are done in a large-N_F limit which allows to resum the fermion bubble-chain to all orders yielding the coefficient of the $\alpha_S^n N_F^{n-1}$ - term exactly. Subsequently, it is converted into the exact coefficient of the $\alpha_S^n \beta_0^{n-1}$ - term by the substitution $N_F \rightarrow N_F - 33/2 = -6\pi\beta_0$, known as the 'Naive Non-Abelianization' (NNA) [5]. The asymptotic character of the resulting perturbative series leads to resummation ambiguities - a singularity in the Borel integral destroys the unambiguous reconstruction of the series and shows up as a factorial divergence of the coefficients of the perturbative expansion. The general uncertainty in the perturbative prediction can be estimated to be of the order of the minimal term in the expansion, or by taking the imaginary part (divided by π) of the Borel integral. Both procedures lead to resummation ambiguities of the form $C \times \left(\Lambda^2/Q^2\right)^r$, with $r = 1$ for the leading IR renormalon, and a numerical coefficient C which either can be taken as it comes out from the NNA calculation, see below, or can be fitted as a free parameter.

Structure functions $F_i, i = L, 2, 3$, which parametrize the unpolarized hadronic scattering tensor for leptons scattering off nucleons, can be generally decomposed in the following manner

$$F_i(x, Q^2) = F_i^{\text{t}-2}(x, Q^2) + \frac{1}{Q^2}h_i^{\text{TMC}}(x, Q^2) + \frac{1}{Q^2}h_i(x, Q^2) + \mathcal{O}(\frac{1}{Q^4}) \quad , \quad (1)$$

where $F_i^{\text{t}-2}$ describes the leading twist-2 contribution. h_i^{TMC} describes the target mass corrections which are directly related to twist-2 matrix elements. h_i contains the genuine twist-4 contribution which we want to estimate using the renormalon model. Note that h_i has dimension 2 (if radiative corrections are neglected) and hence it has to be proportional to a certain mass scale μ^2. In the analysis of the twist-4 contributions [8, 9, 10] it is common to neglect the Q^2 dependence of h_i, which is due to radiative corrections, and to sum target mass corrections and the twist-2 contributions into a leading-twist (LT) structure function thus arriving at the notation

$$F_i(x, Q^2) = F_i^{(LT)}(x, Q^2) + \frac{1}{Q^2}h_i(x) = F_i^{(LT)}(x, Q^2)\left(1 + \frac{C_i(x)}{Q^2}\right) \quad . \quad (2)$$

Now we shall shortly summarize the main features of the renormalon model for higher-twist contributions. In this approach the expression for the higher-twist correction $h_i(x)$ to a structure function F_i has the form of a Mellin convolution [11]

$$h_i = P_i \otimes F_i^{\text{t}-2} \quad (3)$$

of a coefficient $P_i(z)$, and the twist-2 part $F_i^{\text{t}-2}(x)$ of the structure function F_i. The mass scale μ^2 which enters $P_i(z)$ equals to $\mu_R^2 = C_F \Lambda_s^2 e^{-C_\bullet}/(2\pi\beta_0)$ where $C_F = 4/3$, Λ_s is the QCD scale parameter in a given renormalization scheme s, and C_s is determined by the finite part of the fermion loop, $C_{\overline{MS}} = -5/3$. This identification, which follows from the NNA prescription, is RG invariant only in the strict large-N_f limit. One should mention also that due to the two possible contour deformations above or below the pole in the Borel integral, as discussed above, in principle the overall sign of the coefficients $P_i(z)$ cannot be uniquely determined.

So far the renormalon model was used to calculate the power corrections for the non-singlet part of the structure functions F_L and g_1 [11]. The physically equivalent approach based on dispersion relations of [6, 13] was used to calculate the higher-twist contribution to the non-singlet part of the structure functions F_2 and F_3 [14].

In the case of the structure function $F_2(x)$ an experimental analysis of higher-twist corrections exists for proton and deuteron targets [9]. Assuming Q^2 to be low enough, i.e., below the charm threshold, one obtains

$$
\begin{aligned}
C_d &= \frac{1}{F_2^{d,t-2}} \left(\frac{2}{9} P_S \otimes F_2^{0,t-2} + \frac{1}{18} P_{NS} \otimes F_2^{8,t-2} + P_G \otimes G \right) \\
C_{(p,n)} &= \frac{1}{F_2^{(p,n),t-2}} \left[\frac{2}{9} P_S \otimes F_2^{0,t-2} + \frac{1}{6} P_{NS} \otimes (\pm F_2^{3,t-2} + \frac{1}{3} F_2^{8,t-2}) \right. \\
&\quad + \left. P_G \otimes G \right],
\end{aligned}
\tag{4}
$$

where $F_2^{i,t-2}$, $i = 0, 3, 8$ denote corresponding SU(3) combinations of parton densities, and G is the gluon twist-2 structure function of the nucleon. The corresponding expression for the $F_3(x)$ structure function reads

$$
C_3 = \frac{1}{F_3^{t-2}} (P_3 \otimes F_3^{t-2}).
\tag{5}
$$

Equations (4) and (5) define the renormalon model of higher-twist corrections. Comparison of the model predictions with the data shows surprisingly good agreement as far as the x-dependence of the higher-twist correction is concerned, see [15] for details. The mass scale μ^2 predicted by the model is typically by a factor 2 - 3 too low, depending on the observable, so it is necessary to fit a mass scale to each observable separately. Nevertheless, it is amazing that such relatively simple approach can parametrize so well the twist-4 corrections to DIS structure functions.

Acknowledgments

This work has been supported by BMBF, the KBN grant 2 P03B 065 10 and German-Polish exchange program X081.91.

References

[1] Shuryak E.V. and Vainshtein A.I., Nucl. Phys. **B199** (1982) 451.
    Jaffe R.L. and Soldate M., Phys. Rev. **D26** (1982) 49.
    Ellis R.K., Furmanski W. and Petronzio R., Nucl. Phys. **B207** (1982) 1.
[2] Martinelli G. and Sachrajda C.T., Nucl. Phys. **B478** (1996) 660.
[3] Braun, V.M., Kolesnichenko A.V., Nucl. Phys. **283B**, (1987) 723.
    Balitsky I.I., Braun, V.M. and Kolesnichenko A.V., Phys. Lett. **B242** (1990) 245. (E), Phys. Lett. **B318** (1993) 648.
    Stein E., Górnicki P., Mankiewicz L., Schäfer A. and Greiner W., Phys. Lett. **B343** (1995) 369.
    Stein E., Górnicki P., Mankiewicz L. and Schäfer A., Phys. Lett. **B353** (1995) 107.
[4] Neubert M., Phys.Rev. **D51** (1995) 5924.
    Beneke M. and Braun V.M., Nucl. Phys. **B454** (1995) 253.
    Akhoury R. and Zakharov V.I., Phys. Lett. **B357** (1995) 646.
    Lovett-Turner C.N. and Maxwell C.J., Nucl. Phys. **B452** (1995) 188.
    Krasnikov N. V. and Pivovarov A.A. , Mod. Phys. Lett. **A11** (1996) 835.
    Nason P. and Seymour M.H., Nucl. Phys. **B454** (1995) 291.
    Korchemsky P. and Sterman G., Nucl. Phys. **B437** (1995) 415.
[5] Zakharov V. I., Nucl. Phys. **B385** (1992) 452.
    Mueller A. H., Phys. Lett. **B308** (1993) 355.
    Beneke M., Nucl. Phys. **B405** (1993) 424.
    Broadhurst D. J., Z. Phys. **C58** (1993) 339.
[6] Dokshitser Yu. L., Marchesini G. and Webber B. R., Nucl. Phys. **B469** (1996) 93.
[7] Beneke M., Braun V.M. and Magnea L., hep-ph/9701309 (1997).
[8] The New Muon Collaboration, Nucl. Phys. **B371** (1992) 3.
[9] Virchaux M. and Milsztajn A., Phys. Lett. **B274** (1992) 221.
[10] Sidorov A.V., Phys. Lett. **B389** (1996) 379.
[11] Stein E., Meyer-Hermann M., Mankiewicz L. and Schäfer A., Phys. Lett. **B376** (1996) 177.
[12] Meyer-Hermann M., Maul M., Mankiewicz L., Stein E. and Schäfer A., Phys. Lett. **B383** (1996) 463.
[13] Beneke M. and Braun V. M., *Phys. Lett.* **B348** (1995) 513.
    Ball P., Beneke M. and Braun V.M., *Nucl. Phys.* **B452** (1995) 563.
[14] Dasgupta M. and Webber B. R., *Phys. Lett.* **B382** (1996) 273.
[15] Maul M., Stein E., Schäfer A. and Mankiewicz L., Phys. Lett. **B401** (1997) 100.

Power Terms in QCD Hard Processes and Running Coupling

Giuseppe Marchesini

Dipartimento di Fisica, Università di Milano
INFN, Sezione di Milano, Italy

1. INTRODUCTION

A QCD observable depending on a scale Q, such as the e^+e^- annihilation cross section into hadron or the deep inelastic scattering (DIS) scaling violations, can be studied both as expansion in the running coupling $\alpha_S(Q^2)$ and in powers of Λ^2/Q^2, with Λ the QCD physical scale. The first one, the perturbative expansion, is significative only in the short distance regime, i.e. for large Q^2. The second one, the operator product expansion (OPE) [1], depends on large distance non-perturbative effects and is significative for Q^2 not too large in such a way that power corrections are not fully suppressed by the perturbative contributions.

It has been recently proposed [2] that these two frameworks can have a common basis. In this way one could have a unified description for both the short and the large distance regimes. The first associated to free quarks and gluons, the second to confinement.

A QCD process at a hard scale $Q \gg \Lambda$ is determined by the small space-time intervals between the quark and gluon involved. However the cross section of such a process may acquire a contribution from a soft gluon which travels a finite distance 1 fm $\gg 1/Q$ and thus is sensitive to the non-perturbative interaction domain. A soft gluons involves the running coupling in the large distance regime and its effect can be analyzed provided one can introduce and define the coupling at any scale. The effect of these gluons can by studied by the usual perturbative methods and they gives rise to corrections to the usual perturbative QCD terms proportional to powers of Λ^2/Q^2. In the case of Euclidean observables, i.e. observables for which one can perform the Wick rotation, one finds that these powers terms correspond to the OPE terms. This means that a long distance gluons generate powers in Λ^2/Q^2 which correspond to the dimension of the local composite operators with the quantum number of the observable under consideration.

The possibility of a unified description of the perturbative and the non-perturbative contributions is based on the coupling constant and in particular on its possibility of being defined also at small scale. Moreover, the OPE terms correspond to the powers which are coming from the response of the perturbative matrix element to the long distance gluon. It is clear that one should have under control the power terms in the behaviour of the coupling constant in the short distance region, i.e. in the perturbative regime. Possible power terms in the running coupling need not to be related to OPE and they contribute to the power corrections to the observable. This point has been recently arisen [4] by Grunberg and Akhoury and Zakharov.

In this talk I first recall the dispersive method for the study of power corrections [3]. In particular the definition of the running coupling for any value of the scale. Given the interest of this meeting I will recall the application of this analysis to DIS structure function. Then I report some evidence [5] from lattice calculation of the presence of power terms which are not associated to OPE and could be due to the presence of power terms of the running coupling in the short distance region.

2.   DISPERSIVE METHOD AND POWER CORRECTIONS

In [3] it has been suggested a convenient way to define $\alpha_S(k^2)$ for any k^2 which is based on dispersion relations. Inspired by QED and to one-loop accuracy one defines $\alpha_S(k^2)$ in terms of an effective coupling α_S^{eff}

$$\frac{\alpha_S(k^2)}{k^2} = \int_0^\infty \frac{\alpha_S^{\text{eff}}(m^2)}{(m^2 + k^2)^2} \, dm^2 \, . \tag{1}$$

At short distances the couplings are approximately the same $\alpha_S^{\text{eff}} - \alpha_S \sim \alpha_S^3$. This relation has the following important implications, see [3].
i) The running coupling has singularities only in the causal k^2 region. The one-loop calculation of $\alpha_S(k^2)$ with its Landau pole at $k^2 = \Lambda^2$ does not satisfies (1). One can compute $\alpha_S^{\text{eff}}(\mu^2)$ to one-loop and, by using (1), one gets the well known regular function

$$\alpha_S(k^2) = \frac{4\pi}{\beta_0} \left\{ \frac{1}{\ln(k^2/\Lambda^2)} - \frac{\Lambda^2}{k^2 - \Lambda^2} \right\} \, , \tag{2}$$

in which the Landau pole is removed. The asymptotic behaviour at large k^2 contains power corrections of order Λ^2/k^2.

ii) Given (1), a virtual gluon free propagator can be effectively replaced by the exchange of a gluon with mass m with $\alpha_S^{\text{eff}}(mu2)$. This provides a systematic basis for the study of power corrections by massive gluon techniques [2].

iii) The large distance part of $\alpha_S(k^2)$ with $k^2 \lesssim \Lambda^2$ is probed by the part of the effective coupling which has vanishing integer moments. Then only the non-analytic dependence in $m2$ of the distributions is associated to large distances. This provides the connection between power corrections and OPE.

The perturbative answer for an observable $\mathcal{O}_j(Q^2)$, which is collinear and infrared safe, can be formulated in terms of the following m^2 integral

$$\mathcal{O}_j(Q^2) = \int_0^\infty \frac{dm^2}{m^2}\, \alpha_S^{\text{eff}}(m^2) \cdot \dot{\mathcal{F}}_j(m^2/Q^2)\,, \tag{3}$$

with $\dot{\mathcal{F}}_j(m^2/Q^2) \equiv -(d/d\ln m^2)\,\mathcal{F}_j(m^2/Q^2)$ and $\mathcal{F}_j(m^2/Q^2)$ the *characteristic function* obtained by computing the m^2-dependence of the matrix element for a virtual gluon with mass m^2. For an infrared/collinear safe observable the m^2 integral converges and is mainly determined by the region $m^2 \sim Q^2$, the typical hard scale of the process. This region reproduces the standard one-loop perturbative prediction $\mathcal{O}_j^{\text{PT}} = \alpha_S^{\text{eff}}(Q^2)\,\mathcal{F}(1) \approx \alpha_S(Q^2)\,\mathcal{F}(1)$. The genuine non-perturbative component of the answer, $\delta\mathcal{O}^{\text{NP}}$, is triggered by the non-perturbative component of the effective coupling $\delta\alpha_S^{\text{eff}}(m^2)$. Imposing the ITEP/OPE restriction [1] that the non-perturbative part of $\alpha_S(k^2)$ in the ultraviolet region receives no power correction $(k^2)^{-p}$ due to the large-distance interaction domain, one derives from (1) that the integer moments of the non-perturbative effective coupling vanish: $\int_0^\infty \frac{dm^2}{m^2}\delta\alpha_S^{\text{eff}}(m^2)\,(m^2)^i = 0$, $i = 1, 2, \ldots p$. Thus, the power n of the leading power correction Q^{-n} can be triggered by studying the non-analytic $m^2 \to 0$ behaviour of the one-loop matrix element (characteristic function). The magnitude of the power correction is then expressed in terms of the m-moments of an "effective charge" $\delta\alpha_S^{\text{eff}}(m^2)$ describing the intensity of non-perturbative gluon radiation:

$$\delta\mathcal{O}_j^{\text{NP}}(Q^2) = \int_0^\infty \frac{dm^2}{m^2}\delta\alpha_S^{\text{eff}}(m^2)\,\dot{\mathcal{F}}_j(m^2/Q^2) \simeq C_j \frac{A_{n,q}}{Q^n}\,, \tag{4}$$

where

$$A_{2p,q} = \frac{C_F}{2\pi} \int_0^\infty \frac{dm^2}{m^2}\,\delta\alpha_S^{\text{eff}}(m^2) \cdot (m^2)^p \ln^q m^2 \tag{5}$$

(for p half-integer or $q > 0$).

The method can be applied to scaling violation of collinear singular quantities such as the DIS structure function $F_2(x, Q^2)$ and one finds non-perturbative corrections of the form

$$F_2(x, Q^2) = F_2^{\text{PT}}(x, Q^2)\left(1 + \frac{C(x)}{Q^2}\right) \tag{6}$$

where $C(x) \sim A_{2,1}/(1-x)$. The agreement with the data is quite good. Notice that due to the singular behaviour at large x near the phase space the power correction becomes of order one.

3. LATTICE CALCULATION AND POWER TERMS

Here I report same evidence [5] that in the gluon condensate in lattice gauge theory (LGT) there is a contribution proportional to Λ^2. Since in LGT there is no operator of dimension two, this contribution is not connected to OPE. It has been argued [4] that a Λ^2 term in the condensate could be due to the presence in the running coupling at short distance of power terms.

According to OPE the gluon condensate $W \equiv \langle \alpha_S \, Tr \, F^2 \rangle / Q^4$ has the expansion

$$W = W_0 + (\Lambda^4/Q^4) \, W_4 + \cdots \tag{7}$$

In lattice theory the frequencies are bounded by the UV cutoff $Q = \pi/a$ with a the lattice spacing. Here W_4 is the "genuine" gluon condensate. The Λ^2/Q^2 term is absent since there are no gauge invariant operators of dimension two. The perturbative contributions are present only in W_0, which can be represented by the typical integral

$$W_0 \sim \int_{\rho\Lambda^2}^{Q^2} \frac{k^2 \, dk^2}{Q^4} \, \alpha_S(k^2) \,, \tag{8}$$

where k is the softest virtual momentum in the perturbative Feynman diagrams. To make k^2 within the perturbative region one takes $\rho \gg 1$.

If one assumes that $\alpha_S(k^2)$ is given by the beta function at two loops, one finds

$$W_0^{\text{ren}} = \sum_{\ell=1} \alpha_S^\ell \, \{w_\ell^{\text{ren}} + \mathcal{O}(\Lambda^4/Q^4)\}, \quad w_\ell^{\text{ren}} = \mathcal{N}' \, (8\pi/\beta_0)^\ell \, \Gamma(\ell+\beta_1/\beta_0^2) \,, \tag{9}$$

with $\alpha_S = \alpha_S(Q)$ and β_0 and β_1 the first and second beta function coefficients.

In order to identify the Λ^2/Q^2 powers in W one should know the behaviour of $\alpha_S(k^2)$ at large k^2 at the level of powers, i.e. one should know the power corrections as the ones in the example (2).

I recall now the evidence of the presence of a Λ^2/Q^2 term in W. The coefficients w_ℓ^{lat} of the perturbative expansion of W on the lattice with UV coupling $\alpha_{\text{lat}} = \alpha_{\text{lat}}(Q^2)$

$$W^{pert} = \sum_{\ell \geq 1} w_\ell^{\text{lat}} \, \alpha_{\text{lat}}^\ell \,, \tag{10}$$

are known up to eight loops. The first three terms have been computed analytically [6] and eight terms have been computed numerically [7]. It has been shown [7] that the growth of w_ℓ^{lat} with ℓ is consistent with the factorial behaviour in (9). This requires to take into account the differences in the lattice and continuum scales Λ, i.e. the relation between α_{lat} and $\alpha_S = \alpha_S(Q^2)$ in (9).

The evidence that in lattice gauge theory the condensate contains terms of order Λ^2/Q^2 is obtained by studying $W - W_0$ as a function of the lattice coupling $\beta_{\text{lat}} = 3/2\pi\alpha_{\text{lat}}$ in the range $\beta_{\text{lat}} = 6 - 7$ in which lattice artifacts are small. This corresponds to a very small $\alpha_{\text{lat}} \sim 0.1$ for which the perturbative tail is important. The values of W are taken from the Monte Carlo simulation in [8]. W_0 is obtained by taking the partial sum of the first 8 computed coefficients and adding a remainder estimated from W_0 in (8) in which $\alpha_S(k^2)$ is taken at two loops. This estimation is justified by the fact that for large orders the perturbative coefficients in (9) approach the lattice ones. The perturbative part provides over 90% of the full contribution. This is consistent with the fact that the lattice physical value of Λ is very small. From this analysis one finds that the powers correction are proportional to Λ^2/Q^2 rather than to the OPE expected Λ^4/Q^4. The reason for the unexpected behaviour Λ^2/Q^2 could be that our analysis is not complete. The major problem is the finiteness of the lattice size. However we have estimated its effects on the considered contributions to W_0 and they seem to be small.

The Λ^2/Q^2 term can be explained by the presence of highly subleading power corrections at large momentum in the running coupling. For example a Λ^2/k^2 contribution in $\alpha_S(k^2)$ in (8) gives a contribution of order Λ^2/Q^2 coming from the UV region ($k^2 \approx Q^2$). These contributions are then of "perturbative" nature, naturally associated to the contribution W_0 in the OPE, and process independent as the running coupling. An important question is whether these Λ^2/Q^2 are phenomenologically relevant (see [4]).

References

[1] Wilson K. Phys. Rev. D 179 (1969) 1499; Shifman M.A., Vainstein A.I. and Zakharov V.I., Nucl. Phys. B147 (1979) 385,448,519

[2] See for instance: Webber B.R., Phys. Lett. 339B (1994) 148; Korchemsky G.P. and Sterman G., Nucl. Phys. B437 (1995) 415; Akhoury R. and Zakharov V.I., Phys. Lett. 357B (1995) 646; Beneke M., Braun V.M. and Zakharov V.I., Phys. Rev. Lett. 73 (1994) 3058; Ball P., Beneke M. and Braun V.M., Nucl. Phys. B452 (1995) 563.

[3] Dokshitzer Yu.L., Marchesini G. and Webber B.R., Nucl. Phys. B469 (1996) 93

[4] Grunberg G., hep-ph/9705290; Akhoury R. and Zakharov V.I., hep-ph/9705318.

[5] Burgio G., Di Renzo F., Marchesini G. and Onofri E., hep-ph/9706209.

[6] Allés B., Campostrini M., Feo A. and Panagopoulos H., Phys. Lett. 324B (1994) 443 and references therein.

[7] Di Renzo F., Onofri E. and Marchesini G., Nucl. Phys. B457 (1995) 202 and Nucl.Phys.B (to appear), hep-lat/9612016.

[8] Campostrini M., DiGiacomo A. and Günduk V., Phys. Lett. 223 (1989) 393.

Highlights on deep inelastic scattering at HERA

C. Royon

CEA, DAPNIA, Service de Physique des Particules, Centre d'Etudes de Saclay, France

1. Introduction

The new collider HERA, located at DESY, Hamburg, between electrons (positrons) and protons of respectively 27.5 and 820 GeV has provided new interesting results in quantum chromodynamics at low x, where x is the momentum fraction of the interacting parton in the Feynman parton model. We will describe the new results concerning the measurement of the proton structure functions obtained by the H1 and ZEUS collaborations at very low Q^2, the transfered energy squared between the electron and the interacting quark. We will then discuss some new interesting events where there is no deposition of energy in the proton direction. We finally present the new very high Q^2 events observed by both collaborations. This short review represents only a small part of the results obtained by both collaborations.

2. Measurement of the proton structure function

The measurement of the proton structure function has always been of great importance for the understanding of the quark and gluon structure of the proton. In the continuity of the previous fixed target experiments, HERA allows the measurement of the proton structure function in a very wide kinematical domain (a gain of almost two orders of magnitude is reached compared with the previous fixed target experiments): $3.10^{-5} < x < 0.3$, and $0.15 < Q^2 < 5000 GeV^2$. In figures 1a and 1b are shown the new results obtained at low Q^2 by both collaborations [2], as well as the results obtained by

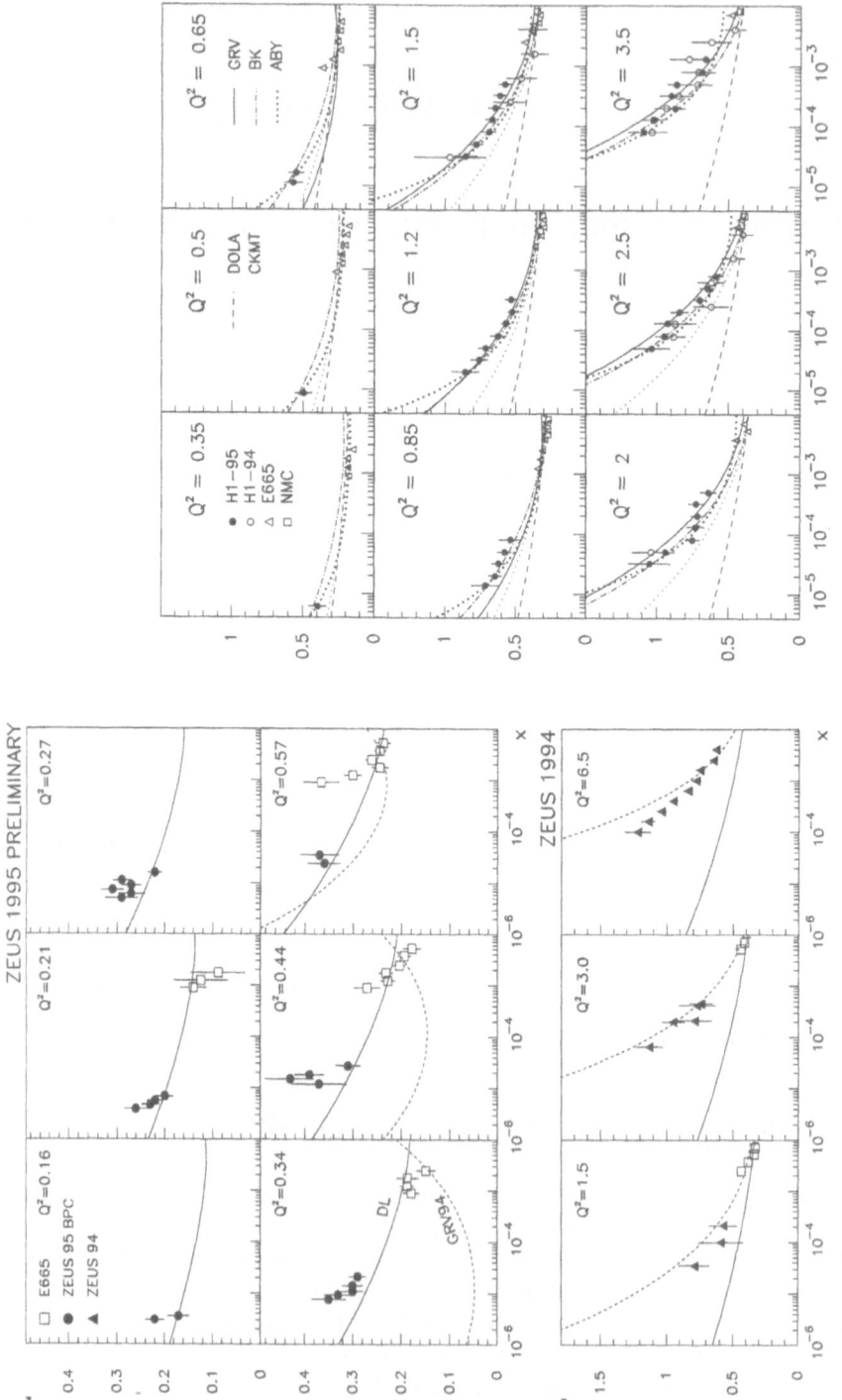

Fig. 1. — Measurement of the proton structure function F_2 from the H1 and ZEUS collaborations at very low Q^2 compared with the fixed target experiments data from NMC, E665, the data are compared with some Regee inspired models (DS Donnachie Landshoff), or based on perturbative QCD (GRV Glück, Reya, Vogt)

the fixed target experiment NMC. It is seen that the rise of F_2 observed at low x for higher values of Q^2 [2] is still valid in this new domain even if it is attenuated. The measurements are compared with parametrizations motivated by Regge theory relating the structure function to Reggeon exchange amplitudes. These Regge parametrisations successfully describe the slow rise of the total cross section with energy in hadron-hadron and γp interactions. We can see that these models can give a correct description of the data only at very low Q^2. On the other hand, some QCD inspired models like GRV (based on the Dokshitzer Gribov Lipatov Altarelli Parisi evolution equation starting at very low scale) can describe the data at higher Q^2 but fail completely when Q^2 decreases. This new low Q^2 domain is thus the intermediate region between photoproduction where Regge theories apply and the standard deep inelastic scattering region where perturbative QCD can be used.

Another important result on the proton structure functions obtained by the H1 and ZEUS experiments is the analysis of the charm content of the proton. For $8.10^{-4} < x < 8.10^{-3}$, it could be shown that the charm fraction of the proton strcuture function $F_2^{c\bar{c}}/F_2$ is equal to $0.237 \pm 0.021^{+0.043}_{-0.039}$, and that $F_2^{c\bar{c}}$ increases steeply with decreasing x.

3. Diffraction at HERA

The data taken in 1992 by the H1 and ZEUS experiments showed some new interesting events with an interval of rapidity around the incident proton direction devoid of any hadronic activity.This means that a colourless exchange ("pomeron") must have occured since there is no colour connection between the remnant proton and the struck quark.

For the first time at HERA, the statistics obtained in 1994 allowed a measurement of the proton diffractive structure function defined in analogy to the standard proton structure function in a wide kinematical domain. In order to describe the exchange of a colourless object, we need two more kinematical variables to determine the kinematics of the event: Q^2 as before, t the Mandelstam variable, β, the momentum fraction of the colourless exchanged object, and $x_p = x/\beta$. x_p is the momentum fraction of the parton inside this object if we assume it has a partonic structure. The obtained results from the H1 experiment are shown in figure 2 [4]. In a Regge interpretation, the x_P dependance can be factorised out from the β and Q^2 dependance. The x_P-term is assumed to describe a reggeon flux associated with the proton (expected to be of the form $F_2^{D(3)} \sim 1/x_P^n$), and the β and Q^2 dependance the structure function of the colourless exchanged object. Experimentally this factorisation breaks out, and there is a clear dependance of n as a function of β, but not as a function of Q^2. A natural interpretation of this result is that we need two separate factorisable components which contribute to the cross section. In this case, there is a contribution to the measured cross section from sub-leading trajectories which may be identified with the exchange of quantum numbers corresponding

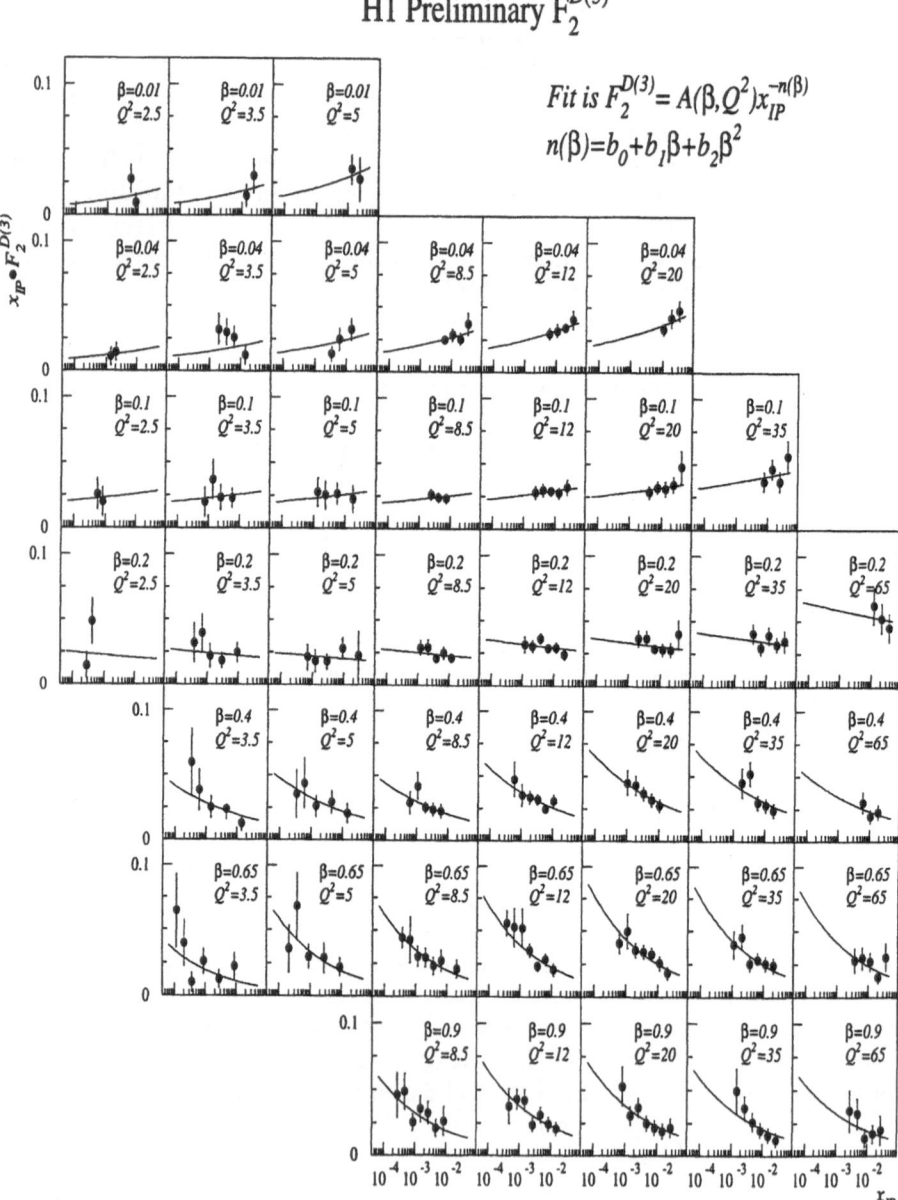

Fig. 2. — Measurement of the diffractive structure function, the result of the phenomenological fit in which two separately factorisable components contribute to the cross section is also shown

to physical meson states. The leading trajectory would be the pomeron, and the next-leading exchanges the trajectories with the quantum numbers of the ρ, ω, a_2, and f_2 mesons. An experimental fit assuming these contributions has been successfully performed by the H1 collaboration. Another possibility also leading to two components would be to use an unified description of total and diffractive structure functions in the framework of the dipole model of the hard BFKL pomeron [1].

4.   High Q^2 events

As we noticed before, it is possible to reach very high values of Q^2 at HERA up to 35000 GeV^2. It is then possible to count the number of events above a given Q^2 threshold and to compare it with the predictions of the standard model. The H1 experiment [3] observes 12 neutral current candidate events for $Q^2 > 15000 GeV^2$ where the expectation from the standard model is $N_{DIS} = 1.77 \pm 0.76$ events. The probability $P(N \geq N_{obs})$ that the DIS model signal fluctuates to $N \geq N_{obs}$ in a random set of experiments is 6.10^{-3}. The ZEUS collaboration [3] has also reported for an excess: for $Q^2 > 35000 GeV^2$ (resp. $x > 0.55$ and $y > 0.25$), they observe 2 (resp. 4) events whereas the standard model expectations are 0.145 ± 0.013 (resp. 0.91 ± 0.08), which corresponds to a probability $P(N \geq N_{obs})$ of 6.10^{-2} (resp. 7.10^{-3}). This excess of events is very interesting and more statistics is needed to say if this is a discovery of a phenomenon beyond the standard model.

References

[1] Navelet H., Peschanski R., Royon Ch., and Wallon S., *Phys.Lett.*, **B385**, (1996) 357, and references therein, Navelet H., Peschanski R. and Royon Ch., *Phys.Lett.*, **B366**, (1996) 329, Bialas A., Peschanski R. and Royon Ch, to appear

[2] H1 Coll., DESY preprint 97-042, to be published in *Nucl.Phys.B*, H1 Coll. *Nucl.Phys.*, **B470**, (1996) 3, ZEUS Coll., *Z.Phys.*, **C72**, (1996), 399, ZEUS Coll., *Z.Phys.*, **C69**, (1996), 607

[3] H1 Coll., *Z.Phys.*, **C774**, (1997) 191, ZEUS Coll., *Z.Phys.*, **C774**, (1997) 207

[4] H1 Coll. to be submitted, H1 Coll., *Phys.Lett.*, **B348**, (1995) 681, ZEUS Coll., *Z.Phys.*, **C70** (1996), 391

α_S: from DIS to LEP

W.J. Stirling[1]

([1]) *Departments of Mathematical Sciences and Physics,*
University of Durham, Durham DH1 3LE, England

1. INTRODUCTION

The strong coupling α_S is a fundamental parameter of the Standard Model. In comparison to parameters like α_{em}, M_Z and $\sin^2 \theta_W$ it is relatively poorly known. However the precision of α_S measurements has improved dramatically in recent years. More than twenty different types of process, from lattice QCD studies to the highest energy colliders, can be used to measure α_S accurately. The most precise determinations now quote uncertainties in $\alpha_S(M_Z^2)$ of less than 5%. There is also a remarkable consistency between the various measurements.

A comprehensive review of α_S measurements, including detailed descriptions of the underlying physics for the most important processes, can be found in Ref. [1]. One year later, several of the measurements quoted in Ref. [1] have been updated, resulting in a slight shift in the overall 'world average' value. The purpose of the present review is to update the discussion on α_S measurements given in Ref. [1], focusing on the new values reported in the last year. For more theoretical details, descriptions of other measurements and a full set of references, the reader is referred to the original review in Ref. [1].

The current situation is summarised in Fig. 1, which updates Table 12.1 of Ref. [1]. Before discussing the new measurements in detail, we begin with some technical preliminaries. In perturbative QCD the dependence of the strong coupling on the renormalisation scale is determined by the β–function:

$$Q^2 \frac{\partial \alpha_S(Q^2)}{\partial Q^2} \;\; = \;\; \beta(\alpha_S(Q^2)),$$

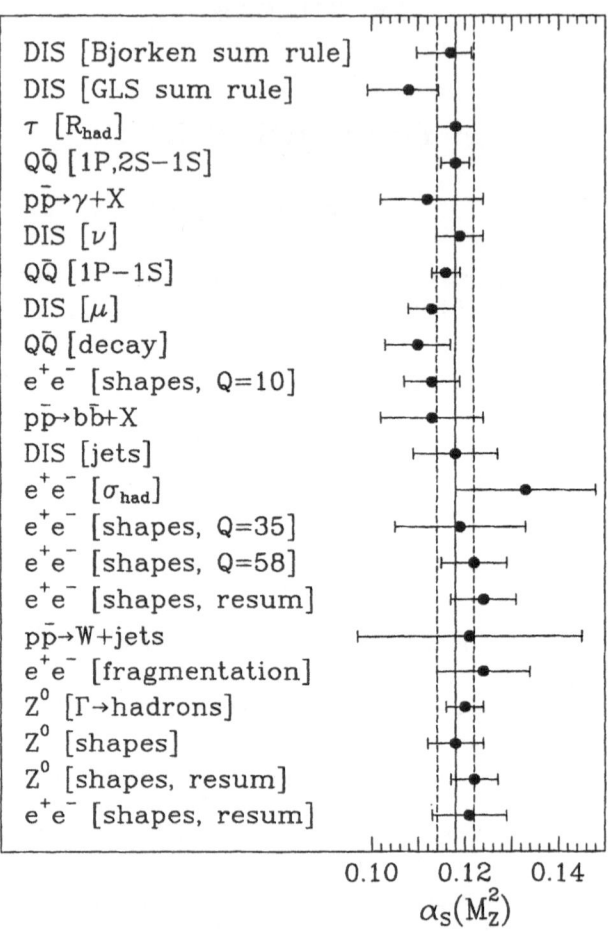

Fig. 1. — Measurements of $\alpha_S(M_Z^2)$, in the $\overline{\text{MS}}$ renormalisation scheme, updated from Ref. [1].

$$\beta(\alpha_S) = -b\alpha_S^2 \left(1 + b'\alpha_S + b''\alpha_S^2 + \ldots\right), \qquad (1)$$

where $b = (33 - 2n_f)/(12\pi)$ etc. The coefficients in the perturbative expansion depend, in general, on the renormalisation scheme (RS), although for massless quarks the first two coefficients, b and b', are RS independent. In essentially all phenomenological applications the $\overline{\text{MS}}$ RS is used; see Ref. [1] for further discussion and explicit expressions for the known β–function coefficients.

At leading order, i.e. retaining only the coefficient b, Eq. (1) can be solved for α_S to give

$$\alpha_S(Q^2) = \frac{\alpha_S(Q_0^2)}{1 + \alpha_S(Q_0^2)\, b\ln(Q^2/Q_0^2)} \tag{2}$$

or

$$\alpha_S(Q^2) = \frac{1}{b\ln(Q^2/\Lambda^2)}. \tag{3}$$

These two expressions are entirely equivalent – they differ only in the choice of boundary condition for the differential equation, $\alpha_S(Q_0^2)$ in the first case and the dimensionful parameter Λ in the second. In fact nowadays Λ is disfavoured as the fundamental parameter of QCD, since its definition is not unique beyond leading order (see below), and its value depends on the number of 'active' quark flavours. Instead, it has become conventional to use the value of α_S in the $\overline{\text{MS}}$ scheme at $Q^2 = M_Z^2$ as the fundamental parameter. The advantage of using M_Z as the reference scale is that it is (a) very precisely measured [2], (b) safely in the perturbative regime, i.e. $\alpha_S(M_Z^2) \ll 1$, and (c) far from quark thresholds, i.e. $m_b \ll M_Z \ll m_t$.

The parameter Λ is, however, sometimes still used as a book-keeping device. At next-to-leading order there are two definitions of Λ which are widely used in the literature:

$$\text{definition 1}: \qquad b\ln\frac{Q^2}{\Lambda^2} = \frac{1}{\alpha_S(Q^2)} + b'\ln\left(\frac{b'\alpha_S(Q^2)}{1 + b'\alpha_S(Q^2)}\right), \tag{4}$$

$$\text{definition 2}: \qquad \alpha_S(Q^2) = \frac{1}{b\ln(Q^2/\Lambda^2)}\left[1 - \frac{b'}{b}\frac{\ln\ln(Q^2/\Lambda^2)}{\ln(Q^2/\Lambda^2)}\right]. \tag{5}$$

The first of these solves Eq. (1) exactly when b'' and higher coefficients are neglected, while the second (the 'PDG' definition [2]) provides an explicit expression for $\alpha_S(Q^2)$ in terms of Q^2/Λ^2 and is a solution of Eq. (1) up to terms of order $1/\ln^3(Q^2/\Lambda^2)$. Note that these two Λ parameters are *different* for the *same* value of $\alpha_S(M_Z^2)$, the difference being about one quarter the size of the current measurement uncertainty:

$$\Lambda_1^{(5)} - \Lambda_2^{(5)} \simeq 15 \text{ MeV} \simeq \frac{1}{4}\delta_{\exp}\Lambda^{(5)}. \tag{6}$$

In this review we will be mainly concerned with measurements from e^+e^- colliders (in practice LEP and SLC) and from deep inelastic scattering. Both processes offer several essentially independent measurements, summarised in Table 1. Note that all of these use the $q\bar{q}g$ vertex to measure α_S, with the high Q^2 scale provided by an electroweak gauge boson, for example a highly virtual γ^* in DIS or an on-shell Z^0 boson at LEP1 and SLC. There are two main theoretical issues which affect these determinations. The first is the effect of unknown higher-order (next-to-next-to-leading order (NNLO) in most

Table I. — Summary of the most important processes for α_S determinations in e^+e^- collisions and in deep inelastic lepton-hadron scattering.

	quantity	perturbation series
e^+e^-	R_{ee}, R_Z, R_τ	$R = R_0[1 + \alpha_S/\pi + \ldots]$
	event shapes, f_3, ...	$1/\sigma d\sigma/dX = A\alpha_S + B\alpha_S^2 + \ldots$
	$D^h(z, Q^2)$	$\partial D^h/\partial \ln Q^2 = \alpha_S D^h \otimes P + \ldots$
ℓN DIS	$F_i(x, Q^2)$	$\partial F_i/\partial \ln Q^2 = \alpha_S F_i \otimes P + \ldots$
		$\int dx F_i(x, Q^2) = A + B\alpha_S + \ldots$
	$\sigma(2+1\ \text{jet})$	$\sigma = A\alpha_S + B\alpha_S^2 + \ldots$

cases) perturbative corrections, which leads to a non-negligible renormalisation scheme dependence uncertainty in the extracted α_S values. This is particularly true for the event shape measurements at e^+e^- colliders. The exceptions here are the total e^+e^- hadronic cross section (equivalently, the Z^0 hadronic decay width) and the DIS sum rules, which are known to NNLO. The second issue concerns the residual impact of $\mathcal{O}(1/Q^n)$ power corrections. For some processes it can be shown that the leading corrections are $\mathcal{O}(1/Q)$ (for example $\mathcal{O}(1/M_Z)$ for the corrections to event shapes at LEP1 and SLC) which can easily be comparable in magnitude to the NLO perturbative contributions. In deep inelastic scattering, the higher-twist power corrections are $\mathcal{O}(1/Q^2(1-x))$ and must be included in scaling violation fits especially at large x. Such power corrections (and their uncertainties) must be taken into account in α_S determinations, either using phenomenological parametrisations or theoretical models.

Before discussing the new high-energy collider measurements of α_S it is important to mention also determinations from lattice QCD, which have very small uncertainties. One of the simplest ways to define α_S on the lattice is to use the average value of the 1×1 Wilson loop (plaquette) operator:

$$\ln W_{1,1} = \frac{4\pi}{3}\alpha_P\left(\frac{3.4}{a}\right)[1 - (1.19 + 0.07n_f)\alpha_P], \qquad (7)$$

where a is the lattice spacing. A variety of choices is available for determining a, i.e. measuring the scale at which α_P has the value measured in (7). Quarkonium level splittings, for example $\Upsilon(S - P)$ and $\Upsilon(1S - 2S)$, are particularly suitable. Subsequently the plaquette α_P can be converted to the standard $\overline{\text{MS}}$

α_S for comparison with other determinations:

$$\alpha_S^{(\overline{\text{MS}},nf)}(Q^2) = \alpha_P^{(n_f)}(e^{5/3}Q^2) \left[1 + \frac{2}{\pi}\alpha_P^{(n_f)} + C_2(n_f)\left(\alpha_P^{(n_f)}\right)^2 + \dots\right]. \quad (8)$$

At present the two-loop coefficient is known only for $n_f = 0$ [3] – the shift in α_S between using $C_2(n_f = 0)$ and $C_2 = 0$ can be used to define a 'conversion' error. Several new lattice α_S values have been obtained recently, see for example Ref. [4], and are included in Fig. 1. As an example of the high precision of these measurements, we quote the value obtained by the NRQCD collaboration [5] using the $\Upsilon(S - P)$ splitting:

$$\alpha_S^{\overline{\text{MS}}}(M_Z^2) = 0.1175 \pm 0.0011(\text{stat.} + \text{sys.}) \pm 0.0013(m_q^{\text{dyn.}}) \pm 0.0019(\text{conv.}), \quad (9)$$

where the first error is due to the lattice statistics and systematics, the second is from the extrapolation in the dynamical quark mass, and the third is the conversion error mentioned above.

In the following sections we will discuss new α_S measurements from LEP/SLC and from deep inelastic scattering. Section 4 presents a new value for the α_S world average.

2. α_S FROM LEP AND SLD

In principle the most reliable determination of α_S at the LEP and SLD e^+e^- colliders comes from the Z^0 hadronic width. In particular we have, for the ratio R_Z,

$$R_Z = \frac{\Gamma(Z^0 \to \text{hads.})}{\Gamma(Z^0 \to \ell^+\ell^-)} = R_0 \left[1 + \frac{\alpha_S}{\pi} + C_2\left(\frac{\alpha_S}{\pi}\right)^2 + C_3\left(\frac{\alpha_S}{\pi}\right)^3 + \dots\right] \quad (10)$$

with $R_0 = 3\sum_q(v_q^2 + a_q^2)/(v_\ell^2 + a_\ell^2)$. The perturbative coefficients are known up to third order, see Ref. [1] for explicit expressions and references, and as a result the prediction is very stable with respect to variations in the renormalisation scale. In practice, since R_0 depends on the weak mixing angle and other electroweak parameters, it is more appropriate to perform a global fit to all relevant electroweak quantities, for example the (LEP and SLD) Z^0 partial widths and decay asymmetries, $p\bar{p}$ collider measurements of M_W and m_t, etc. Such analyses are performed regularly by the LEP Electroweak Working Group, and the results of a recent (1996) fit [6] are summarised in Table 2. An additional theory error from unknown higher-order corrections of $\delta\alpha_S(M_Z^2) = \pm0.002$ has been estimated, see Ref. [6] and references therein. The resulting α_S value,

$$\alpha_S(M_Z^2) = 0.120 \pm 0.003(\text{fit}) \pm 0.002(\text{theory}), \quad (11)$$

is displayed in Fig. 1.

Table II. — Values for the Standard Model parameters obtained from a global fit to LEP, SLD, $p\bar{p}$ and νN data, from Ref. [6].

parameter	fit value
m_t [GeV]	172 ± 6
M_H [GeV]	149^{+148}_{-82}
$\alpha_S(M_Z^2)$	0.120 ± 0.003
$\sin^2 \theta_{\text{eff}}^{\text{lept}}$	0.23167 ± 0.00023
$1 - M_W^2/M_Z^2$	0.2235 ± 0.0006
M_W [GeV]	80.352 ± 0.033

The other high-precision determination of α_S at LEP and SLC comes from *event shapes*, quantities which measure the relative contribution of the $\mathcal{O}(\alpha_S)$ $e^+e^- \to q\bar{q}g$ process to the total hadronic cross section, see Table 1. A typical example is the thrust distribution:

$$\frac{1}{\sigma}\frac{d\sigma}{dT} = \alpha_S A_1(T) + \alpha_S^2 A_2(T) + \ldots + \mathcal{O}\left(\frac{1}{E_{\text{cm}}}\right). \tag{12}$$

Such quantities are known in perturbation theory to $\mathcal{O}(\alpha_S^2)$, and the theoretical predictions in the $T \to 1$ region can be improved by resumming the leading logarithmic $A_n \sim \ln^{(2n-1)}(1-T)/(1-T)$ contributions to all orders, as discussed in Ref. [1]. Another important recent theoretical development has been an improved understanding of the leading $\mathcal{O}(1/E)$ power corrections [7], which at LEP can be as numerically important as the next-to-leading perturbative corrections.

Event shapes have yielded α_S measurements over a wide range of e^+e^- collision energies, the most recent measurements being at the LEP2 energies $\sqrt{s} = 161$ and 172 GeV. Although the statistical precision of these measurements cannot match that obtained at the Z^0 pole, the results are consistent with the Q^2 evolution of α_S predicted by Eq. (1). For example, Fig. 2 shows the α_S values determined by the L3 collaboration [8] from event shape measurements at LEP1 and LEP2 energies. The solid line is the evolution predicted by perturbative QCD. Figure 1 contains a new 'LEP1.5' average value for α_S obtained from event shapes at $\sqrt{s} = 133$ GeV, taken from the 1996 review by Schmelling [9]:

$$\alpha_S(Q^2 = (133 \text{ GeV})^2) = 0.114 \pm 0.007 \quad \Rightarrow \quad \alpha_S(M_Z^2) = 0.121 \pm 0.008. \tag{13}$$

Another updated value in Fig. 1 is that obtained from the scaling violations of the fragmentation function measured in $e^+e^- \to hX$ over a range of collision energies, the analogue of the scaling violations of structure functions in DIS. The new value (an ALEPH/DELPHI average taken from Ref. [9]) corresponds to

$$\alpha_S(M_Z^2) = 0.124 \pm 0.010. \tag{14}$$

Finally, the CLEO collaboration have published [10] a new value for α_S obtained from the relative decay rate of the $\Upsilon(1S)$ into a single hard photon:

$$\frac{\Gamma^{\gamma gg}}{\Gamma^{ggg}} = \frac{4}{5}\frac{\alpha}{\alpha_S(\mu^2)}\left[1 - (2.6 - 2.1\ln(m_b^2/\mu^2))\frac{\alpha_S(\mu^2)}{\pi} + \ldots\right]. \quad (15)$$

The new value,

$$\alpha_S(M_{\Upsilon(1S)}^2) = 0.163 \pm 0.002(\text{stat.}) \pm 0.014(\text{sys.})$$

$$\Rightarrow \quad \alpha_S(M_Z^2) = 0.110 \pm 0.001(\text{stat.}) \pm 0.007(\text{sys.}), \quad (16)$$

is included in Fig. 1.

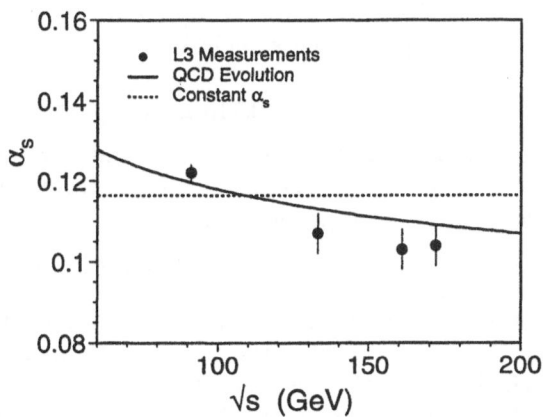

Fig. 2. — Measurements of α_S from event shapes at LEP1 and LEP2 from the L3 collaboration [8]. The errors correspond to experimental uncertainties.

3. α_S FROM DEEP INELASTIC SCATTERING

The traditional method of measuring α_S in deep inelastic scattering is from the strength of the structure function scaling violations predicted by the DGLAP equations:

$$Q^2\frac{\partial q^{NS}}{\partial Q^2} = \frac{\alpha_S(Q^2)}{2\pi}P^{qq} \otimes q^{NS}$$

$$Q^2 \frac{\partial q^S}{\partial Q^2} = \frac{\alpha_S(Q^2)}{2\pi} \left(P^{qq} \otimes q^S + 2n_f P^{qg} \otimes g \right)$$

$$Q^2 \frac{\partial g}{\partial Q^2} = \frac{\alpha_S(Q^2)}{2\pi} \left(P^{gq} \otimes q^S + P^{gg} \otimes g \right), \tag{17}$$

where q^{NS} and q^S are respectively non-singlet and singlet combinations of quark distribution functions. The fixed target and HERA structure function data, spanning a large range in x and Q^2, are all consistent with NLO DGLAP evolution, and yield α_S values which are in broad agreement. As an example, Fig.3 [11] shows the χ^2 values for various DIS data sets as a function of the $\alpha_S(M_Z^2)$ value in the evolution equations. With one exception, all data sets exhibit a χ^2 minimum in the $\alpha_S = 0.11 - 0.13$ range. In fact the 'best fit' value for these data sets is $\alpha_S(M_Z^2) = 0.118$, exactly the world average value (see Section 4 below).

It is difficult to extract a proper error on α_S from such global fit analyses. This requires a rigorous treatment of systematic errors and inclusion of higher-twist contributions in the fit. Several groups have performed such analyses. For example. the Milsztajn–Virchaux analysis of the SLAC/BCDMS ($eN, \mu N$) data [12] yields

$$\alpha_S(M_Z^2) = 0.113 \pm 0.005, \tag{18}$$

where the error includes statistical, systematic and scale dependence uncertainties. Recently the CCFR collaboration have reported [13] a new value of α_S from their $F_2^{\nu N}, x F_3^{\nu N}$ high-precision data (see Fig. 3):

$$\alpha_S(M_Z^2) = 0.119 \pm 0.002(\text{exp.}) \pm 0.001(\text{HT}) \pm 0.004(\text{scale}). \tag{19}$$

The second error is from an estimate of the higher-twist contribution using the model of Ref. [14], and the third is the scale dependence uncertainty implemented as in Ref. [12]. Note that the value in (19) is somewhat larger than the earlier (1993) CCFR value of $\alpha_S = 0.111 \pm 0.004$. The change is due to new energy calibrations of the detector [13].

Deep inelastic scattering structure functions satisfy a variety of *sum rules*, corresponding to the conservation of various nucleon quantum numbers. In general the parton model values of the sums have $\mathcal{O}(\alpha_S)$ corrections, which can be used to extract α_S from measurements of structure function integrals at fixed Q^2. Two sums rules which have been used to obtain precision measurements are the Gross–Llewellyn Smith and Bjorken sum rules (see Ref. [1] for more discussion and references):

$$\text{GLS}: \quad \int_0^1 dx(F_3^{\nu p} + F_3^{\bar\nu p}) = 6\left[1 + \frac{\alpha_S}{\pi} + \ldots\right] + \Delta_{\text{HT}}, \tag{20}$$

$$\text{BjS}: \quad \int_0^1 dx(g_1^p - g_1^n) = \frac{1}{6}\frac{g_A}{g_V}\left[1 - \frac{\alpha_S}{\pi} + \ldots\right] + \Delta_{\text{HT}}, \tag{21}$$

where Δ_{HT} represents $\mathcal{O}(1/Q^2)$ higher-twist contributions. A new analysis [15] of polarised structure function measurements has produced an update of the

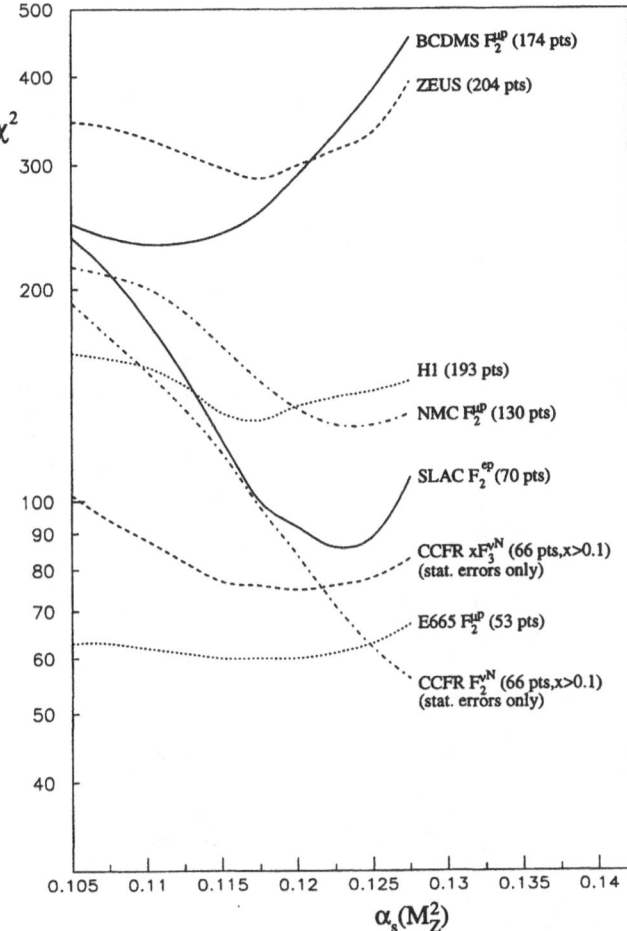

Fig. 3. — χ^2 values for various DIS data sets obtained in a global fit to these and other hard scattering data [11].

α_S value from the Bjorken sum rule. In Ref. [15] Padé Summation is used to reduce the theoretical error from the choice of renormalisation scheme in the calculation of the perturbation series on the right-hand side of (21). The resulting theoretical error in $\alpha_S(M_Z^2)$ is estimated at ±0.002:

$$\alpha_S(M_Z^2) = 0.117^{+0.004}_{-0.007}(\text{exp.}) \pm 0.002(\text{theory}).\tag{22}$$

This new value is included in Fig. 1.

Finally, α_S can be obtained from jet fractions and event shapes in DIS, see Table 1. For example, NLO theoretical predictions are currently being used at

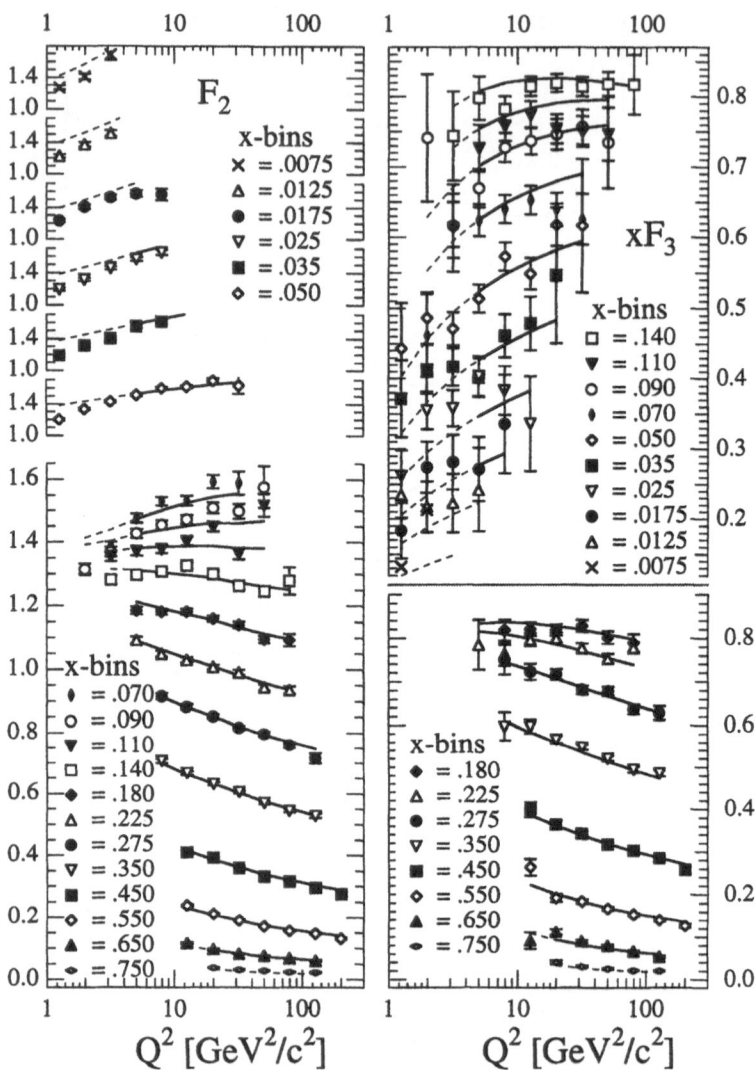

Fig. 4. — Measurements of the structure function $F_2^{\nu N}$ from the CCFR collaboration together with a NLO QCD fit, from Ref. [13].

HERA to extract α_S from the relative rate of '2+1' jet production at high Q^2, the analogue of f_3 in e^+e^- annihilation. No new results have been published since the review in Ref. [1].

4. SUMMARY

The average value([1]) of the measurements presented in Fig. 1 is

$$\text{WORLD AVERAGE:} \quad \alpha_S(M_Z^2) = 0.118 \pm 0.004. \qquad (23)$$

Following Ref. [1], the error here is defined as 'the uncertainty equal to that of a typical measurement by a reliable method'. In view of the recent improvements in the lattice, Z^0 hadronic width, and DIS (νN) determinations, it seems appropriate to decrease the uncertainty of ± 0.005 in Ref. [1] to ± 0.004. The central value in (23) has increased by $+0.002$ from that given in Ref. [1]. This is due primarily to (a) increases of $+0.003$ and $+0.004$ in the central values of the two lattice determinations, and (b) an increase in the CCFR νN DIS scaling violation central value of $+0.008$. In view of the remarkable consistency of all the measurements, and in particular of those with the smallest uncertainties, it seems unlikely that future 'world average' values of α_S will deviate significantly, if at all, from the current value given in (23).

Acknowledgments

I am grateful to my co-authors Keith Ellis and Bryan Webber for their help in preparing this review. This work was supported in part by the EU Programme "Human Capital and Mobility", Network "Physics at High Energy Colliders", contract CHRX-CT93-0537 (DG 12 COMA).

References

[1] Ellis R.K., Stirling W.J. and Webber B.R., QCD and Collider Physics (Cambridge University Press, Cambridge, 1996). See also http://www.hep.phy.cam.ac.uk/theory/webber/QCDbook.
[2] Particle Data Group: Barnett R.M. et al., Phys. Rev. **D54** (1996) 77.
[3] Lüscher M. and Weisz P., Phys. Lett. **B349** (1995) 165.
[4] Shigemitsu J., Nucl. Phys. Proc. Suppl. **53** (1997) 16.
[5] NRQCD collaboration: Davies C.T.H. et al., preprint hep-lat/9703010 (1997).
[6] LEP Electroweak Working Group report CERN-PPE/96-183 (1996).
[7] Webber B.R., Phys. Lett. **B339** (1994) 148.
Dokshitzer Yu.L. and Webber B.R., preprint Cavendish-HEP-97/2 (1997); Phys. Lett. **B352** (1995) 451.
Dokshitzer Yu.L., Marchesini G. and Webber B.R., Nucl. Phys. **B469** (1996) 93.
Korchemsky G.P., Sterman G., preprint hep-ph/9505391; Nucl. Phys.

([1]) obtained by χ^2 minimisation, as described in Ref. [1]

B437 (1995) 415.

Nason P. and Seymour M.H., *Nucl. Phys.* **B454** (1995) 291.

Akhoury R. and Zakharov V.I., *Phys. Lett.* **B357** (1995) 646; *Nucl. Phys.* **B465** (1996) 295.

Braun V.M., presented at 1996 Annual Divisional Meeting (DPF 96) of the Division of Particles and Fields of the American Physical Society, Minneapolis, August 1996, preprint hep-ph/9610212 (1996).

Beneke M., presented at the 28th International Conference on High-Energy Physics (ICHEP 96), Warsaw, Poland, July 1996, preprint hep-ph/9609215 (1996).

[8]   L3 collaboration: Acciarri M. *et al.*, preprint CERN-PPE/97-74 (1997).

[9]   Schmelling M., presented at the 28th International Conference on High-Energy Physics (ICHEP 96), Warsaw, Poland, July 1996, preprint hep-ex/9701002 (1997).

[10] CLEO collaboration: Nemati B. *et al.*, *Phys. Rev.* **D55** (1997) 5273.

[11] Martin A.D., Roberts R.G. and Stirling W.J., preprint in preparation.

[12] Milsztajn A. and Virchaux M., *Phys. Lett.* **B274** (1992) 221.

[13] CCFR collaboration: Seligman W.G. *et al.*, preprint hep-ex/9701017 (1997).

[14] Dasgupta M. and Webber B.R., preprint hep-ph/9704297 (1997).

[15] Ellis J. *et al.*, *Phys. Rev.* **D54** (1996) 6986.

On Small-x Resummations for the Evolution of Deep-Inelastic Structure Functions *

J. Blümlein [1] and A. Vogt [2]

([1]) *DESY–Zeuthen,*
Platanenallee 6, D–15735 Zeuthen, Germany

([2]) *Institut für Theoretische Physik, Universität Würzburg,*
Am Hubland, D–97074 Würzburg, Germany

A brief survey is given of recent results on the resummation of leading small-x terms for unpolarized and polarized, non–singlet and singlet structure function evolution.

1. INTRODUCTION

The evolution kernels of both non–singlet and singlet parton densities contain large logarithmic contributions for small fractional momenta x. In all–order resummations of these terms in the limit $x \to 0$ [1–5], one naturally faces the problem of factorization and renormalization scheme dependence. Therefore these resummations have to be performed in the frame of the corresponding renormalization group equations. In the following we will briefly discuss the resulting small-x resummations for the anomalous dimensions relevant to the various deep-inelastic scattering (DIS) processes and their quantitative consequences. For full accounts the reader is referred to refs. [6, 7] and refs. [8, 9] for the non–singlet and singlet evolutions, respectively. A recent review can be found in ref. [10].

For unpolarized deep-inelastic processes the leading small-x contributions to the gluonic anomalous dimensions behave like [1]

(*) Presented by A. Vogt

$$\left(\frac{\alpha_s}{N-1}\right)^k \quad \leftrightarrow \quad \frac{1}{x}\,\alpha_s^k\,\ln^{k-1}x\ . \tag{1}$$

Here N denotes the Mellin variable. The corresponding quark anomalous dimensions, being one power down in $\ln x$, have been derived in ref. [2]. Recently also first partial results have been obtained for the subleading terms of the gluon–gluon splitting functions [3]. The leading contributions to all anomalous dimensions for the non–singlet [4] and polarized singlet [5] evolutions are given by

$$N\left(\frac{\alpha_s}{N^2}\right)^k \quad \leftrightarrow \quad \alpha_s^k\,\ln^{2k-2}x\ . \tag{2}$$

The resummation of these terms can be completely derived by means of perturbative QCD. Its effect on the behaviour of the various DIS structure functions, however, is necessarily determined as well by the behaviour of the input parton densities at an initial scale Q_0^2, and is therefore not predictable within perturbative QCD but has to be determined by experiment. Thus the resummation effect can only be studied via the evolution of structure functions over some range in Q^2, which moreover probes the anomalous dimensions at all $z \geq x$ via the Mellin convolution with the parton densities.

In leading (LO) and next–to–leading order (NLO) QCD the complete anomalous dimensions are known. Hence the effect of the all–order resummation of the most singular parts of the splitting functions as $x \to 0$ concerns only orders higher than α_s^2. Due to the Mellin convolution also terms which are less singular as $x \to 0$ may contribute substantially at these higher orders as well. In some cases the existence of such terms is, in fact, enforced by conservation laws. Fermion–number conservation, e.g., implies for the non–singlet '$-$'-evolution

$$\int_0^1 dz \sum_{k=1}^\infty \alpha_s^k P_k^-(z) = 0\ . \tag{3}$$

Correspondingly energy–momentum conservation holds for the unpolarized singlet evolution. Even in the polarized singlet case, however, where no conservation laws constrain the anomalous dimensions, the LO and NLO results exhibit terms which are less singular by one power in N and have about the same coefficient but with opposite sign, cf. ref. [8]. Since such contributions and further corrections are not yet known to all orders, it is reasonable to estimate their possible impact by corresponding modifications of the resummed anomalous dimensions $\Gamma(N, \alpha_s)$. Possible examples studied within refs. [6–11] are:

$$
\begin{aligned}
\text{A}: \quad & \Gamma(N,\alpha_s) \to \Gamma(N,\alpha_s) - \Gamma(1,\alpha_s) \\
\text{B}: \quad & \Gamma(N,\alpha_s) \to \Gamma(N,\alpha_s)(1-N) \\
\text{C}: \quad & \Gamma(N,\alpha_s) \to \Gamma(N,\alpha_s)(1-2N+N^2) \\
\text{D}: \quad & \Gamma(N,\alpha_s) \to \Gamma(N,\alpha_s)(1-2N+N^3)\ ,
\end{aligned}
\tag{4}
$$

where $N \to N-1$ for the case of eq. (1). Clearly the presently known resummed terms are only sufficient for understanding the small-x evolution, if the difference of the results obtained by these prescriptions are small.

2.   RESUMMATION OF DOMINANT TERMS FOR $X \to 0$

2.1.   Unpolarized Non–Singlet Structure Functions

The numerical effects due to the resummation of the $O(\alpha_s \ln^2 x)$ terms (2) have been studied in refs. [6, 7] for the structure functions $F_2^{ep} - F_2^{en}$ and $xF_3^{\nu N}(x, Q^2)$ over a wide range of x and Q^2. The resummed terms beyond NLO lead to corrections on the level of 1% and below even at extremely small x. K–factors of about 10 as claimed in ref. [12] are not confirmed. Even moderate less-singular terms can alter the resummation correction by a factor of about 3.

2.2.   Unpolarized Singlet Structure Functions

The quantitative impact of the resummation of the leading small-x terms in the gluonic [1] and quarkonic anomalous dimensions [2] has been studied for the singlet quark (Σ) and gluon (g) distributions and the structure function F_2^{ep} in refs. [11] and [9, 10]. The latter analysis confirms the results of the former one. Related investigations were carried out in refs. [13–15].

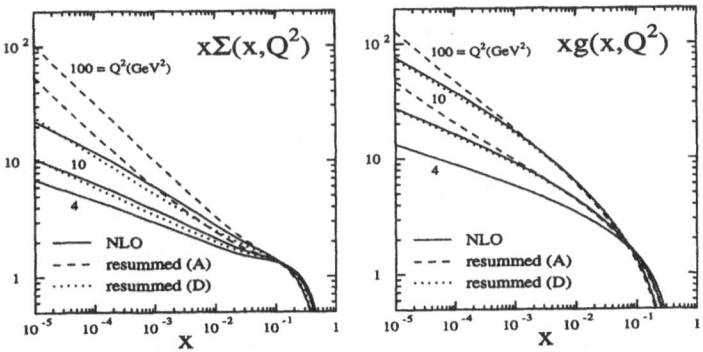

Fig. 1. — The evolution of the (DIS scheme) gluon and singlet quark densities with the resummed kernels [1, 2] as compared to the NLO results. Two prescriptions for implementing the momentum sum rule have been applied, cf. eq. (4).

In Fig. 1 the evolution of the MRS(A') initial distribution [16] at $Q_0^2 = 4$ GeV^2 is displayed. The effect of the resummation is very large. For the quarks and hence F_2 it is entirely dominated by the quarkonic pieces of ref. [2]. Two examples of additional less singular terms are shown. Their vital importance is obvious from the fact that the choice of (D) in eq. (4) leads to even an overcompensation of the enhancement caused by the leading small-x terms.

2.3.   Polarized Non–Singlet Structure Functions

This case was investigated in refs. [6, 7] numerically for the structure function combination $g_1^{ep} - g_1^{en}$ for two different parametrizations of the non-perturbative initial distributions, the results were recently confirmed in ref. [17]. The evolution of the interference structure function $g_{5,\gamma Z}^{ep}(x, Q^2)$ can be found in ref. [10]. As in the unpolarized case the corrections obtained are of the order of 1% with respect to the NLO results in the kinematic ranges experimentally accessible in the foreseeable future. Huge K–factors of about 10 or larger expected for this case in ref. [18] are not present. Again less singular terms in the anomalous dimensions are only marginally suppressed.

2.4. Polarized Singlet Structure Functions

Resummation relations for amplitudes related to the singlet anomalous dimensions of polarized DIS have recently been given [5]. Explicit analytical and numerical results for the evolution kernels beyond NLO have been derived on this basis in ref. [8], including an all-order symmetry relation among the elements of the anomalous dimension matrix and a discussion of a supersymmetric limit.

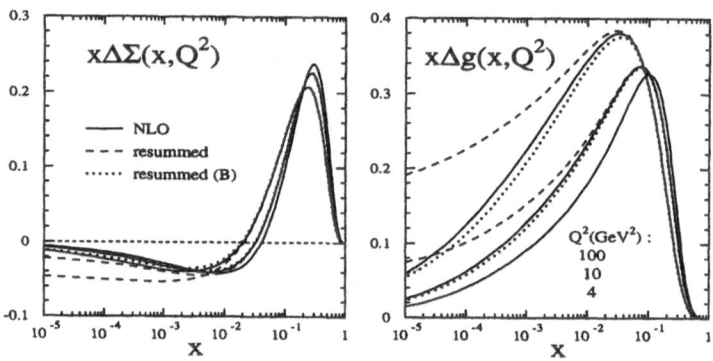

Fig. 2. — The resummed evolution of the polarized gluon and singlet densities as compared to the NLO results [8]. The possible impact of unknown less singular terms is illustrated by the curves (B). The initial distributions are taken from ref. [19].

Numerical results for the evolution of $g_1^{ep,en}(x, Q^2)$ and the parton densities have been given for different input distributions [8]. Fig. 2 shows an example. The situation is rather similar to the unpolarized case: the leading resummation effects are exceedingly large but unstable against less singular terms as $x \to 0$.

2.5. QED Non–Singlet Radiative Corrections

The resummation of the $O(\alpha \ln^2 x)$ terms may yield non–negligible contributions to QED corrections. This has been shown recently [20] for the case of initial state radiation for DIS at large y. There the effect reaches around 10% of the differential Born cross section. On the other hand, the corresponding corrections to $\sigma(e^+e^- \to \mu^+\mu^-)$ near the Z peak are completely irrelevant [20].

3. CONCLUSIONS

The resummations of the leading small-x terms in both unpolarized and polarized, non–singlet and singlet anomalous dimensions have been investigated recently. At NLO the results agree with those found for the most singular terms as $x \to 0$ in fixed order calculations. Since the coefficient functions are known up to $O(\alpha_s^2)$ in the $\overline{\text{MS}}$ scheme, predictions for the most singular terms of three–loop anomalous dimensions have been derived [2, 6–8].

For non–singlet structure functions the corrections due to the $\alpha_s (\alpha_s \ln^2 x)^l$ contributions are about 1% or smaller in the kinematic ranges probed so far and possibly accessible at HERA including polarization [6, 7]. The non–singlet QED corrections in deep-inelastic scattering resumming the $O(\alpha \ln^2 x)$ terms can reach values of about 10% at $x \approx 10^{-4}$ and $y > 0.9$ [20].

In the singlet case very large corrections are obtained for both unpolarized and polarized parton densities and structure functions [8–11]. As in the non–singlet cases possible less singular terms in higher order anomalous dimensions, however, which are in some cases required by conservation laws, are hardly suppressed against the presently resummed leading terms in the evolution: even a full compensation of the resummation effects cannot be excluded.

To draw firm conclusions on the small-x evolution of singlet structure functions also the next less singular terms have to be calculated. Since contributions even less singular than these ones may still cause relevant corrections, it appears to be indispensable to compare the corresponding results to those of future fixed order three–loop calculations.

References

[1] Balitsky Y. and Lipatov L., *Sov. J. Nucl. Phys.* **28** (1978) 822.

[2] Catani S. and Hautmann F., *Nucl. Phys.* **B427** (1994) 475.

[3] Camici G. and Ciafaloni M., *Phys. Lett.* **B386** (1996) 341.

[4] Kirschner R. and Lipatov L., *Nucl. Phys.* **B213** (1983) 122.

[5] Bartels J., Ermolaev B., and Ryskin M., *Z. Phys.* **C72** (1997) 627.

[6] Blümlein J. and Vogt A., *Phys. Lett.* **B370** (1996) 149.

[7] Blümlein J. and Vogt A., *Acta Phys. Polonica* **B27** (1996) 1309.

[8] Blümlein J. and Vogt A., *Phys. Lett.* **B386** (1996) 350.

[9] Blümlein J. and Vogt A., DESY 96-096.

[10] Blümlein J., Riemersma S., and Vogt A., *Nucl. Phys. B (Proc. Suppl.)* **51C** (1996) 30.

[11] Ellis K., Hautmann F. and Webber B., *Phys. Lett.* **B348** (1995) 582.

[12] Ermolaev B., Manayenkov S. and Ryskin M., *Z. Phys.* **C69** (1996) 259.

[13] Ball R. and Forte S., *Phys. Lett.* **B351** (1995) 313; **B358** (1995) 365.

[14] Forshaw J., Roberts R. and Thorne R., *Phys. Lett.* **B356** (1995) 79; Thorne R., *Phys. Lett.* **B392** (1997) 463; hep-ph/9701241.

[15] Bojak I. and Ernst M., *Phys. Lett.* **B397** (1997) 296; hep-ph/9702282.

[16] Martin A., Roberts R. and Stirling J., *Phys. Lett.* **B354** (1995) 155

[17] Kiyo J., Kodaira J. and Tochimura H., hep-ph/9701365.

[18] Bartels J., Ermolaev B. and Ryskin M., *Z. Phys.* **C70** (1996) 273.

[19] Glück M., Reya E., Stratmann M. and Vogelsang W., *Phys. Rev.* **D53** (1996) 4775.

[20] Blümlein J., Riemersma S. and Vogt A., hep-ph/9611214, *Z. Phys.* C (1997), in press.

CHAPTER VIII

Phenomenological Applications

LECTURE 28

The Light-Cone Fock State Expansion and QCD Phenomenology*

Stanley J. Brodsky

Stanford Linear Accelerator Center
Stanford University, Stanford, California 94309

---

1. INTRODUCTION

The concept of the "number of constituents" of a relativistic bound state, such as a hadron in quantum chromodynamics, is not only frame-dependent, but its value can fluctuate to an arbitrary number of quanta. Thus when a laser beam crosses a proton at fixed "light-cone" time $\tau = 3Dt + z/c = 3Dx^0 + x^z$, an interacting photon can encounter a state with any given number of quarks, anti-quarks, and gluons in flight (as long as $n_q - n_{\bar{q}} = 3D3$). The probability amplitude for each such n-particle state of on-mass shell quarks and gluons in a hadron is given by a light-cone Fock state wavefunction $\psi_{n/H}(x_i, \vec{k}_{\perp i}, \lambda_i)$, where the constituents have longitudinal light-cone momentum fractions

$$x_i = 3D\frac{k_i^+}{p^+} = 3D\frac{k^0 + k_i^z}{p^0 + p^z} \, , \quad \sum_{i=3D1}^{n} x_i = 3D1 \, , \tag{1}$$

relative transverse momentum

$$\vec{k}_{\perp i} \, , \quad \sum_{i=3D1}^{n} \vec{k}_{\perp i} = 3D\vec{0}_{\perp} \, , \tag{2}$$

and helicities λ_i. The ensemble $\{\psi_{n/H}\}$ of such light-cone Fock wavefunctions is a key concept for hadronic physics, providing a conceptual basis for representing

---

(*) Work supported by the Department of Energy, contract DE–AC03–76SF00515.

physical hadrons (and also nuclei) in terms of their fundamental quark and gluon degrees of freedom.[1]

The light-cone Fock expansion is defined in the following way: one first constructs the light-cone time evolution operator $P^- = 3DP^0 - P^z$ and the invariant mass operator $H_{LC} = 3DP^-P^+ - P_\perp^2$ in light-cone gauge $A^+ = 3D0$ from the QCD Lagrangian. The total longitudinal momentum $P^+ = 3DP^0 + P^z$ and transverse momenta $\vec{P}_\perp$ are conserved, i.e. are independent of the interactions. The matrix elements of H_{LC} on the complete orthonormal basis $\{|n>\}$ of the free theory $H_{LC}^0 = 3DH_{LC}(g = 3D0)$ can then be constructed. The matrix elements $\langle n | H_{LC} | m \rangle$ connect Fock states differing by 0, 1, or 2 quark or gluon quanta, and they include the instantaneous quark and gluon contributions imposed by eliminating dependent degrees of freedom in light-cone gauge.

In practice it is essential to introduce an ultraviolet regulator in order to limit the total range of $\langle n | H_{LC} | m \rangle$, such as the "global" cutoff in the invariant mass of the free Fock states:

$$\mathcal{M}_n^2 = 3D \sum_{i=3D1}^{n} \frac{k_\perp^2 + m^2}{x} < \Lambda_{\text{global}}^2 . \tag{3}$$

One can also introduce a "local" cutoff to limit the change in invariant mass $|\mathcal{M}_n^2 - \mathcal{M}_m^2| < \Lambda_{\text{local}}^2$ which provides spectator-independent regularization of the sub-divergences associated with mass and coupling renormalization.

The natural renormalization scheme for the coupling is $\alpha_V(Q)$, the effective charge defined from the scattering of two infinitely-heavy quark test charges. The renormalization scale can then be determined from the virtuality of the exchanged momentum, as in the BLM and commensurate scale methods. [2, 3, 4]

In the discretized light-cone method (DLCQ) [5, 6] the matrix elements $\left\langle n | H_{LC}^{\Lambda} | m \right\rangle$, are made discrete in momentum space by imposing periodic or anti-periodic boundary conditions in $x^- = 3Dx^0 - x^z$ and $\vec{x}_\perp$. Upon diagonalization of H_{LC}, the eigenvalues provide the invariant mass of the bound states and eigenstates of the continuum. The projection of the hadronic eigensolutions on the free Fock basis define the light-cone wavefunctions. For example, for the proton,

$$\begin{aligned} |p\rangle &= 3D \sum_n \langle n|p\rangle \, |n\rangle \\ &= 3D \, \psi_{3q/p}^{(\Lambda)}(x, \vec{k}_{\perp i}, \lambda_i) \, |uud\rangle \\ &\quad + \psi_{3qg/p}^{(\Lambda)}(x_i, \vec{k}_{\perp i}, \lambda_i) \, |uudg\rangle + \cdots \end{aligned} \tag{4}$$

The light-cone formalism has the remarkable feature that the $\psi_{n/H}^{(\Lambda)}(x_i, \vec{k}_{\perp i}, \Lambda_c)$ are invariant under longitudinal boosts; i.e., they are independent of the total momentum P^+, $\vec{P}_\perp$ of the hadron. Given the $\psi_{n/H}^{(\Lambda)}$, we can construct any electromagnetic or electroweak form factor from the diagonal overlap of the LC wavefunctions.[7] Similarly, the matrix elements of the currents that define quark and gluon structure functions can be computed from the integrated squares of the LC wavefunctions.[8]

In general, any hadronic amplitude such as quarkonium decay, heavy hadron decay, or any hard exclusive hadron process can be constructed as the convolution of the light-cone Fock state wavefunctions with quark-gluon matrix elements [9]

$$\mathcal{M}_{\text{Hadron}} = 3D \prod_H \sum_n \int \prod_{i=3D1}^n d^2 k_\perp \prod_{i=3D1}^n dx\, \delta\left(1 - \sum_{i=3D1}^n x_i\right) \delta\left(\sum_{i=3D1}^n \vec{k}_{\perp i}\right)$$

$$\times \psi_{n/H}^{(\Lambda)}(x_i, \vec{k}_{\perp i}, \Lambda_i)\, \mathcal{M}_{q,g}^{(\Lambda)} . \tag{5}$$

Here $\mathcal{M}_{q,g}^{(\Lambda)}$ is the underlying quark-gluon subprocess scattering amplitude, where the (incident or final) hadrons are replaced by quarks and gluons with momenta $x_i p^+$, $x_i \vec{p}_\perp + \vec{k}_{\perp i}$ and invariant mass above the separation scale $\mathcal{M}_n^2 > \Lambda^2$. The LC ultraviolet regulators thus provide a factorization scheme for elastic and inelastic scattering, separating the hard dynamical contributions with invariant mass squared $\mathcal{M}^2 > \Lambda_{\text{global}}^2$ from the soft physics with $\mathcal{M}^2 \leq \Lambda_{\text{global}}^2$ which is incorporated in the nonperturbative LC wavefunctions. The DGLAP evolution of parton distributions can be derived by computing the variation of the Fock expansion with respect to $\Lambda_{\text{global}}^2$. [9]

The simplest, but most fundamental, characteristic of a hadron in the light-cone representation, is the hadronic distribution amplitudes, [9] defined as the integral over transverse momenta of the valence (lowest particle number) Fock wavefunction; e.g. for the pion

$$\phi_\pi(x_i, Q) \equiv \int d^2 k_\perp \, \psi_{q\bar{q}/\pi}^{(Q)}(x_i, \vec{k}_{\perp i}, \lambda) \tag{6}$$

where the global cutoff Λ_{global} is identified with the resolution Q. The distribution amplitude controls leading-twist exclusive amplitudes at high momentum transfer, and it can be related to the gauge-invariant Bethe-Salpeter wavefunction at equal light-cone time $\tau = 3Dx^+$. The $\log Q$ evolution of the hadron distribution amplitude s $\phi_H(x_i, Q)$ can be derived from the perturbatively-computable tail of the valence light-cone wavefunction in the high transverse momentum regime.[9]

Light-cone quantization methods have had remarkable success in solving quantum field theories in one-space and one-time dimension—virtually any (1+1) quantum field theory can be solved using DLCQ. A beautiful example is "collinear" QCD: a variant of $QCD(3+1)$ defined by dropping all of interaction terms in H_{LC}^{QCD} involving transverse momenta. [10] Even though this theory is effectively two-dimensional, the transversely-polarized degrees of freedom of

the gluon field are retained as two scalar fields. Antonuccio and Dalley [11] have
used DLCQ to solve this theory. The diagonalization of H_{LC} provides not only
the complete bound and continuum spectrum of the collinear theory, including
the gluonium states, but it also yields the complete ensemble of light-cone Fock
state wavefunctions needed to construct quark and gluon structure functions
for each bound state. Although the collinear theory is a drastic approximation
to physical $QCD(3+1)$, the phenomenology of its DLCQ solutions demonstrate general gauge theory features, such as the peaking of the wavefunctions
at minimal invariant mass, color coherence and the helicity retention of leading
partons in the polarized structure functions at $x \to 1$.

2.   APPLICATIONS OF LIGHT-CONE METHODS TO QCD PHE-
NOMENOLOGY

Regge behavior. The light-cone wavefunctions $\psi_{n/H}$ of a hadron are not independent of each other, but rather are coupled via the equations of motion.
Recently Antonuccio, Dalley and I [12] have used the constraint of finite "mechanical" kinetic energy to derive "ladder relations" which interrelate the light-
cone wavefunctions of states differing by 1 or 2 gluons. We then use these
relations to derive the Regge behavior of both the polarized and unpolarized
structure functions at $x \to 0$, extending Mueller's derivation of the BFKL hard
QCD pomeron from the properties of heavy quarkonium light-cone wavefunctions at large N_C QCD. [13]

High momentum transfer exclusive reactions. Given the solution for the
hadronic wavefunctions $\psi_n^{(\Lambda)}$ with $\mathcal{M}_n^2 < \Lambda^2$, one can construct the wavefunction in the hard regime with $\mathcal{M}_n^2 > \Lambda^2$ using projection operator techniques. [9]
The construction can be done perturbatively in QCD since only high invariant
mass, far off-shell matrix elements are involved. One can use this method to derive the physical properties of the LC wavefunctions and their matrix elements
at high invariant mass. Since $\mathcal{M}_n^2 = 3D \sum_{i=3D1}^{n} \left(\frac{k_\perp^2 + m^2}{x} \right)_i$, this method also
allows the derivation of the asymptotic behavior of light-cone wavefunctions
at large $k_\perp$, which in turn leads to predictions for the fall-off of form factors
and other exclusive matrix elements at large momentum transfer, such as the
quark counting rules for predicting the nominal power-law fall-off of two-body
scattering amplitudes at fixed θ_{cm}. The phenomenological successes of these
rules can be understood within QCD if the coupling $\alpha_V(Q)$ freezes in a range
of relatively small momentum transfer. [14].

Analysis of diffractive vector meson photoproduction. The light-cone Fock
wavefunction representation of hadronic amplitudes allows a simple eikonal
analysis of diffractive high energy processes, such as $\gamma^*(Q^2)p \to \rho p$, in terms of
the virtual photon and the vector meson =46ock state light-cone wavefunctions
convoluted with the $gp \to gp$ near-forward matrix element.[15] One can easily
show that only small transverse size $b_\perp \sim 1/Q$ of the vector meson wavefunction
is involved. The hadronic interactions are minimal, and thus the $\gamma^*(Q^2)N \to$

ρN reaction can occur coherently throughout a nuclear target in reactions such as without absorption or shadowing. The $\gamma^* A \to \phi A$ process thus provides a natural framework for testing QCD color transparency. [16]

Structure functions at large x_{bj}. The behavior of structure functions where one quark has the entire momentum requires the knowledge of LC wavefunctions with $x \to 1$ for the struck quark and $x \to 0$ for the spectators. This is a highly off-shell configuration, and thus one can rigorously derive quark-counting and helicity-retention rules for the power-law behavior of the polarized and unpolarized quark and gluon distributions in the $x \to 1$ endpoint domain. It is interesting to note that the evolution of structure functions is minimal in this domain because the struck quark is highly virtual as $x \to 1$; *i.e.* the starting point Q_0^2 for evolution cannot be held fixed, but must be larger than a scale of order $(m^2 + k_\perp^2)/(1 - x)$. [8]

Intrinsic gluon and heavy quarks. The main features of the heavy sea quark-pair contributions of the Fock state expansion of light hadrons can also be derived from perturbative QCD, since $\mathcal{M}_n^2$ grows with m_Q^2. One identifies two contributions to the heavy quark sea, the "extrinsic" contributions which correspond to ordinary gluon splitting, and the "intrinsic" sea which is multiconnected via gluons to the valence quarks. The intrinsic sea is thus sensitive to the hadronic bound state structure.[17] The maximal contribution of the intrinsic heavy quark occurs at $x_Q \simeq m_{\perp Q}/\sum_i m_\perp$ where $m_\perp = 3D\sqrt{m^2 + k_\perp^2}$; *i.e.* at large x_Q, since this minimizes the invariant mass $\mathcal{M}_n^2$. The measurements of the charm structure function by the EMC experiment are consistent with intrinsic charm at large x in the nucleon with a probability of order $0.6 \pm 0.3\%$. [18] Similarly, one can distinguish intrinsic gluons which are associated with multi-quark interactions and extrinsic gluon contributions associated with quark substructure. [20] One can also use this framework to isolate the physics of the anomaly contribution to the Ellis-Jaffe sum rule.

Rearrangement mechanism in heavy quarkonium decay. It is usually taken for granted that a heavy quarkonium state such as the J/ψ decays to light hadrons via the annihilation of the heavy quark constituents to gluons. However, as Karliner and I [19] have recently shown, the transition $J/\psi \to \rho\pi$ can also occur by the rearrangement of the $c\bar{c}$ from the J/ψ into the $|q\bar{q}c\bar{c}>$ intrinsic charm Fock state of the ρ or π. On the other hand, the overlap rearrangement integral in the decay $\psi' \to \rho\pi$ will be suppressed since the intrinsic charm Fock state radial wavefunction of the light hadrons will evidently not have nodes in its radial wavefunction. This observation provides a natural explanation of the long-standing puzzle why the J/ψ decays prominently to two-body pseudoscalar-vector final states, whereas the ψ' does not.

Asymmetry of Intrinsic heavy quark sea. As Ma and I have noted, [21] the higher Fock state of the proton $|uuds\bar{s}>$ should resemble a $|K\Lambda>$ intermediate state, since this minimizes its invariant mass $\mathcal{M}$. In such a state, the strange quark has a higher mean momentum fraction x than the $\bar{s}$. [22, 21] Similarly, the helicity intrinsic strange quark in this configuration will be anti-aligned with the helicity of the nucleon. [21] This $Q \leftrightarrow \overline{Q}$ asymmetry is a remarkable, striking feature of the intrinsic heavy-quark sea.

Direct measurement of the light-cone valence wavefunction. Diffractive multi-jet production in heavy nuclei provides a novel way to measure the shape of the LC Fock state wavefunctions. For example, consider the reaction [23, 24]

$$\pi A \to \text{Jet}_1 + \text{Jet}_2 + A' \tag{7}$$

at high energy where the nucleus A' is left intact in its ground state. The transverse momenta of the jets have to balance so that $\vec{k}_{\perp i} + \vec{k}_{\perp 2} = 3D\vec{q}_\perp < \mathcal{R}_A^{-1}$, and the light-cone longitudinal momentum fractions have to add to $x_1 + x_2 \sim 1$ so that $\Delta p_L < \mathcal{R}_A^{-1}$. The process can then occur coherently in the nucleus. Because of color transparency; *i.e.* the cancellation of color interactions in a small-size color-singlet hadron; the valence wavefunction of the pion with small impact separation will penetrate the nucleus with minimal interactions, diffracting into jet pairs. [23] The $x_1 = 3Dx$, $x_2 = 3D1 - x=$ dependence of the di-jet distributions will thus reflect the shape of the pion distribution amplitude; the $\vec{k}_{\perp 1} - \vec{k}_{\perp 2}$ relative transverse momenta of the jets also gives key information on the underlying shape of the valence pion wavefunction. The QCD analysis can be confirmed by the observation that the diffractive nuclear amplitude extrapolated to $t = 3D0$ is linear in nuclear number A, as predicted by QCD color transparency. The integrated diffractive rate should scale as $A^2/\mathcal{R}_A^2 \, sim A^{4/3}$. A diffractive experiment of this type is now in progress at Fermilab using 500 GeV incident pions on nuclear targets. [25]

Data from CLEO for the $\gamma\gamma^* \to \pi^0$ transition form factor favor a form for the pion distribution amplitude close to the asymptotic solution [9] $\phi_\pi^{asympt}(x) = 3D\sqrt{3}f_\pi x(1 - x)$ to the perturbative QCD evolution equation.[26, 27, 14] It will be interesting to see if the diffractive pion to di-jet experiment also favors the asymptotic form.

It would also be interesting to study diffractive tri-jet production using proton beams $pA \to \text{Jet}_1 + \text{Jet}_2 + \text{Jet}_3 + A'$ to determine the fundamental shape of the 3-quark structure of the valence light-cone wavefunction of the nucleon at small transverse separation. Conversely, one can use incident real and virtual photons: $\gamma^* A \to \text{Jet}_1 + \text{Jet}_2 + A'$ to confirm the shape of the calculable light-cone wavefunction for transversely-polarized and longitudinally-polarized virtual photons. Such experiments will open up a remarkable, direct window on the amplitude structure of hadrons at short distances.

References

[1] For a review and further references, see S. J. Brodsky and D.G. Robertson SLAC-PUB-95-7056, to be published in the proceedings of ELFE (European Laboratory for Electrons) Summer School on Confinement Physics, Cambridge, England. e-Print Archive: hep-ph/9511374

[2] S. J. Brodsky, G. P. Lepage, and P. B. Mackenzie, Phys. Rev. D **28**, 1983, 228.

[3] H. J. Lu and S. J. Brodsky, Phys. Rev. D **48**, 1993, 3310.

[4] S. J. Brodsky, G. T. Gabadadze, A. L. Kataev and H. J. Lu, Phys. Lett. **372B**, 1996, 133.

[5] H. C. Pauli and S. J. Brodsky, Phys. Rev. **D32**, 1985, 2001.

[6] S. J. Brodsky and H. C. Pauli, SLAC-PUB-5558, published in Schladming 1991, Proceedings.

[7] S. J. Brodsky and S. D. Drell, Phys. Rev. **D22** 1980, 2236.

[8] S. J. Brodsky and G. P. Lepage, in *Perturbative Quantum Chromodynamics*, A. H. Mueller, Ed. (World Scientific, 1989).

[9] G. P. Lepage and S. J. Brodsky, Phys. Rev. **D22**,1980, 2157.

[10] S. Dalley, and I. R. Klebanov, Phys. Rev **D47**, 1993, 2517.

[11] F. Antonuccio and S. Dalley, Phys. Lett. **B376**, 1996, 154 e-Print Archive: hep-ph/9512106, and references therein.

[12] F. Antonuccio, S.J. Brodsky, and S. Dalley SLAC-PUB-7472, e-Print Archive: hep-ph/9705413

[13] A. H. Mueller, Nucl. Phys. **B415**, 1994, 373.

[14] S. J. Brodsky, C.-R. Ji, A. Pang, and D. G. Robertson, SLAC–PUB–7473. e-Print Archive: hep-ph/9705221.

[15] S. J. Brodsky, L. Frankfurt, J. F. Gunion, A. H. Mueller, and M. Strikman, Phys. Rev. **D50**, 1994, 3134. e-Print Archive: hep-ph/9402283

[16] S. J. Brodsky and A.H. Mueller, Phys. Lett. **206B**, 1988, 685.

[17] S. J. Brodsky, P. Hoyer, C. Peterson, and N. Sakai, Phys. Lett. **93B**, 1980, 451.

[18] B. W. Harris, J. Smith, and R. Vogt, Nucl. Phys. **B461**, 1996, 181. e-Print Archive: hep-ph/9508403

[19] S. J. Brodsky and M. Karliner SLAC-PUB-7463, e-Print Archive: hep-ph/9704379

[20] S. J. Brodsky and I. A. Schmidt, Phys. Lett.**B234**, 1990, 144.

[21] S. J. Brodsky and B-Q Ma, Phys. Lett. **B381**, 1996, 317. e-Print Archive: hep-ph/9604393

[22] A. I. Signal and A. W. Thomas, Phys. Lett. **191B**, 1987, 205.

[23] G. Bertsch, S.J. Brodsky, A.S. Goldhaber, and J.F. Gunion, Phys. Rev. Lett. **47**, 1981, 297.

[24] L. Frankfurt, G. A. Miller, and M. Strikman, Phys. Lett. **B304**, 1993,1. e-Print Archive: hep-ph/9305228

[25] Fermillab E791 Collaboration, in progress.

[26] P. Kroll and M. Raulfs, Phys. Lett. **B387**, 1996, 848.

[27] I. V. Musatov and A. V. Radyushkin, e-Print Archive: hep-ph/9702443

The Analog of t'Hooft Pions with Adjoint Fermions*

Stephen S. Pinsky

Department of Physics,
The Ohio State University,
Columbus, OH 43210

1. Introduction

In this paper we discuss and solve a class of $1 + 1$ dimensional matrix field theories in the light-cone Hamiltonian approach that are obtained from a dimensional reduction of $3 + 1$ dimensional Yang-Mills theory. Recently a similar procedure has been used to formulate a conjectured for M theory [1].

The strategy in formulating the model is to retain all of the essential degrees of freedom of higher dimensional QCD. We start by considering QCD_{3+1} coupled to Dirac adjoint fermions[2]. The virtual creation of fermion-antifermion pairs is not suppressed in the large-N limit – in contrast to the case for fermions in the fundamental representation [3, 4, 5] – and so one may study the structure of boundstates beyond the valence quark (or quenched) approximation.

The gauge group of the theory is actually $SU(N)/Z_N$, which has nontrivial topology and vacuum structure. For the particular gauge group $SU(2)$ this has been discussed elsewhere [6]. While this vacuum structure may in fact be relevant for a discussion of condensates, for the purposes of this calculation it will be ignored.

In the first section we formulate the $3 + 1$ dimensional $SU(N)$ Yang-Mills theory and then perform dimensional reduction to obtain a $1 + 1$ dimensional

(*) This work is done in collaboration with F. Antonuccio, Max Planck Institute Heidelberg, Germany

matrix field theory. The light-cone Hamiltonian is then derived for the light-cone gauge $A_- = 0$ following a discussion of the physical degrees of freedom of the theory. We then discuss the exact massless solutions of the boundstate integral equations. These massless states have constant wavefunctions in momentum space and are therefore fundamental excitations of the the theory.

2. SU(N) Yang-Mills Coupled to Adjoint Fermions: Definitions

We first consider $3+1$ dimensional $SU(N)$ Yang-Mills coupled to a Dirac spinor field whose components transform in the adjoint representation of $SU(N)$:

$$\mathcal{L} = \mathrm{Tr}\left[-\frac{1}{4}F_{\mu\nu}F^{\mu\nu} + \frac{i}{2}(\bar{\Psi}\gamma^\mu \overset{\leftrightarrow}{D}_\mu \Psi) - m\bar{\Psi}\Psi\right], \tag{1}$$

where $D_\mu = \partial_\mu + ig[A_\mu, \ ]$ and $F_{\mu\nu} = \partial_\mu A_\nu - \partial_\nu A_\mu + ig[A_\mu, A_\nu]$. We also write $A_\mu = A_\mu^a \tau^a$ where τ^a is normalized such that $\mathrm{Tr}(\tau^a\tau^b) = \delta_{ab}$. The projection operators[1] Λ_L, Λ_R permit a decomposition of the spinor field $\Psi = \Psi_L + \Psi_R$, where

$$\Lambda_L = \frac{1}{2}\gamma^+\gamma^-, \quad \Lambda_R = \frac{1}{2}\gamma^-\gamma^+ \quad \text{and} \quad \Psi_L = \Lambda_L\Psi, \quad \Psi_R = \Lambda_R\Psi. \tag{2}$$

Inverting the equation of motion for Ψ_L, we find

$$\Psi_L = \frac{1}{2iD_-}\left[i\gamma^i D_i + m\right]\gamma^+\Psi_R \tag{3}$$

where $i = 1, 2$ runs over transverse space. Therefore Ψ_L is not an independent degree of freedom.

Dimensional reduction of the $3+1$ dimensional Lagrangian (1) is performed by assuming (at the classical level) that all fields are independent of the transverse coordinates $x^\perp = (x^1, x^2)$: $\partial_\perp A_\mu = 0$ and $\partial_\perp \Psi = 0$. In the resulting $1+1$ dimensional field theory, the transverse components $A_\perp = (A_1, A_2)$ of the gluon field will be represented by the $N \times N$ complex matrix fields $\phi_\pm$:

$$\phi_\pm = \frac{A_1 \mp iA_2}{\sqrt{2}}. \tag{4}$$

Here, ϕ_- is just the Hermitian conjugate of ϕ_+. When the theory is quantized, $\phi_\pm$ will correspond to ± 1 helicity bosons (respectively).

The components of the Dirac spinor Ψ are the $N \times N$ *complex* matrices $u_\pm$ and $v_\pm$, which are related to the left and right-moving spinor fields according to

$$\Psi_R = \frac{1}{2^{\frac{1}{4}}}\begin{pmatrix} u_+ \\ 0 \\ 0 \\ u_- \end{pmatrix} \quad \Psi_L = \frac{1}{2^{\frac{1}{4}}}\begin{pmatrix} 0 \\ v_+ \\ v_- \\ 0 \end{pmatrix} \tag{5}$$

---

[1] We use the conventions $\gamma^\pm = (\gamma^0 \pm \gamma^3)/\sqrt{2}$, and $x^\pm = (x^0 \pm x^3)/\sqrt{2}$.

Adopting the light-cone gauge $A_- = 0$ allows one to explicitly rewrite the left-moving fermion fields $v_\pm$ in terms of the right-moving fields $u_\pm$ and boson fields $\phi_\pm$, by virtue of equation (3). We may therefore eliminate $v_\pm$ dependence from the field theory. Moreover, Gauss' Law

$$\partial_-^2 A_+ = g\left(i[\phi_+, \partial_-\phi_-] + i[\phi_-, \partial_-\phi_+] + \{u_+, u_+^\dagger\} + \{u_-, u_-^\dagger\} \right) \qquad (6)$$

permits one to remove any explicit dependence on A_+, and so the remaining *physical* degrees of freedom of the field theory are represented by the helicity $\pm\frac{1}{2}$ fermions $u_\pm$, and the helicity ± 1 bosons $\phi_\pm$. There are no ghosts in the quantization scheme adopted here. In the light-cone frame the Poincaré generators P^- and P^+ for the reduced $1+1$ dimensional field theory are given by

$$P^+ = \int_{-\infty}^{\infty} dx^- \text{Tr}\left[2\partial_-\phi_- \cdot \partial_-\phi_+ + \frac{i}{2}\sum_h \left(u_h^\dagger \cdot \partial_- u_h - \partial_- u_h^\dagger \cdot u_h \right) \right] \qquad (7)$$

$$P^- = \int_{-\infty}^{\infty} dx^- \text{Tr}\left[m_b^2 \phi_+\phi_- - \frac{g^2}{2} J^+ \frac{1}{\partial_-^2} J^+ + \frac{tg^2}{2}[\phi_+, \phi_-]^2 + \sum_h F_h^\dagger \frac{1}{i\partial_-} F_h^\dagger \right] \qquad (8)$$

where the sum $\sum_h$ is over $h = \pm$ helicity labels, and

$$J^+ = i[\phi_+, \partial_-\phi_-] + i[\phi_-, \partial_-\phi_+] + \{u_+, u_+^\dagger\} + \{u_-, u_-^\dagger\} \qquad (9)$$

$$F_\pm = \mp sg[\phi_\pm, u_\mp] + \frac{m}{\sqrt{2}}u_\pm \qquad (10)$$

We have generalized the couplings by introducing the variables t and s, which do not spoil the $1+1$ dimensional gauge invariance of the reduced theory; the variable t will determine the strength of the quartic-like interactions, and the variable s will determine the strength of the Yukawa interactions between the fermion and boson fields, and appears explicitly in equation (10). The dimensional reduction of the original $3+1$ dimensional theory yields the canonical values $s = t = 1$.

Renormalizability of the reduced theory also requires the addition of a bare coupling m_b, which leaves the $1+1$ dimensional gauge invariance intact. In all calculations, the renormalized boson mass $\tilde{m}_b$ will be set to zero.

Canonical quantization of the field theory is performed by decomposing the boson and fermion fields into Fourier expansions at fixed light-cone time $x^+ = 0$:

$$u_\pm = \frac{1}{\sqrt{2\pi}}\int_{-\infty}^{\infty} dk\, b_\pm(k)e^{-ikx^-} \quad \text{and} \quad \phi_\pm = \frac{1}{\sqrt{2\pi}}\int_{-\infty}^{\infty} \frac{dk}{\sqrt{2|k|}} a_\pm(k)e^{-ikx^-} \qquad (11)$$

where $b_\pm = b_\pm^a \tau^a$ etc. We also define

$$b_\pm(-k) = d_\mp^\dagger(k), \quad a_\pm(-k) = a_\mp^\dagger(k), \qquad (12)$$

where $d_\pm$ correspond to antifermions. Note that $(b_\pm^\dagger)_{ij}$ should be distinguished from $b_{\pm ij}^\dagger$, since in the former the quantum conjugate operator $\dagger$ acts on (color) indices, while it does not in the latter. The latter formalism is sometimes customary in the study of matrix models. The precise connection between the usual gauge theory and matrix theory formalism may be stated as follows:

$$b_{\pm ji}^\dagger = b_\pm^{a\dagger} \tau_{ji}^{a*} = b_\pm^{a\dagger} \tau_{ij}^a = (b_\pm^\dagger)_{ij}$$

The commutation and anti-commutation relations (in matrix formalism) for the boson and fermion fields take the following form in the large-N limit ($k, \tilde{k} > 0$; $h, h' = \pm$):

$$\left[a_{hij}(k), a_{h'kl}^\dagger(\tilde{k}) \right] = \{ b_{hij}(k), b_{h'kl}^\dagger(\tilde{k}) \} = \{ d_{hij}(k), d_{h'kl}^\dagger(\tilde{k}) \} = \delta_{hh'} \delta_{jl} \delta_{ik} \delta(k - \tilde{k}),$$

$$(13)$$

where we have used the relation $\tau_{ij}^a \tau_{kl}^a = \delta_{il} \delta_{jk} - \frac{1}{N} \delta_{ij} \delta_{kl}$. All other (anti)commutators vanish.

The Fock space of physical states is generated by the color singlet states, which have a natural 'closed-string' interpretation. They are formed by a color trace of the fermion, antifermion and boson operators acting on the vacuum state $|0\rangle$. Multiple string states couple to the theory with strength $1/N$, and so may be ignored.

3. The Light-Cone Hamiltonian

The current-current term $J^+ \frac{1}{\partial_-^2} J^+$ in equation (8) in momentum space, for example, takes the form

$$
\begin{aligned}
P_{J^+ \cdot J^+}^- = {} & \frac{g^2}{2\pi} \int_{-\infty}^{\infty} dk_1 dk_2 dk_3 dk_4 \frac{\delta(k_1 + k_2 - k_3 - k_4)}{(k_3 - k_1)^2} \frac{\mathrm{Tr}}{2} \Bigg[\\
& \sum_{h,h'} : \{ b_h^\dagger(k_1), b_h(k_3) \} :: \{ b_{h'}^\dagger(k_2), b_{h'}(k_4) \} : \\
& + \frac{(k_1 + k_3)(k_2 + k_4)}{4\sqrt{|k_1||k_2||k_3||k_4|}} : [a_+^\dagger(k_1), a_+(k_3)] :: [a_+^\dagger(k_2), a_+(k_4)] : \\
& + \frac{(k_2 + k_4)}{2\sqrt{|k_2||k_4|}} \sum_h : \{ b_h^\dagger(k_1), b_h(k_3) \} :: [a_+^\dagger(k_2), a_+(k_4)] : \\
& + \frac{(k_3 + k_1)}{2\sqrt{|k_1||k_3|}} \sum_{h'} : [a_+^\dagger(k_1), a_+(k_3)] :: \{ b_{h'}^\dagger(k_2), b_{h'}(k_4) \} : \Bigg] \quad (14)
\end{aligned}
$$

The explicit form of remaining terms of the Hamiltonian (14) in terms of the operators $b_\pm$, $d_\pm$ and $a_\pm$ is straightforward to calculate, but too long to be written down here. It should be stressed, however, that several $2 \to 2$ parton processes are suppressed by a factor $1/N$, and so are ignored in the large-N

limit. No terms involving $1 \leftrightarrow 3$ parton interactions are suppressed in this limit, however.

One can show that this Hamiltonian conserves total helicity h, which is an additive quantum number. Moreover, the number of fermions *minus* the number of antifermions is also conserved in each interaction, and so we have an additional quantum number $\mathcal{N}$. States with $\mathcal{N} = even$ will be referred to as *boson* boundstates, while the quantum number $\mathcal{N} = odd$ will refer to *fermion* boundstates. The cases $\mathcal{N} = 0$ and 3, are analogous to conventional mesons and baryons (respectively).

4. Exact Solutions

For the special case $s = m = \tilde{m}_b = 0$, the only surviving terms in the Hamiltonian (8) are the current-current interactions $J^+ \frac{1}{\partial^2} J^+$ and the four gluon interaction $[\phi_+, \phi_-]^2$. The current-current interaction includes four gluon interaction, four Fermion interaction and two gluon, two Fermion interactions. This theory has infinitely many massless boundstates, and the partons in these states are either fermions or antifermions. States with bosonic $a_\pm$ quanta are always massive. One also finds that the massless states are pure, in the sense that the number of partons is a fixed integer, and there is no mixing between sectors of different parton number. In particular, for each integer $n \geq 2$, one can always find a massless boundstate consisting of a superposition of only n-parton states. A striking feature is that the wavefunctions of these states are *constant*, and so these states are natural generalizations of the constant wavefunction solution appearing in t'Hooft's model [7].

We present an explicit examples below of such a constant wavefunction solution involving a three fermion state with total helicity $+\frac{3}{2}$, which is perhaps the simplest case to study. We apply P^- to the state and show that the result vanishes. Massless states with five or more partons appear to have more than one wavefunction which are non-zero and constant, and in general the wavefunctions are unequal. It would be interesting to classify all states systematically, and we leave this to future work. One can, however, easily count the number of massless states. In particular, for $\mathcal{N} = 3$, $h = +\frac{3}{2}$ states, there is one three-parton state, 2 five-parton states, 14 seven-parton states and 106 nine-parton states that yield massless solutions.

Let us now consider the action of the light-cone Hamiltonian P^- on the three-parton state

$$|b_+ b_+ b_+\rangle = \int_0^\infty dk_1 dk_2 dk_3 \ \delta(\sum_{i=1}^3 k_i - P^+)$$

$$f_{b_+ b_+ b_+}(k_1, k_2, k_3) \frac{1}{N^{3/2}} \text{Tr}[b_+^\dagger(k_1) b_+^\dagger(k_2) b_+^\dagger(k_3)]|0\rangle \qquad (15)$$

The quantum number $\mathcal{N}$ is 3 in this case, and ensures that the state $P^-|b_+ b_+ b_+\rangle$

must have at least three partons. In fact, one can deduce the following:

$$P^- \mid b_+ b_+ b_+ \rangle = \int_0^\infty dk_1 dk_2 dk_3 \; \delta(\sum_{i=1}^{3} k_i - P^+) \times$$

$$\left\{ -\frac{g^2 N}{2\pi} \int_0^\infty d\alpha d\beta \frac{\delta(\alpha + \beta - k_1 - k_2)}{(\alpha - k_1)^2} \left[f_{b_+ b_+ b_+}(\alpha, \beta, k_3) - f_{b_+ b_+ b_+}(k_1, k_2, k_3) \right] \right.$$

$$\frac{1}{N^{3/2}} \text{Tr} \left[b_+^\dagger(\alpha) b_+^\dagger(\beta) b_+^\dagger(k_3) \right] \mid 0\rangle$$

$$+ \frac{g^2 N}{2\pi} \int_0^\infty d\alpha d\beta d\gamma \sum_h \frac{\delta(\alpha + \beta + \gamma - k_1)}{(\alpha + \beta)^2} f_{b_+ b_+ b_+}(\alpha + \beta + \gamma, k_2, k_3) \frac{1}{N^{5/2}}$$

$$\text{Tr} \left[\{ b_h^\dagger(\alpha), d_{-h}^\dagger(\beta) \} b_+^\dagger(\gamma) b_+^\dagger(k_2) b_+^\dagger(k_3) - \{ b_h^\dagger(\alpha), d_{-h}^\dagger(\beta) \} b_+^\dagger(k_2) b_+^\dagger(k_3) b_+^\dagger(\gamma) \right] \mid 0\rangle$$

$$+ \frac{g^2 N}{4\pi} \int_0^\infty d\alpha d\beta d\gamma \sum_h \frac{\delta(\alpha + \beta + \gamma - k_1)}{\sqrt{\alpha\beta}(\alpha + \beta)^2} f_{b_+ b_+ b_+} \frac{1}{N^{5/2}} (\alpha + \beta + \gamma, k_2, k_3)$$

$$\text{Tr} \left[[a_h^\dagger(\alpha), a_{-h}^\dagger(\beta)] b_+^\dagger(\gamma) b_+^\dagger(k_2) b_+^\dagger(k_3) - [a_h^\dagger(\alpha), a_{-h}^\dagger(\beta)] b_+^\dagger(k_2) b_+^\dagger(k_3) b_+^\dagger(\gamma) \right] \mid 0\rangle$$

$$+ \quad \text{cyclic permutations} \Bigg\} \tag{16}$$

The five-parton states above correspond to virtual fermion-antifermion and boson-boson pair creation. The expression (16) vanishes if the wavefunction $f_{b_+ b_+ b_+}$ is constant.

One of the exact seven particle wave functions is

$$\int_0^\infty dk_1 dk_2 dk_3 dk_4 dk_5 dk_6 dk_7 \; \delta(\sum_{i=1}^{7} k_i - P^+) \frac{1}{N^{7/2}} \frac{1}{\sqrt{2}}$$

$$\text{Tr}[b_+^\dagger(k_1) b_+^\dagger(k_2) b_+^\dagger(k_3) b_-^\dagger(k_4) b_-^\dagger(k_5) d_+^\dagger(k_6) d_+^\dagger(k_7) - \tag{17}$$

$$b_+^\dagger(k_1) b_+^\dagger(k_2) b_+ \dagger(k_3) b_-^\dagger(k_4) d_-^\dagger(k_5) b_+^\dagger(k_6) d_+^\dagger(k_7)] \mid 0\rangle$$

5.   Conclusions

We have presented a non-perturbative Hamiltonian formulation of a class of $1 + 1$ dimensional matrix field theories, which may be derived from a classical dimensional reduction of QCD_{3+1} coupled to Dirac adjoint fermions. We choose to adopt the light-cone gauge $A_- = 0$, and solve the boundstate integral equations in the large-N limit. Different states may be classified according to total helicity h, and the quantum number $\mathcal{N}$, which defines the number of fermions minus the number of antifermions in a state.

For a special choice of couplings we find an infinite number of pure massless states of arbitrary length. The wavefunctions of these states are always constant, and are the analogs of the t' Hooft pion. Sometimes the wavefunctions involves several (possibly different) constants. In this model we have both fermion and boson states of this type. The state explicitly shown here are all fermions.

The techniques employed here are not specific to the choice of field theory, and are expected to have a wide range of applicability, particularly in the light-cone Hamiltonian formulation of supersymmetric field theories[8, 9].

Acknowledgments

The work was supported in part by a grant from the US Department of Energy. Travel support was provided in part by a NATO collaborative grant.

References

[1] Banks T., Fischler W., Shenker S.H. and Susskind L. Phys. Rev. **D55** 5112, 1997 hep-th/9610043; Susskind L. "Another Conjecture about M Theory" hep-th/9704080

[2] Antonuccio F. and Pin S. Phys. Lett. **B397** 42, 1997 hep-th/9612021

[3] Antonuccio F. and Dalley S., Phys. Lett. **B376** (1996) 154-162.

[4] Antonuccio F. and Dalley S., Nucl. Phys. **B461** (1996) 275-301.

[5] Burkardt M. and van de Sande B., Phys. Rev. **D53** 4628, 1995, hep-th/9510104; Dalley S. and van de Sande B. Color-Delectric Gauge Theory on a Transverse Lattice heop-ph/9704408

[6] Pinsky S.S. and Robertson D.G. Phys. Lett. B379 (1996) 169 ; Pinsky S. and Mohr R., " The Condensate for $SU(2)$ Int. J. Mod. Phys. **A12** (1997) 1063 ; Pinsky S.S., "(1+1)-Dimensional Yang-Maills Theory Coupled to Adjoint Fermions on the Light Front" hep-th/9612073

[7] 't Hooft G., Nucl. Phys. **B75**, 461 (1974).

[8] Matsumura Y. and Sakai N., Phys. Rev. **D52** (1995) 2446 hep-th/9504150.

[9] Hashimoto A. and Klebanov I.R., Mod. Phys. Lett. **A10** (1995) 2639 hep-th/9507062.

Physical Coupling Schemes and QCD Exclusive Processes

David G. Robertson

Department of Physics
The Ohio State University
Columbus, OH 43210, USA

One of the most important problems in making reliable predictions in perturbative QCD is dealing with the dependence of the truncated perturbative series on the choice of renormalization scale μ and scheme s for the QCD coupling $\alpha_s(\mu)$. Consider a physical quantity $\mathcal{O}$, computed in perturbation theory and truncated at next-to-leading order (NLO) in α_s:

$$\mathcal{O} = \alpha_s(\mu) \left[1 + (A_1 + B_1 n_f) \frac{\alpha_s(\mu)}{\pi} + \cdots \right] , \qquad (1)$$

where n_f is the effective number of quark flavors. The finite-order expression depends on both μ and the choice of scheme used to define the coupling. In fact, Eq. (1) can be made to take on essentially any value by varying μ and the renormalization scheme, which are *a priori* completely arbitrary. The scale/scheme problem is that of choosing μ and the scheme s in an "optimal" way, so that an unambiguous theoretical prediction, ideally including some plausible estimate of theoretical uncertainties, can be made.([1])

([1]) The precise meaning of "optimal" in this context is connected to the minimization of remainders for the truncated series. As is well known, perturbation series in QCD are asymptotic, and thus there is an optimum number of terms that should be computed for a given observable. In general, very little is known about the remainders in pQCD; however, if we assume that pQCD series are sign-alternating, then the remainder can be estimated by the first neglected (or last included) term. This term

For any given observable there is no rigorously correct way to make this choice in general. However, a particular prescription may be supported to a greater or lesser degree by general theoretical arguments and, *a posteriori*, by its success in practical applications. From these perspectives, a particularly successful method for choosing the renormalization scale is that proposed by Brodsky, Lepage and MacKenzie [1]. In the BLM procedure, the renormalization scales are chosen such that all vacuum polarization effects from fermion loops are absorbed into the running couplings. A principal motivation for this choice is that it reduces to the correct prescription in the case of Abelian gauge theory. Furthermore, the BLM scales are physical in the sense that they typically reflect the mean virtuality of the gluon propagators. Another important advantage of the method is that it "pre-sums" the large and strongly divergent terms in the pQCD series which grow as $n!(\alpha_s \beta_0)^n$, i.e., the infrared renormalons associated with coupling constant renormalization.

Dependence on the renormalization scheme can be avoided by considering relations between physical observables only. By the general principles of renormalization theory, such a relation must be independent of any theoretical conventions, in particular the choice of scheme in the definition of α_s. A relation between physical quantities in which the BLM method has been used to fix the renormalization scales is known as a "commensurate scale relation" (CSR) [2]. An important example is the generalized Crewther relation [2, 3], in which the radiative corrections to the Bjorken sum rule for deep inelastic lepton-proton scattering at a given momentum transfer Q are predicted from measurements of the e^+e^- annihilation cross section at a commensurate energy scale $\sqrt{s} \propto Q$.

In this talk I summarize recent applications of the BLM procedure to obtain CSRs relating QCD exclusive amplitudes to other observables, in particular the heavy quark potential $V(Q^2)$ [4]. As we shall see, the heavy quark potential can be used to define a physical coupling scheme which is quite natural for perturbative calculations. It may also be useful in the context of a nonperturbative formulation of QCD based on light-cone quantization.

BLM SCALE FIXING

The term involving n_f in Eq. (1) arises solely from quark loops in vacuum polarization diagrams. In QED these are the only contributions responsible for the running of the coupling, and thus it is natural to absorb them into the definition of the coupling. The BLM procedure is the analog of this approach in QCD. Specifically, we rewrite Eq. (1) in the form

$$\mathcal{O} = \alpha_s(\mu) \left[1 + \left(\frac{3\beta_0 B_1}{2} \right) \frac{\alpha_s(\mu)}{\pi} \right] \left[1 + \left(A_1 + \frac{33 B_1}{2} \right) \frac{\alpha_s(\mu)}{\pi} \right] , \quad (2)$$

---

can take on essentially any value, however, by simply varying the scale and scheme, and thus its minimization is meaningless without invoking additional criteria.

correct to order α_s^2, where $\beta_0 = 11 - 2n_f/3$ is the lowest-order QCD beta function. The first term in square brackets can then be absorbed by a redefinition of the renormalization scale in the leading-order coupling, using

$$\alpha_s(\mu^*) = \alpha_s(\mu) \left[1 - \frac{\beta_0 \alpha_s(\mu)}{2\pi} \ln(\mu^*/\mu) + \cdots \right] . \qquad (3)$$

That is, the BLM procedure consists of defining the prediction for $\mathcal{O}$ at this order to be

$$\mathcal{O} = \alpha_s(\mu^*) \left[1 + \left(A_1 + \frac{33B_1}{2} \right) \frac{\alpha_s(\mu^*)}{\pi} + \cdots \right] , \qquad (4)$$

where

$$\mu^* \equiv \mu e^{3B_1} . \qquad (5)$$

Note that knowledge of the NLO term in the expansion is necessary to fix the scale at LO. The scale occurring in the highest term in the expansion will in general be unknown. A natural prescription is to set this scale to be the same as that in the next-to-highest-order term.

A very important feature of this prescription is that μ^* is actually independent of μ. (This follows from considering the μ dependence of B_1.) Thus pQCD predictions using the BLM procedure are unambiguous.

The same basic idea can be extended to higher orders, by systematically shifting n_f dependence into the renormalization scales order by order. The result is that a generic perturbative expansion

$$\frac{\alpha_s(\mu)}{\pi} + (A_1 + B_1 n_f) \left(\frac{\alpha_s(\mu)}{\pi} \right)^2 + (A_2 + B_2 n_f + C_2 n_f^2) \left(\frac{\alpha_s(\mu)}{\pi} \right)^3 + \cdots \quad (6)$$

is replaced by a series of the form

$$\frac{\alpha_s(\mu^*)}{\pi} + \tilde{A}_1 \left(\frac{\alpha_s(\mu^{**})}{\pi} \right)^2 + \tilde{A}_2 \left(\frac{\alpha_s(\mu^{***})}{\pi} \right)^3 + \cdots . \qquad (7)$$

In general a different scale appears at each order in perturbation theory. In addition, the coefficients $\tilde{A}_n$ are constructed to be independent of n_f, and so the form of the expansion is unchanged as momenta vary across quark mass thresholds. All effects due to quark loops in vacuum polarization diagrams are automatically incorporated into the effective couplings.

As discussed above, one motivation for this prescription is that it reduces to the correct result in the case of QED. In addition, when combined with the idea of commensurate scale relations (see below), the BLM method can be shown to be consistent with the generalized renormalization group invariance of Stückelberg and Peterman, in which one considers "flow equations" both in μ and in the parameters that define the scheme.

To avoid scheme dependence, it is convenient to introduce physical effective charges, defined via some convenient observable, for use as an expansion parameter. An expansion of a physical quantity in terms of such a charge is a relation between observables and therefore must be independent of theoretical conventions, such as the renormalization scheme, to any fixed order of perturbation theory. In practice, a CSR relating two observables A and B is obtained by applying BLM scale-fixing to their respective perturbative predictions in, say, the $\overline{MS}$ scheme, and then algebraically eliminating $\alpha_{\overline{MS}}$.

A particularly useful scheme is furnished by the heavy quark potential $V(Q^2)$, which can be identified as the two-particle-irreducible amplitude for the scattering of an infinitely heavy quark and antiquark at momentum transfer $t = -Q^2$. The relation

$$V(Q^2) = -\frac{4\pi C_F \alpha_V(Q)}{Q^2} \, , \tag{8}$$

with $C_F = (N_C^2 - 1)/2N_C = 4/3$, then defines the effective charge $\alpha_V(Q)$. This coupling provides a physically-based alternative to the usual $\overline{MS}$ scheme. Another useful charge is provided by the total $e^+e^- \to X$ cross section, via the definition $R(s) \equiv 3\Sigma e_q^2 \left(1 + \alpha_R(\sqrt{s})/\pi\right)$. The CSR relating α_V and α_R is

$$\alpha_V(Q_V) = \alpha_R(Q_R) \left(1 - \frac{25}{12}\frac{\alpha_R}{\pi} + \cdots\right) , \tag{9}$$

where the ratio of commensurate scales is $Q_R/Q_V = e^{23/12 - 2\zeta_3} \simeq 0.614$ [4].

Physical couplings like α_V are of course renormalization-group-invariant, i.e. $\mu \partial \alpha_V / \partial \mu = 0$. However, the dependence of $\alpha_V(Q)$ on Q is controlled by an equation which is formally identical to the usual RG equation. Since α_V is dimensionless we must have

$$\alpha_V = \alpha_V \left(\frac{Q}{\mu}, \alpha_s(\mu)\right) . \tag{10}$$

Then $\mu \partial \alpha_V / \partial \mu = 0$ implies

$$Q\frac{\partial}{\partial Q}\alpha_V(Q) = \beta_s(\alpha_s)\frac{\partial \alpha_V}{\partial \alpha_s} \equiv \beta_V(\alpha_V) , \tag{11}$$

where

$$\beta_s = \mu\frac{\partial}{\partial \mu}\alpha_s(\mu) . \tag{12}$$

This is formally a change of scheme, so that the first two coefficients $\beta_0 = 11 - 2n_f/3$ and $\beta_1 = 102 - 38n_f/3$ in the perturbative expansion of β_V are the standard ones.

EXCLUSIVE AMPLITUDES AND $V(Q^2)$

Exclusive processes are particularly challenging to compute in QCD because of their sensitivity to the unknown nonperturbative bound state dynamics of the hadrons. However, there is an extraordinary simplification which occurs when the hadrons are forced to absorb a large momentum transfer Q: one can separate the nonperturbative long-distance physics associated with hadron structure from the short-distance quark-gluon hard scattering amplitudes responsible for the dynamical reaction. A meson form factor, for example, factorizes to leading order in $1/Q$ in the form

$$F_M(Q^2) = \int_0^1 dx \int_0^1 dy \phi_M(x,\tilde{Q}) T_H(x,y,Q^2) \phi_M(y,\tilde{Q}) , \qquad (13)$$

where $\phi_M(x,\tilde{Q})$ is the meson distribution amplitude, which encodes the nonperturbative dynamics of the bound valence Fock state up to the resolution scale $\tilde{Q}$, and T_H is the leading-twist perturbatively-calculable amplitude for the subprocess $\gamma^* q(x)\bar{q}(1-x) \to q(y)\bar{q}(1-y)$, in which the incident and final mesons are replaced by valence quarks collinear up to the resolution scale $\tilde{Q}$. Contributions from nonvalence Fock states and the correction from neglecting the transverse momenta in the subprocess amplitude from the nonperturbative regime are higher twist, i.e., power-law suppressed. The transverse momenta in the perturbative domain lead to the evolution of the distribution amplitude in $\tilde{Q}$ and to NLO corrections in α_s. For further details and references see [5].

It is straightforward to obtain CSRs relating exclusive amplitudes to, e.g., the heavy quark potential. For the pion form factor, for example, we find [4]

$$F_\pi(Q^2) = \int_0^1 dx \phi_\pi(x) \int_0^1 dy \phi_\pi(y) \frac{4\pi C_F \alpha_V(Q_V^*)}{(1-x)(1-y)Q^2} \left(1 + C_V \frac{\alpha_V(Q_V^*))}{\pi}\right) , \qquad (14)$$

where $C_V = -1.91$ and $Q_V^{*2} = (1-x)(1-y)Q^2$ is the virtuality of the exchanged gluon in the underlying hard scattering amplitude. Eq. (14) represents a general connection between the form factor of a bound-state system and the irreducible kernel that describes the scattering of its constituents.

If we expand the QCD coupling about a fixed point [4], and assume that the pion distribution amplitude has the asymptotic form $\phi_\pi(x) = \sqrt{3} f_\pi x(1-x)$, with $f_\pi \simeq 93$ MeV, then the integral over the effective charge in Eq. (14) can be performed explicitly. We thus find

$$Q^2 F_\pi(Q^2) = 16\pi f_\pi^2 \alpha_V(e^{-3/2}Q) \left(1 - 1.91\frac{\alpha_V}{\pi}\right) \qquad (15)$$

for the asymptotic distribution amplitude. In addition

$$Q^2 F_{\gamma\pi}(Q^2) = 2f_\pi \left(1 - \frac{5}{3}\frac{\alpha_V(e^{-3/2}Q)}{\pi}\right) \qquad (16)$$

for the $\gamma \to \pi^0$ transition form factor. A further prediction resulting from the factorized form of these results is that the normalization of the ratio

$$R_\pi(Q^2) \equiv \frac{F_\pi(Q^2)}{4\pi Q^2 |F_{\pi\gamma}(Q^2)|^2} \qquad (17)$$

$$= \alpha_V(e^{-3/2}Q) \left(1 + 1.43\frac{\alpha_V}{\pi}\right) \qquad (18)$$

is formally independent of the form of the pion distribution amplitude. The NLO correction given here assumes the asymptotic distribution amplitude.

A striking feature of these results is that the physical scale controlling the form factor in the α_V scheme is very low: $e^{-3/2}Q \simeq 0.22Q$, reflecting the characteristic momentum transfer experienced by the spectator valence quark in lepton-meson elastic scattering. In order to compare these expressions to data, therefore, we require an Ansatz for α_V at low scales. In Ref. [4] we consider a parameterization of the form

$$\alpha_V(Q) = \frac{4\pi}{\beta_0 \ln\left(\frac{Q^2+4m_g^2}{\Lambda_V^2}\right)}, \tag{19}$$

which effectively freezes α_V to a constant value for $Q^2 \leq 4m_g^2$. A primary motivation for this is the observation that the data for exclusive amplitudes such as form factors, two-photon processes such as $\gamma\gamma \to \pi^+\pi^-$, and photoproduction at fixed $\theta_{c.m.}$ are consistent with the nominal scaling of the leading-twist QCD predictions for momentum transfers down to a few GeV. This can be immediately understood if α_V is slowly varying at low momentum. The scaling of the exclusive amplitude then follows that of the subprocess amplitude T_H with effectively fixed coupling.

The parameters Λ_V and m_g^2 are determined by fitting to a lattice determination of $V(Q^2)$ [6] and to a value of α_R advocated in [7] using Eq. (9). We find $\Lambda_V \simeq 0.16$ GeV and $m_g^2 \simeq 0.2$ GeV^2. With these values, the prediction for $F_{\gamma\pi}$ is in excellent agreement with the data for Q^2 in the range 2–10 GeV^2. We also reproduce the scaling and normalization of the $\gamma\gamma \to \pi^+\pi^-$ cross section at large momentum transfer. However, the normalization of the space-like pion form factor $F_\pi(Q^2)$ obtained from electroproduction experiments is somewhat higher than that predicted by Eq. (15). This discrepancy may actually be due to systematic errors introduced by the extrapolation of the $\gamma^*p \to \pi^+n$ data to the pion pole. What is at best measured in electroproduction is the transition amplitude between a mesonic state with an effective space-like mass $m^2 = t < 0$ and the physical pion. It is theoretically possible that the off-shell form factor $F_\pi(Q^2, t)$ is significantly larger than the physical form factor because of its bias towards more point-like $q\bar{q}$ valence configurations in its Fock state structure. These and related issues are discussed elsewhere [8].

In any case, we find no compelling argument for significant higher-twist contributions in the few-GeV regime from the hard scattering amplitude or the endpoint regions, since such corrections would violate the observed scaling behavior of the data.

SUMMARY

As we have emphasized, the α_V scheme is quite natural when analyzing QCD processes perturbatively. By definition, it automatically incorporates quark (as well as the corresponding gluon) vacuum polarization contributions into the coupling; thus the coefficients in a perturbative expansion do not change as momenta vary across quark mass thresholds. It is directly connected to one of the most useful observables in QCD, the heavy quark potential, which is accessible on the lattice as well as phenomenologically via the spectrum of heavy quarkonium. Finally, the scale-setting problem in QCD appears much less mysterious from this point of view: the scale appropriate for each appearance of α_V in a Feynman diagram is just the momentum transfer of the corresponding exchanged gluon.[2] This prescription is equivalent to the BLM procedure.

It may also prove useful in the context of nonperturbative calculations based on the light-cone formalism. A light-cone Hamiltonian expressed in terms of α_V, so that it reproduces covariant perturbation theory with α_V appearing inside the momentum integrals, should be very well suited to studying, e.g., heavy quark systems: the effects of the light quarks and higher Fock state gluons that renormalize the coupling are already contained in α_V. At high momentum scales α_V should be computable via perturbation theory, while at low scales a semi-phenomenological Ansatz would be necessary. As we have discussed, exclusive processes can provide a valuable window for determining α_V in the low-energy domain.

It is a pleasure to thank S. J. Brodsky, C.-R. Ji and A. Pang for the very enjoyable collaboration which led to this work. I also thank the organizers, in particular P. Grangé, for putting together such a stimulating and enjoyable meeting. This work was supported in part by a grant from the U.S. Department of Energy.

References

[1] Brodsky S.J., Lepage G.P. and Mackenzie P.B., Phys. Rev. D **28**, 228 (1983).

[2] Lu H.J. and Brodsky S.J., Phys. Rev. D **48**, 3310 (1993).

[3] Brodsky S.J., Gabadadze G.T., Kataev A.L. and Lu H.J., Phys. Lett. **372B**, 133 (1996).

[4] Brodsky S.J., Ji C.-R., Pang A. and Robertson D.G., "Optimal renormalization scale and scheme for exclusive processes," SLAC–PUB–7473, hep-ph/9705221.

[5] Brodsky S.J. and Lepage G.P., in *Perturbative Quantum Chromodynamics*, Mueller A.H. Ed. (World Scientific, 1989).

[6] Davies C.T.H. *et. al.*, Phys. Rev. D **52**, 6519 (1995).

[7] Mattingly A.C. and Stevenson P.M., Phys. Rev. D **49**, 437 (1994).

[8] Brodsky S.J. and Robertson D.G., manuscript in preparation.

[9] Lu H.J., Ph. D. thesis, Stanford University (1992).

---

[2] There are complications which arise when gluon self-couplings are present. These are discussed in [9].

CHAPTER IX

Condensates and Chiral Symmetry Breaking

A 3+1 Dimensional LF Model with Spontaneous χSB

M. Burkardt and H. El-Khozondar

Department of Physics
New Mexico State University
Las Cruces, NM 88003-0001, U.S.A.

---

1. INTRODUCTION

Light-front (LF) quantization provides an intuitive (physical basis!) description of hadron structure that stays close to the relevant degrees of freedom in high energy scattering: many high-energy scattering processes probe hadrons along a light-like direction, since particles at very high energies travel close to the light-cone. More detailed discussions of this main motivation for studying LF field theories can be found in Refs. [1, 2] as well as in the lecture by Stan Brodsky [3].

If the main application for LF field theory is supposed to be the phenomenology of high-energy scattering processes, must one worry about the structure the vacuum in the LF formalism? At first one might think that the answer to this question is no. However, since the naive (no 0-modes) LF vacuum is known to be trivial, one might worry, for example, whether deep-inelastic structure functions can be correctly calculated on the LF in a theory like QCD where the vacuum is known to have a nontrivial structure and where one knows that this nontrivial vacuum structure plays an important role for phenomenology.

It is well known that LF Hamiltonians allow for a richer counter-term structure [4], and spontaneous symmetry breaking in normal coordinates can manifest itself as explicit symmetry breaking counter-terms in the corresponding LF Hamiltonian. In other words, the vacuum structure is shifted from states to fields. Thus, one can account for a nontrivial vacuum structure in the renormalization procedure. Some immediate questions that arise in this context are

- Can a LF Hamiltonian, with a trivial vacuum, have the same "physics" (in the sense of physical spectrum or deep inelastic structure function) as an equal time Hamiltonian with nontrivial vacuum?

- What are implications for renormalization, i.e. how does one have to renormalize in order to obtain the same physics?

- What is the structure of the effective interaction for non-zero-modes

Of course, the general answer (i.e. QCD_{3+1}) is difficult to find, but above questions have been studied in simple examples ([1]):

QED/QCD_{1+1} [6, 7, 8], Yukawa$_{1+1}$ [9], scalar theories (in any number of dimensions) [10], perturbative QED/QCD_{3+1} [11] and "mean field models": Gross-Neveu/NJL-model [12].

The goal of such toy model studies is to build intuition which one can hopefully apply to QCD_{3+1} (using trial and error). However, while these models have been very useful for studying nonperturbative renormalization in 1+1 dimensional LF field theories, it is not clear to what extend these results can be generalized to sufficiently nontrivial theories in 3+1 dimensions.

2. A 3+1-DIMENSIONAL TOY MODEL

One would like to study a 3+1 dimensional model which goes beyond the mean field approximation (NJL !), but on the other hand being too ambitious results in very difficult or unsolvable models. ([2]) We decided to place the following constraints on our model:

- Most importantly, the model should be 3+1 dimensional, but we do not require full rotational invariance.

- The model should have spontaneous χSB (but not just mean field)

- Finally, it should be solvable both on the LF and using a conventional technique (to provide a reference calculation).

Given these constraints, the most simple model that we found is described by the Lagrangian

$$\mathcal{L} = \bar{\psi}_k \left[\delta^{kl} \left(i\not{\partial} - m \right) - \frac{g}{\sqrt{N_c}} \vec{\gamma}_\perp \vec{A}_\perp^{kl} \right] \psi_l - \frac{1}{2} \vec{A}_\perp^{kl} \left(\square + \lambda^2 \right) \vec{A}_\perp^{kl}, \tag{1}$$

([1]) For a more complete list of examples and references on this topic, see Ref. [2].

([2]) For example, demanding Lorentz invariance, chiral symmetry and asymptotic freedom leaves QCD as the most simple model.

$$\Sigma \ =$$

Fig. 1. — Typical Feynman diagram contributing to the fermion self-energy in the large N_C limit of the model. No crossed "gluon" lines are allowed.

where k, l are "color" indices ($N_c \to \infty$), $\perp = x, y$ and where a cutoff is imposed on the transverse momenta. A fermion mass was introduced to avoid pathologies associated with the strict $m = 0$ case. χSB can be studied by considering the $m \to 0$ limit of the model.

The reasons for this bizarre choice of model [Eq. (1)] are as follows. If one wants to study spontaneous breaking of chiral symmetry, then one needs to have a chirally invariant interaction to start with, which motivates a vector coupling between fermions and bosons. However, we restricted the vector coupling to the $\perp$ component of a vector field since otherwise one has to deal with couplings to the bad current j^- ([3]) In a gauge theory, such couplings can be avoided by choice of gauge, but we preferred not to work with a gauge theory, since this would give rise to additional complications from infrared divergences. Furthermore, we used a model with "color" degrees of freedom and considered the limit where the number of colors is infinite, because such a model is solvable, both on and off the LF. No interaction among the bosons was included because this would complicate the model too much. Finally, we used a cutoff on the transverse momenta because such a cutoff can be used both on the LF as well as in normal coordinates and therefore one can compare results from these two frameworks already for finite values of the cutoff.

3. DYSON-SCHWINGER SOLUTION OF THE MODEL

Because we are considering the limit $N_C \to \infty$, of Eq. (1), the iterated rainbow approximation (Fig. 1) for the fermion self-energy Σ becomes exact, yielding

$$
\begin{aligned}
\Sigma(p^\mu) &= ig^2 \int \frac{d^4 k}{(2\pi)^4} \vec{\gamma}_\perp S_F(p^\mu - k^\mu) \vec{\gamma}_\perp \frac{1}{k^2 - \lambda^2 + i\varepsilon} \\
&= \not{p}_L \Sigma_L(\vec{p}_L^2, \vec{p}_\perp^2) + \Sigma_0(\vec{p}_L^2, \vec{p}_\perp^2),
\end{aligned}
\tag{2}
$$

([3]) j^- is bilinear in the constrained component of the fermion field, which makes it very difficult to renormalize this component of the current in the LF framework.

with:

$$S_F^{-1} = \not{p}_L \left[1 - \Sigma_L(\vec{p}_L^2, \vec{p}_\perp^2)\right] + \not{p}_\perp - \left[m + \Sigma_0(\vec{p}_L^2, \vec{p}_\perp^2)\right]. \qquad (3)$$

These equations can be solved by iteration. From the self-consistently obtained solution of the Dyson-Schwinger (DS) equation (2) one can extract the physical mass of the fermion. For sufficiently large coupling constant, the physical mass for the fermion remains finite in the limit $m \to 0$, proving the spontaneous breakdown of chiral symmetry in the model.

4. LF SOLUTION OF THE MODEL

Since we wanted to investigate the applicability of the effective LF Hamiltonian formalism, we formulated above model without explicit zero-mode degrees of freedom. In principle, the calculation should thus be straightforward, using standard numerical techniques, such as DLCQ [13]. However, in this approach it is hard to take full advantage of the large N_C limit so that it is difficult to compare the obtained spectrum with the results from solving the DS equation. Instead, we use the following 2-step procedure to obtain a formal solution for the LF formulation

1. First, we derive a self-consistent Green's function equation which is equivalent to the DLCQ calculation. The Green's function calculation was originally derived by starting from the covariant calculation and performing k^- integrations first (throwing away zero modes in k^+). In order to convince even the skeptics that this procedure is equivalent to DLCQ, we demonstrate numerically that, for finite and fixed DLCQ parameter K, the spectrum obtained by diagonalizing the DLCQ matrix and the spectrum obtained by solving the Green's function equation self-consistently (^{4}) are identical.

2. In the next step we compare the self-consistent Green's function equation with the DS equation. In order to facilitate the comparison with the LF calculation, we rewrite the DS equation (2), using a spectral representation for the fermion propagator S_F. In the resulting DS equation with the spectral density, we combine energy denominators, using Feynman parameter integral and perform the longitudinal momentum integral covariantly.

Details of this procedure can be found in Ref. [14]. The main results from the comparison between LF and DS equations are as follows

• The LF Green's function equation and the DS equation are identical (and thus have identical solutions) if and only if one introduces an additional (in addition to the self-induced inertias) counterterm to the kinetic mass term for the fermion.

(^{4}) Replacing integrals by finite sums in order to account for the finite DLCQ parameter K.

- For fixed transverse momentum cutoff, this additional kinetic mass term is finite.

- The value of the vertex mass in the LF Hamiltonian is the same as the value of the current mass in the DS equation.

- In the chiral limit, mass generation for the (physical) fermion occurs through the kinetic mass counter term

5. IMPLICATIONS FOR RENORMALIZATION

We have studied a 3+1 dimensional model with spontaneous breaking of chiral symmetry both in a LF framework as well as in a Dyson-Schwinger framework. Our work presents an explicit 3+1 dimensional example demonstrating that there is no conflict between chiral symmetry breaking and trivial LF vacua provided the renormalization is properly done.

The effective interaction (after integrating out 0-modes) can be summarized by a few simple terms — which are already present in the canonical Hamiltonian. The current quark mass in the covariant formulation and the "vertex mass" in the LF formulation are the same if one does not truncate the Fock space and if one uses the same cutoff on and off the LF. This is perhaps surprising, since the vertex mass multiplies the only term in the canonical Hamiltonian which explicitly breaks (LF-) chiral symmetry. Thus one might think that chiral symmetry breaking would manifest itself through a nonzero vertex mass. If one does not truncate Fock space, this is <u>not</u> what happens in this model! (^{5}) χSB, in the sense of physical mass generation for the fermion, manifests itself through a "kinetic mass" counterterm.

Even though we determined the kinetic mass counter term by directly comparing the LF and DS calculation, several methods are conceivable which avoid reference to a non-LF calculation in order to set up the LF problem. One possible procedure would be to impose parity invariance for physical observables as a constraint [9].

M.B. would like to acknowledge Michael Frank and Craig Roberts for helpful discussions on the Schwinger-Dyson solution to the model. We would like to thank Brett vande Sande for carefully reading the manuscript. This work was supported by the D.O.E. under contract DE-FG03-96ER40965 and in part by TJNAF.

References

[1] Burkardt M., Advances Nucl. Phys. **23**, 1 (1996).

(^{5}) Further studies show that a renormalization of the vertex mass arises from a Tamm-Dancoff truncation but not from integrating out zero-modes [15].

[2] Brodsky S., Pauli H-C and Pinsky S., submitted to Physics Reports, hep-ph/9705477.

[3] Brodsky S.J., these proceedings, hep-ph/9706236.

[4] Wilson K. G. et al., Phys. Rev. D **49**, 6720 (1994).

[5] 't Hooft G., Nucl. Phys. B **75**, 461 (1974).

[6] Lenz F. et. al., A.. Phys. **208**, 1 (1990).

[7] Hornbostel K., Phys. Rev. D **45** (1992) 3781;
Robertson D.G., Phys. Rev. D **47**, 2549 (1993).

[8] Zhitnitsky A., Phys. Lett. B **165** , 405 (1985);
Burkardt M., Phys. Rev. D **53**, 933 (1996).

[9] Burkardt M., Phys. Rev. D **54**, 2913 (1996).

[10] Prokhvatilov E.V. and Franke V. A., Sov. J. Nucl. Phys. **49** (1989) 688;
Burkardt M., Phys. Rev. D **47**, 4628 (1993); Prokhvatilov E. V., Naus H.
W. L. and Pirner H.-J., Phys. Rev. D **51**, 2933 (1995); Vary J.P., Fields
T.J. and Pirner H.-J., Phys. Rev. D **53**, 7231 (1996).

[11] Burkardt M. and Langnau A., Phys. Rev. D **44**, 3857 (1991).

[12] Dietmaier C. et al., Z. Phys. A **334**, 220 (1989);
Itakura K. and Maedan S., Prog. Theor. Phys. **97**, 635 (1997).

[13] Pauli H.-C. and Brodsky S. J., Phys. Rev. D **32**, 1993 (1985); *ibid* 2001
(1985).

[14] Burkardt M. and El-Khozondar H., Phys. Rev. D **55**, 6514 (1997).

[15] Burkardt M., hep-ph/9705224.

Chiral Symmetry and Light-Cone Wave Functions

T. Heinzl([1])

([1]) Institut für Theoretische Physik,
Universität Regensburg, 93040 Regensburg,
Germany

1. INTRODUCTION

The purpose of this contribution is to use a particular non-perturbative approach to strong interaction physics, namely light-cone (LC) quantization [1] for the study of spontaneous breaking of chiral symmetry (SBCS). In this approach, one aims at diagonalizing the LC Hamiltonian, which amounts to solving for the bound states of the underlying field theory, in particular their masses and LC wave functions. For a recent review on the subject the reader is referred to [2].

Before we consider any such bound state equation, however, let us consider shortly some peculiar features of spontaneous symmetry breaking in the LC framework. The text-book definition of spontaneous symmetry breaking is the statement that the dynamics of a particular system is invariant under a symmetry, but the ground state is not. In other words: the generator Q of the symmetry, $Q(t) = \int d^4x\, j^0(x)\delta(x^0 - t)$, commutes with the Hamiltonian (and is thus conserved), but does not annihilate the ground state. From this one can conclude that there must be condensates, i.e. non-vanishing vaccum expectation values that are order parameters of the symmetry breaking.

How do these things look like within LC quantisation? It has been known since a long time that the relevant charges, called 'light-like charges', $Q(\tau) = \int d^4x\, j^+(x)\delta(x^+ - \tau)$([1]), always annihilate the vacuum [3]. This is in agreement

---

([1]) Our LC conventions are: $a^\pm = a^0 \pm a^3$, with a an arbitrary four-vector.

with the central feature of LC quantisation that the vacuum is 'trivial', thus void of particles and structure. The question that immediately arises, however, is the following: How does one define and/or determine condensates within LC quantisation, if the vacuum is 'trivial'?

A clue to a possible answer is gained by considering the well known relation of Gell-Mann, Oakes and Renner (GOR) [4],

$$m_\pi^2 = -\frac{1}{f_\pi^2} \langle 0|\bar{\psi}\psi|0\rangle m_q + O(m_q^2) \,, \tag{1}$$

which relates the observable pion mass m_π to the current quark mass m_q and the quark condensate, $\langle 0|\bar{\psi}\psi|0\rangle$. The GOR relation entails Goldstone's theorem, the vanishing of the pion mass in the chiral limit ($m_q = 0$), and allows for a definition of the condensate as the slope of m_π as a function of the quark mass m_q. As in [5], we will use the GOR relation as a paradigm in what follows.

2. D = 1+1: 'T HOOFT AND SCHWINGER MODEL

In the first part of this contribution we will concentrate on the 't Hooft [6, 7] and Schwinger models [8, 9] (tHM and SM). The tHM is QCD in 1+1 dimensions with N_C, the number of colors, going to infinity. The (massive) SM is QED_{1+1}. The features of the models are:

(i) Their spectrum does not contain elementary fermions (charge screening and/or confinement) due to the linear Coulomb potential in $d = 1 + 1$.

(ii) One has SBCS in the tHM and anomalous chiral symmetry breaking in the SM.

(iii) The lowest bound state, which we call the 'pion' for brevity, can be symbolically written as 'pion' = $q\bar{q} + q\bar{q}q\bar{q} + \ldots$, the contributions from the non-valence sector (four or more particles) being either zero (tHM) or small (SM) [10].

(iv) For the 'pion' mass-squared one has an analogue of the GOR relation of the form

$$m_\pi^2 = \alpha + M_1 m + M_2 m^2 + M_3 m^3 + \ldots \,, \tag{2}$$

where m is the fermion mass and the parameter α the anomaly contribution. All masses are measured in units of a basic scale μ_0. For the tHM, $\mu_0^2 = g^2 N_C/2\pi$ and $\alpha = 0$ (no anomaly), so that the 'pion' is a Goldstone boson [11]. For the SM, $\mu_0^2 = e^2/\pi$, the mass-squared of the boson of the massless SM, and $\alpha = 1$. The axial anomaly thus implies that, for $m = 0$, $m_\pi = \alpha = 1$.

The present knowledge of the coefficients M_i in (2) is as follows. For the SM with vanishing θ-parameter, the first two coefficients have been calculated via bosonization techniques,

$$M_1 = -4\pi\langle 0|\bar{\psi}\psi|0\rangle = 2e^\gamma = 3.5621 \,, \tag{3}$$
$$M_2 = 4\pi^2\langle 0|\bar{\psi}\psi|0\rangle^2 A = 3.3874 \,, \tag{4}$$

where A is a numerical constant and $\langle 0|\bar{\psi}\psi|0\rangle$ the condensate of the massless SM [12]. The first-order result (3) was originally obtained in [13], the second-order one, (4), recently in [14, 15]. Note that both M_1 and M_2 are related to the condensate in a rather straightforward manner.

The coefficient M_1 has also been obtained by solving the light-front bound state equation (LFBSE) in the valence sector, with the result [16]

$$M_1 = 2\pi/\sqrt{3} = 3.6276 , \tag{5}$$

which differs from (3) by 2%. The same methods applied to the tHM yield the same value for M_1 [7].

In order to have a chance of resolving the 2% discrepancy for the SM, we will perform a high-precision analysis of the LFBSE by using a variational method together with perturbation theory in m. The calculations in the 2-particle (valence) sector will be done analytically; higher particle sectors will be included numerically. In order to truncate the system of coupled equations we use the light-front Tamm-Dancoff (LFTD) approximation [17] and allow for a maximum number of six particles.

The valence sector LFBSEs for the tHM and SM can be written in a unified way [5] as

$$m_\pi^2 \phi(x) = \frac{m^2 - 1}{x(1 - x)}\phi(x) - \mathcal{P}\int_0^1 dy\, \frac{\phi(y)}{(x - y)^2} + \alpha \int_0^1 dx\, \phi(x) , \tag{6}$$

which we will call the 't Hooft-Bergknoff equation hereafter. $\phi(x)$ is the LC wave function of the 'pion', the amplitude for finding a quark in the 'pion' with longitudinal momentum fraction x. The 't Hooft-Bergknoff equation (6) cannot be solved analytically, and one has to resort to approximations. We use a variational procedure with trial functions $\phi[v]$ depending on a set $\{v\}$ of variational parameters,

$$\phi[v] = x^\beta (1 - x)^\beta + \sum_{i=1}^{4} v_i x^{\beta+i}(1 - x)^{\beta+i} . \tag{7}$$

The first term on the r.h.s. corresponds to 't Hooft's original ansatz which is extended by the following sum. The set of variational parameters is thus $\{v\} = \{\beta, v_1, \ldots, v_4\} \equiv \{\beta, a, b, c, d\}$. As a second ingredient, we use mass perturbation theory, i.e. the 'pion' mass and all variational parameters are expanded as power series in m. A high intrinsic precision is guaranteed by the fact that all calculations are performed analytically using the computer algebraic language MAPLE. This can be done as there are analytic expressions for the matrix elements (integrals) appearing in the variational expression for the 'pion' mass.

Our best values for the expansion coefficients M_i obtained within the valence approximation (2PTD) and including up to six particles (6PTD) are displayed in Tables I and II, where we also compare with related work.

Table I. — The expansion coefficients M_i for the 't Hooft model ($\alpha = 0$).

	M_1	M_2	M_3
variational 2PTD	$2\pi/\sqrt{3} = 3.627599$	3.581055	0.061793
numerical 2PTD	$3.62758 \pm 2 \cdot 10^{-5}$	$3.5829 \pm 3 \cdot 10^{-4}$	$0.064 \pm 2 \cdot 10^{-3}$
Li et al. [18]	3.64 ± 0.03	3.60 ± 0.06	0.04 ± 0.04

Table II. — The expansion coefficients M_i for the Schwinger model ($\alpha = 1$).

	M_1	M_2	M_3
variational 2PTD	$2\pi/\sqrt{3} = 3.627599$	3.308608	0.348204
numerical 2PTD	$3.6268 \pm 8 \cdot 10^{-4}$	3.32 ± 0.02	0.28 ± 0.06
numerical 6PTD	$3.6267 \pm 4 \cdot 10^{-4}$	3.22 ± 0.02	0.5 ± 0.1
DLCQ [19]	3.7 ± 0.02	3.5 ± 0.3	–
lattice [20]	3.5 ± 0.2	3.7	0.02
bosonization [14, 15]	$2e^\gamma = 3.562146$	3.387399	–

Let us summarize our findings [21]:

1. To order m one obtains the 't Hooft-Bergknoff result, $M_1 = 2\pi/\sqrt{3}$. This value does *not* change upon enlarging the space of variational parameters. As M_1 is independent of the anomaly α it is the same in both the tHM and SM. To leading order, the 'pion' mass is thus entirely determined by 't Hooft's ansatz; extensions of this ansatz only affect the higher order coefficients M_2 and M_3.

2. The higher order coefficients depend on α and are thus no longer universal. For the SM, the coefficient M_2 is smaller than the bosonization result (4). The difference is again 2%. This discrepancy in M_1 and M_2 for the SM thus does not become resolved by refining the solution in the valence sector.

3. The convergence of the variational procedure is quite good and is used as a measure of the errors. For the two-particle LFTD approximation they are less than $5 \cdot 10^{-6}$.

4. The endpoint behaviour of the wave functions is stable under the improvements a, b, c, d. The latter, however, lead to a slight enhancement of the wave function in the intermediate-x region.

A natural explanation of the 2% discrepancy for the SM seems to be the neglect of higher particle sectors. It is only for the tHM that these are strictly absent as they are suppressed by $1/N_C$ and thus vanish for $N_C \rightarrow \infty$. In the Schwinger model they vanish as $m \rightarrow 0$, but the exact m-dependence of these contributions is unknown. We have calculated the 'pion' mass-squared numerically including up to six particles with the following results (Table II): (i) the 't Hooft-Bergknoff value for M_1 is not affected, and (ii), M_2 becomes slightly smaller, which is the *wrong* tendency compared to (4). It might therefore be that one has to include a very large number of higher Fock states to exactly obtain the bosonization results.

3. D = 3+1: PERSPECTIVES

In his 1974 solution of the tHM, 't Hooft did not use the LC formalism in the manner which nowadays might be called standard, namely to derive the canonical light-front Hamiltonian and set up the LFBSE by projecting the Schrödinger equation on the different sectors of Fock space. Instead, he started from covariant equations, namely the Schwinger-Dyson equation (SDE) for the quark self-energy, and the Bethe-Salpeter equation (BSE) for the bound state amplitude which needs the quark self-energy as an input. The LFBSE was then obtained by projecting the BSE onto hypersurfaces of equal LC time, thereby spoiling explicit covariance.

In a condensed notation the SDE can be written as $\Sigma = S_0^{-1} - S^{-1}$, where Σ denotes the quark self-energy, $S^{-1} = \not{p} - m_0 - \Sigma$ and $S_0^{-1} = \not{p} - m_0$ the inverse of the full and free propagator, respectively. The BSE can be written as $\chi_{BS} = S_1 S_2 K \chi_{BS}$ with S_1, S_2 the full propagators of quark and anti-quark, K the Bethe-Salpeter kernel, defined through the interaction, and χ_{BS} the Bethe-Salpeter amplitude. From the latter, one obtains the LC wave function ϕ_{LC} via integration over the energy variable p^-, $2\pi \phi_{LC}(\underline{p}) = \int dp^- \chi_{BS}(p)$, where $\underline{p} = (p^+, \mathbf{p}_\perp)$.

The program outlined above, deriving the LFBSE from the covariant equations, if applied to QCD, is of course ambitious but should still be feasible in view of the wealth of results already obtained in studying the SDEs of QCD [22]. In a first step, a simpler approach will be pursued here, namely a calculation within a particularly suited framework, the Nambu-Jona-Lasinio (NJL) model [23] (for a recent review, see [24]).

The NJL model has a chirally invariant four fermion interaction of the form $(\bar{\psi}\Gamma\psi)^2$. It is not renormalizeable and needs a cutoff as a parameter that has to be fixed by phenomenology. Above a critical coupling, $g > g_c$, a fermion mass is dynamically generated so that chiral symmetry is spontaneously broken. The dynamical fermion mass is proportional to the fermionic condensate in the vacuum.

These results are obtained via a self-consistent mean field solution of the SDE resulting in the gap equation (for $m_0 = 0$),

$$\Sigma \equiv m = 2igtrS(0) = \frac{igm}{2\pi^4} \int \frac{d^4k}{k^2 - m^2} = -2g\langle 0|\bar{\psi}\psi|0\rangle_m . \qquad (8)$$

This equation determines the dynamically generated mass m. The integral on the r.h.s. is of course divergent and has to be cut off (see below). $\langle 0|\bar{\psi}\psi|0\rangle_m$ is the condensate of the dynamical fermions which are viewed as quasi-particles or 'constituent quarks'. Once the physical fermion mass m is known, it can be plugged into the BSE which in ladder approximation reads

$$\phi_{LC}(\underline{k}) = \int \frac{dk^-}{2\pi} iS(k) iS(k - P) \int \frac{d^3\tilde{k}}{(2\pi)^3} \int \frac{d\tilde{k}^-}{2\pi} K(k, \tilde{k}) \chi_{BS}(\tilde{k}) , \qquad (9)$$

with P the bound state four-momentum. On the l.h.s., the projection onto $x^+ = 0$ (i.e. the k^--integration) has already been carried out. On the r.h.s., the two integrations over k^- and $\tilde{k}^-$ still have to be performed. Whether this can easily be done depends of course crucially on the kernel K, which in principle is a function of both energy variables. For the NJL model, however, one has $K(k, \tilde{k}) = 2\gamma_5 \otimes \gamma_5 - \gamma_\mu \gamma_5 \otimes \gamma^\mu \gamma_5$, i.e. the kernel is momentum independent due to the four-point contact interaction! Thus, the $\tilde{k}^-$-integration immediately yields $\phi_{\rm LC}$, and the k^--integration can be performed via residue technique and is completely determined by the poles of the propagators, $S(k)$ and $S(k - P)$. As a result, one finds a non-vanishing result only if $0 \leq k^+ \leq P^+$, and *one* of the two particles is put on-shell, e.g. $k^2 = m^2$.

The upshot of all this is nothing but the LFBSE, which reads explicitly

$$
\begin{aligned}
\phi_{\rm LC}(x, \mathbf{k}_\perp) &= -\frac{2gC_\Lambda}{x(1-x)} \frac{(\hat{\slashed{k}} + m)\gamma_5(\hat{\slashed{k}} - \slashed{P} + m)}{M^2 - M_0^2(x)} \\
&+ \frac{gD_\Lambda}{x(1-x)} \frac{(\hat{\slashed{k}} + m)\slashed{P}\gamma_5(\hat{\slashed{k}} - \slashed{P} + m)}{M^2 - M_0^2(x)} .
\end{aligned}
\tag{10}
$$

Here we have defined the longitudinal momentum fraction $x = k^+/P^+$, the on-shell momentum $\hat{k} = (\hat{k}^-, \mathbf{k}_\perp, k^+)$ with $\hat{k}^- = (k_\perp^2 + m^2)/k^+$, the bound state mass squared, $M^2 = P^2$, which is the eigenvalue to be solved for, and the free kinetic energy (or invariant mass), $M_0^2(x) = (k_\perp^2 + m^2)/x(1 - x)$. The (cutoff dependent) expressions C_Λ and $P_\mu D_\Lambda$ denote the $\tilde{k}$-integrals over $\gamma_5 \phi_{\rm LC}$ and $\gamma_\mu \gamma_5 \phi_{\rm LC}$, respectively, and thus are mere normalization constants. The cutoff Λ is chosen such that transverse *and* longitudinal momenta are cut off, namely $M_0^2 < \Lambda^2$ ('invariant mass cutoff') and $m^2/\Lambda^2 < x$ ('small-x cutoff'). In this way, one maintains rotational invariance (in terms of the ordinary three-vector $\mathbf{k}$), and parity invariance.

The crucial observation now is that, due to the simplicity of the kernel, the solution of the LFBSE (10) is already given by its r.h.s., which does not contain $\phi_{\rm LC}(x, \mathbf{k}_\perp)$ anymore!

As a first check of this bound state wave function we look for a massless pion in the chiral limit. To this end we decompose the LC wave function according to

$$
\phi_{\rm LC} = \phi_{\rm S} + \phi_{\rm P}\gamma_5 + \phi_{\rm A}^\mu \gamma_\mu \gamma_5 + \phi_{\rm V}^\mu \gamma_\mu + \phi_{\rm T}^{\mu\nu} \sigma_{\mu\nu} .
\tag{11}
$$

Multiplying (10) with γ_5, taking the trace and integrating over $\tilde{k}$ one finds for $M = 0$ (corresponding to a Goldstone pion),

$$
1 = \frac{g}{2\pi^2} \int_{m^2/\Lambda^2}^1 dx \int_0^\infty dk_\perp^2 \, \theta \left(x(1-x)\Lambda^2 - k_\perp^2 - m^2 \right) .
\tag{12}
$$

This, however, is just the gap equation (8), if one performs the k^--integration with the appropriate cutoff [25]. Thus, there is a Goldstone pion obtained as the solution of the LFBSE exactly if the gap equation holds. This fact is well known from the covariant treatment and to derive it from the LFBSE provides additional evidence for the self-consistency of the suggested procedure. The next steps will be to explicitly calculate the pion wave function and distribution amplitude, as well as the pion decay constant (which should fix the cutoff) and the pion electromagnetic form factor.

Acknowledgements

It is a pleasure to thank the organizers of this workshop, in particular P. Grangé, for all their efforts; the participants for providing a stimulating atmosphere; and K. Harada and C. Stern for their collaboration on part of the presented work.

References

[1] Dirac P.A.M., *Rev. Mod. Phys.* **21** (1949) 392.

[2] Burkardt M., *Advances Nucl. Phys.* **23** (1996) 1.

[3] Jersak J. and Stern J., *Nucl. Phys.* **B7** (1968) 413.

[4] Gell-Mann M., Oakes R.J. and Renner B., *Phys. Rev.* **175** (1968) 2195.

[5] Heinzl T., *Phys. Lett.* **B388** (1996) 129.

[6] 't Hooft G., *Nucl. Phys.* **B75** (1974) 461.

[7] 't Hooft G., in: *New Phenomena in Subnuclear Physics*, Zichichi A., ed., (Plenum, New York, 1977), p. 261

[8] Schwinger J., *Phys. Rev.* **128** (1962) 2425.

[9] Coleman S., *Ann. Phys. (N.Y.)* **101** (1976) 239.

[10] Mo Y. and Perry R.J., *J. Comp. Phys.* **108** (1993) 159.

[11] Zhitnitsky A., *Sov. J. Nucl. Phys.* **43** (1986) 999.

[12] Marinari E., Parisi G. and Rebbi C., *Nucl. Phys.* **B190** (1981) 734.

[13] Banks T., Kogut J. and Susskind L., *Phys. Rev.* **D13** (1976) 1042.

[14] Adam C., *Phys. Lett.* **B382** (1996) 383.

[15] Fields T.J., Pirner H.-J. and Vary J., *Phys. Rev.* **D53** (1996) 7231.

[16] Bergknoff H., *Nucl. Phys.* **B122** (1977) 215.

[17] Harindranath A., Perry R.J. and Wilson K.G., *Phys. Rev. Lett.* **65** (1990) 2969.

[18] Li M., Birse M. and Wilets L., *J. Phys.* **G13** (1987) 915.

[19] Eller T., Pauli H.-C. and Brodsky S.J., *Phys. Rev.* **D35** (1987) 1493.

[20] Crewther D.P. and Hamer C.J., *Nucl. Phys.* **B170** (1980) 352.

[21] Harada K., Heinzl T. and Stern C., hep-th/9705159

[22] Roberts C.D. and Williams A.G., *Prog. Part. Nucl. Phys.* **33** (1994) 477.

[23] Nambu Y. and Jona-Lasinio G., *Phys. Rev.* **122** (1961) 345.

[24] Hatsuda T. and Kunihiro T., *Phys. Rep.* **247** (1994) 221.

[25] Dietmaier C., Heinzl T., Schaden M. and Werner E., *Z. Phys.* **A334** (1989) 215.

QCD at Finite Extension[*]

F. Lenz and M. Thies

Institute for Theoretical Physics III,
University of Erlangen-Nürnberg,
Staudtstr. 7, 91058 Erlangen,
Germany

The study of QCD at finite extension, i.e., in a geometry where the system is of finite extent (L_3) in one direction (x_3), is of interest for various reasons. First, with L_3 a parameter is introduced which helps to control infrared ambiguities. This is of particular importance when using axial like gauges which are appropriate for such a geometry (for an early discussion of ambiguities in the axial gauge, see Ref. [1]).

Second, by covariance, QCD at finite extension is equivalent to finite temperature QCD. This equivalence of finite extension and finite temperature of relativistic field theories has been noted in Ref. [2] and used e.g. in a discussion of the finite temperature quark propagators [3]. By rotational invariance in the Euclidean, the value of the partition function of a system with finite extension L_3 in 3 direction and β in 0 direction is invariant under the exchange of these two extensions,

$$Z(\beta, L_3) = Z(L_3, \beta) , \tag{1}$$

provided bosonic (fermionic) fields satisfy periodic (antiperiodic) boundary conditions in both time and 3 coordinate. As a consequence of (1), energy density and pressure are related by

$$\epsilon(\beta, L_3) = -p(L_3, \beta) . \tag{2}$$

(*) Talk presented by F. Lenz; work supported by Bundesministerium für Bildung und Forschung

For a system of non-interacting particles this relation connects energy density or pressure of the Stefan Boltzmann law with the corresponding quantities measured in the Casimir effect.

In QCD, covariance also implies by Eq. (2) that at zero temperature a confinement-deconfinement transition occurs when compressing the QCD vacuum (i.e. decreasing L_3). From lattice gauge calculations [4] we infer that this transition occurs at a critical extension $L_3^c \approx 0.8$ fm in the absence of quarks and at $L_3^c \approx 1.3$ fm when quarks are included. For extensions smaller than L_3^c, the energy density and pressure reach values which are typically 80% of the corresponding "Casimir" energy and pressure. The order parameter which characterizes the phases of QCD and in particular the realization of the center symmetry (in the absence of quarks) is the vacuum expectation value of the Polyakov loop operator at finite temperature [5] and correspondingly of the operator

$$W\left(x_\perp\right) = N_c^{-1}\mathrm{tr}\, P \exp\left\{ig \int_0^{L_3} dx_3 A_3\left(x\right)\right\} \qquad (3)$$

at finite extension ($x_\perp = (x_0, x_1, x_2)$).

When compressing the system beyond the typical length scales of strong interaction physics (e.g. beyond a typical hadron radius R), two further limits are of interest. Again as supported by results from lattice calculations [6], correlation functions at transverse momenta or energies $|p| \ll 1/L_3$ are dominated by the zero "Matsubara wave-numbers" in 3-direction and can therefore be expected to be given by the dimensionally reduced QCD_{2+1}. On the other hand we may consider excitations with large momenta p in the 3 direction such that the Lorentz-contracted extension of a hadron (of mass m) is small in comparison to L_3,

$$mR/p \ll L_3 \ll R . \qquad (4)$$

For sufficiently high momenta, spectrum and structure of hadrons should not be affected by the finite extension, although the vacuum on which these excitations are built is not the confining ground state with broken chiral symmetry, but that of the quark gluon plasma with the chiral symmetry restored. This situation of large momentum excitations at small extension is closely related to the light cone or infinite momentum frame limit in which, because of the off-diagonal metric, a finite extension along the light-cone space axis x^- actually describes an interval of vanishing invariant length [7].

For the theoretical treatment of QCD at finite extension an axial type gauge is particularly appropriate. Within the canonical formalism, the Gauss law can be resolved explicitly [8]. It is quite obvious that the process of eliminating the 3-component of the gauge field cannot be carried through completely; Polyakov loops winding around the compact 3-direction are gauge invariant variables and thus must be present also after gauge fixing. The formal procedure shows that these variables are the only remainders of A_3 and properly determines their kinematics. These results are confirmed in the path integral approach, where the axial type gauges have the distinct advantage that the Faddeev Popov

determinant can be evaluated in closed form (cf. also Ref. [9]). The result for
the SU(2) Yang Mills theory is summarized in the generating functional

$$Z[J, j_3] = \int D[A_\mu] d[a_3] \exp \left\{ iS_{\text{YM}}[A_\mu, a_3] + iS_{\text{gf}}[A_\mu^3] \right.$$
$$\left. +i \int d^4x \left(A_\mu^a J^{a,\mu} + a_3 j^3 \right) \right\} , \tag{5}$$

where in the standard Yang Mills action the 3 component of the gluon field
$A_3(x)$ is replaced by the neutral (i.e., diagonal) 2-dimensional field $a_3(x_\perp)$.
At finite extension L_3, this two dimensional field can be eliminated neither
in QED nor in QCD. In QED it describes transverse photons polarized in 3
direction and propagating in the 1-2 plane, while in SU(2) QCD, $a_3(x_\perp)$ are
the eigenvalues of the (untraced) "Polyakov loop" variable $W(x_\perp)$ of Eq. (3).
After an appropriate choice of coordinates in color space, $W(x_\perp)$ can be written
as

$$W(x_\perp) = \cos(gL_3 a_3(x_\perp)/2) . \tag{6}$$

The presence of the two dimensional field $a_3(x_\perp)$ makes a further gauge fixing
of the transverse ($\mu = 0, 1, 2$), neutral, x_3 independent gauge fields necessary,
which is achieved by the additional contribution to the action in Eq. (5).
The following choice of the gauge fixing term is particularly convenient for
perturbative calculations,

$$S_{\text{gf}}[A_\mu^3] = -\frac{1}{2L_3^2} \left(\int_0^{L_3} dx_3 \partial^\mu A_\mu^3(x) \right)^2 . \tag{7}$$

The integration measure for the a_3 functional integral is given by

$$d[a_3] = \prod_{y_\perp} \sin^2(gL_3 a_3(y_\perp)/2) \, \Theta(a_3(y_\perp)) \, \Theta(2\pi/gL_3 - a_3(y_\perp)) \, da_3(y_\perp) \tag{8}$$

and accounts for the fact that $gL_3 a_3/2$ appears in the parametrization of the
group manifold S_3 as the first polar angle. As a consequence the corresponding
part of the Haar measure enters with its finite range of integration. In the
canonical formulation, this Faddeev Popov determinant arises as a Jacobian
modifying the kinetic energy of the Polyakov loop variables $a_3(x_\perp)$ [8]. In
QED the same procedure yields the standard flat measure for $a_3(x_\perp)$. Non-flat
measures for the Polyakov loop variables appear in gauge fixed formulations of
QCD irrespective of space-time dimension. However the consequences of the
compactness of these variables depend in general on the dimension. For 1+1
dimensional QCD with adjoint fermions e.g. it has been shown [10] that only
by accounting properly for the non-flatness of the measure the symmetries of
the system are correctly described.

    In most of the approaches to finite temperature QCD in the temporal gauge,
one ignores the finite range of integration and accounts for the $\sin^2$-factor in

the volume element by introducing ghost fields $c_a(x)$,

$$\prod_{y_\perp} \sin^2(gL_3a_3(y_\perp)/2) = \int d[c]d[c^\dagger] \exp\left\{ i \int d^4x c_a^\dagger(x) \left(-i\partial_3\delta^{ab}\right.\right.$$

$$\left.\left. + ig\epsilon^{ab3}a_3(x_\perp)\right) c_b(x)\right\}. \tag{9}$$

In perturbative treatments, such ghosts have no effect, as is most easily seen in dimensional regularization, and therefore the QCD measure is effectively replaced by the flat measure appropriate for QED. Debye screening is obtained as a result of this procedure but also, as in QED, a shift symmetry (Z) of the effective potential for the Polyakov loop variables a_3 not present in the original theory [11].

We have not been able to justify the above procedure. We find that the functional integral over a_3 cannot be approximated by a Gaussian integral with corresponding perturbative corrections. Rather, the relevant limit is that of a vanishing instead of an infinite range of integration, as is easily seen when considering the propagator for non-interacting Polyakov loop variables $W(x_\perp)$ in the discretized Euclidean formulation (with lattice spacing ℓ and unit vectors **e**),

$$\langle \Omega | T\left[W(x_\perp)W(0)\right] |\Omega\rangle = \frac{1}{4}\delta_{\mathbf{x},0} + O\left(\frac{\ell}{g^2L_3}\right)\left(\delta_{\mathbf{x},0} + \sum_{\pm\mathbf{e}}\delta_{\mathbf{x},\mathbf{e}}\right)$$

$$\approx \frac{\ell^3}{4}\delta^{(3)}(x_\perp). \tag{10}$$

Hopping terms are suppressed by powers of ℓ/g^2L_3, and the propagator becomes local in the continuum limit. We note that this property is due to the "macroscopic" nature of the variable $a_3(x_\perp)$. When describing a single link variable on the lattice, the factor ℓ/L_3 does not appear and the discretized form of the free propagator is obtained. The winding of a_3 around the circle in 3 direction apparently provides an infinite inertia. (Suppression of the hopping term as a consequence of the Haar measure for the Polyakov loop variables is characteristic for 3+1 dimensional QCD; in 2+1 dimensions e.g. hopping is controlled by $1/g^2L_3$ and therfore dependent on extension and coupling constant.) Propagation of excitations induced by $a_3(x_\perp)$ can consequently only arise by coupling to the other microscopic degrees of freedom. Formally this suggests the Polyakov loop variables a_3 to be integrated out by disregarding the contribution of the free a_3 action, but keeping the coupling to the other degrees of freedom. Also disregarding possible effects from singular field configurations, the following effective action is obtained

$$S_{\text{eff}}[A_\mu] = S_{\text{YM}}[A_\mu, A_3 = 0] + S_{\text{gf}}[A_\mu^3] + M^2/2 \sum_{a=1,2}\int d^4x A_\mu^a(x) A^{a,\mu}(x). \tag{11}$$

In this effective action, the Polyakov loop variables have left their signature in the geometrical mass term of the charged gluons

$$M^2 = \left(\pi^2/3 - 2\right)/L_3^2 \tag{12}$$

and in the change to antiperiodic boundary conditions

$$A_\mu^{1,2}\left(x_\perp, L_3\right) = -A_\mu^{1,2}\left(x_\perp, 0\right) , \tag{13}$$

while the neutral gluons remain massless and periodic. The antiperiodic boundary conditions reflect the mean value of the Polyakov loop variables, the geometrical mass their fluctuations; notice that in both of these corrections, the coupling constant has dropped out. We emphasize that periodic boundary conditions for the gluon fields are imposed in the representation (5) of the generating functional. Otherwise the equivalence between QCD at finite temperature and finite extension would break down. The antiperiodic boundary conditions in (11) describe the appearence of Aharonov-Bohm fluxes in the elimination of the Polyakov loop variables. Periodic charged gluon fields may be used if the differential operator ∂_3 is replaced by

$$\partial_3 \to \partial_3 + \frac{i\pi}{2L_3}[\tau_3, \qquad \cdot \tag{14}$$

As for a quantum mechanical particle on a circle, such a magnetic flux is technically most easily accounted for by an appropriate change in boundary conditions – without changing the original periodicity requirements. With regard to the rather unexpected physical consequences, the space-time independence of this flux is important, since it induces global changes in the theory. These global changes are missed if the Polyakov loops are treated as Gaussian variables. In such treatments underlying most of the finite temperature formulations, no such Aharonov-Bohm flux arises and concomitantly the center symmetry cannot be maintained.

The role of the order parameter is taken over by the neutral color current in 3-direction,

$$u\left(x_\perp\right) = \int_0^{L_3} dx_3 \epsilon^{ab3} A_\mu^a\left(x\right) \partial_3 A^{b,\mu}\left(x\right) , \tag{15}$$

which determines the vacuum expectation value of the Polyakov loops

$$\langle\Omega|W\left(x_\perp\right)|\Omega\rangle \quad \propto \quad \langle\Omega|u\left(x_\perp\right)|\Omega\rangle \tag{16}$$

and the corresponding correlation function

$$\langle\Omega|T\left[W\left(x_\perp\right)W\left(0\right)\right]|\Omega\rangle \quad \propto \quad \langle\Omega|T\left[u\left(x_\perp\right)u\left(0\right)\right]|\Omega\rangle . \tag{17}$$

Formulated in these variables, the center symmetry ([1]) acts as "charge conjugation"

$$C: \qquad A_\mu^3 \to -A_\mu^3 , \quad A_\mu^1 + iA_\mu^2 \to A_\mu^1 - iA_\mu^2 . \tag{18}$$

([1]) Note that unlike at finite temperature, the center symmetry is an ordinary symmetry at finite extension. These considerations can be generalized to SU(3) and provide new perspectives on the controversial issue of Z(N) bubbles, see e.g. [12]

The treatment of the Polyakov loop variables required by the finite range of integration leads to a picture of QCD which is quite different from that based on the treatment of the a_3 functional integral as an approximate Gaussian one. Among the most important differences, we note that the perturbative vacuum corresponding to the effective action (11) is an eigenstate of the center symmetry. In the perturbative ground state, no color currents are present, hence

$$\langle \Omega_{\text{pt}} | W\left(x_\perp\right) | \Omega_{\text{pt}} \rangle = 0 \ . \tag{19}$$

Loosely speaking we can take this as an indication for an infinite free energy of a static quark. A more precise characterization of the realization of confinement can be obtained from properties of the associated correlation function (17) which, in the Euclidean, determines the static quark-antiquark interaction energy [5]. Due to the locality property (10), this static quark-antiquark potential is given directly by the $a = b = 3, \mu = \nu = 3$ component of the vacuum polarization tensor $\Pi_{\mu\nu}^{ab}$ and not by the zero mass propagator of the Gaussian approximation to the Polyakov loop action. Up to an irrelevant factor we have in the Euclidean

$$\exp\left\{-L_3 V\left(r\right)\right\} = \langle \Omega | T\left[u(x_\perp^E)u\left(0\right)\right] | \Omega \rangle \ . \tag{20}$$

This identity implies a linearly rising potential at large distances, if the vacuum expectation value of the Polyakov loop operator vanishes and if the spectrum of states excited by u exhibits a gap ΔE,

$$V\left(r\right) \to \sigma r = \Delta E r/L_3 \ . \tag{21}$$

Thus in this axial like gauge, confinement is connected to a shift in the spectrum of gluonic excitations to excitation energies

$$E \geq \sigma L_3 \ . \tag{22}$$

After a gradual decrease of this threshold with decreasing L_3, the whole spectrum of excitations becomes suddenly available when at the deconfinement transition the string tension vanishes. At the perturbative level already a linearly rising potential is obtained, however with a string tension decreasing with increasing extension ($\propto L_3^{-2}$). The change from this value of the string tension to the physical one together with the emergence of the proper QCD scale is beyond a perturbative treatment also after elimination of the Polyakov loop variables.

For small distances, Coulomb-like behaviour requires this particular component of the vacuum polarization tensor to possess an essential singularity at infinite momentum

$$\int d^3x e^{ipx} \langle \Omega | T\left[u\left(x\right)u\left(0\right)\right] | \Omega \rangle \to e^{-\sqrt{g^2 L_3 p/\pi}} \tag{23}$$

which, in the Euclidean, implies a density of intermediate states contributing to the vacuum polarization which increases exponentially with the square root of the excitation energy. A perturbative evaluation of the vacuum polarization therefore has to yield increasingly high powers of $L_3 p$ with increasing order.

With respect to perturbative treatments, we remark that the difficulties encountered in finite temperature perturbation theory [13] are not expected to arise in this perturbative, confining phase. With the charged gluons satisfying antiperiodic boundary conditions, the infrared properties of QCD at finite extension should resemble those of QED. Moreover, the finite extension gluon propagator is well defined and, unlike the continuum axial gauge propagator, does not need a prescription how to handle spurious double poles [14].

A final remark concerns the perturbative phase with its signatures of confinement. This phase is most likely not relevant for QCD at extensions smaller than L_3^c. Not only do we expect the center symmetry to be broken at small extensions but also dimensional reduction to QCD_{2+1} to happen. On the other hand, in the process of dimensional reduction, charged gluons decouple from the low-lying excitations due to their antiperiodic boundary conditions; the small extension or high temperature limit in this phase is QED_{2+1}. Thus in the deconfinement phase transition arising when compressing the QCD vacuum, a change to periodic boundary conditions as well as the disappearence of the geometrial mass must occur. This change in boundary conditions results in a change in Casimir energy density and pressure which for non-interacting gluons (and neglecting the effects of the geometrical mass) is given by

$$\Delta\epsilon = -\pi^2/12 L_3^4 , \qquad \Delta p = 3\Delta\epsilon . \qquad (24)$$

This estimate is of the order of magnitude of the change in the energy density across the confinement-deconfinement transition when compressing the system,

$$\Delta\epsilon = -0.45/L_3^4 , \qquad (25)$$

deduced from the finite temperature lattice calculation of Ref. [15]. If QCD in the high temperature or small extension quark gluon plasma phase is to be described perturbatively, one has to abandon the non-perturbative resolution which is implicit to the temporal or axial like gauge with its characteristic $N-1$ Polyakov loops as zero modes. Starting point has to be QCD at $g = 0$, and for this $\text{U}(1)^{N^2-1}$ theory an axial gauge with $N^2 - 1$ photons as zero modes is e.g. appropriate. This procedure breaks the center symmetry and the high temperature limit results.

In summary we have investigated QCD at finite extension in an axial like gauge. Novel aspects in the theoretical description have emerged due to the peculiar dynamics of the Polyakov loops; these variables are described by functional integrals which do not have a natural Gaussian limit. By elimination of these variables an effective description has been derived. It yields a QCD ground state which already at the perturbative level exhibits certain characteristics of the confining vacuum and permits to connect confinement with properties of the spectrum of gluonic excitations.

Acknowledgments

We thank V.L. Eletskii and A.C. Kalloniatis for valuable discussions.

References

[1] Schwinger J., *Phys. Rev.* **130** (1963) 402.
[2] Toms D. J., *Phys. Rev. D* **21** (1980) 928.
[3] Koch V., Shuryak E. V., Brown G. E. and Jackson A. D., *Phys. Rev. D* **46** (1992) 3169.
[4] For a recent review, see Kanayo K., *Nucl. Phys. B (Proc. Suppl.)* **47** (1996) 144.
[5] Svetitsky B., *Phys. Rep.* **132** (1986) 1.
[6] Reisz T., *Z. Phys. C* **53** (1992) 169.
[7] Lenz F., Thies M., Levit S. and Yazaki, K., *Ann. Phys.* **208** (1991) 1.
[8] Lenz F., Naus H. W. L. and Thies M., *Ann. Phys.* **233** (1994) 317;
   Lenz F., Moniz E. J. and Thies M., *Ann. Phys.* **242** (1995) 429.
[9] Reinhardt H., *Phys. Rev. D* **55** (1997) 2331.
[10] Lenz F., Shifman M. and Thies M., *Phys. Rev. D* **51** (1995) 7060.
[11] Weiss N., *Phys. Rev. D* **24** (1981) 475.
[12] Smilga A. V., *Ann. Phys.* **234** (1994) 1;
   Bhattacharya T., Gocksch A. and Korthals Altes C., *Nucl. Phys. B* **383** (1992) 497.
[13] Linde A. D., *Phys. Lett. B* **96** (1980) 289.
[14] Landshoff P. V., *Phys. Lett. B* **169** (1986) 69;
   Leibbrandt G. and Staley M., *Nucl. Phys. B* **428** (1994) 469.
[15] Engels J., Karsch F. and Redlich K., *Nucl. Phys. B* **435** (1995) 295.

Technicalities of the Zero Modes in the Light-Cone Representation

G. McCartor

Department of Physics
Southern Methodist University
Dallas, Texas 75275
USA

---

1. INTRODUCTION

In this talk I shall review the role of the light-cone zero modes in accommo-dating such features as degenerate vacua, condensates and anomalies in the light-cone representation. In addition to discussing the mechanisms by which the zero modes represent these features, I shall discuss the (still not completely solved) problem of calculating the dynamical operators in the presence of these zero modes. I shall illustrate my remarks, first with the best studied case — the Schwinger model — then with QED in 3+1 dimensions and QCD in 1+1 dimensions. I shall also remark on the possible relation of my remarks to an inconsistency in published solutions to the 't Hooft model.

To begin with I wish to recall the oft given argument that in the light-cone representation, the perturbative vacuum is the (unique) physical vacuum. That argument, combined with the known fact that there are systems with degenerate ground states, provided the original motivation for the studies I shall be discussing. While it can be stated several equivalent ways, the argument amounts to the following: The degree to which the operator P^+ can be modified from its value in free theory, P^+_{FREE} to some other value due to interactions is severely limited by the Heisenberg relation

$$\partial_-\phi = \frac{i}{2}[P^+, \phi]$$

Since we initialize our fields to be isomorphic to free fields on the initial value surface $x^+ = 0$, we have immediately that on that surface

$$\partial_- \phi = \frac{i}{2}[P^+_{FREE}, \phi]$$

If the set of fields initialized on $x^+ = 0$, here generically called ϕ, forms an irreducible set, the only possible modification to P^+_{FREE} would be to add a multiple of the identity which would then get removed in a vacuum adjustment. On the other hand, the Fock space for fields initialized on $x^+ = 0$ contains only one state, the perturbative vacuum, for which

$$P^+_{FREE}|s\rangle = 0$$

Thus we conclude that the perturbative vacuum is the only state annihilated by the (interacting) P^+ and thus must be the unique physical vacuum.

The catch in the argument is that since $x^+ = 0$ is a characteristic, one can not always completely specify a physics problem with data on that surface and so the fields specified on $x^+ = 0$ may not form an irreducible set. If there are fields — the zero modes, $\phi_z = \phi_z(x^+)$ — which are functions only of x^+, then P^+ will commute with them and any vector of the form

$$\mathcal{F}(\phi_z)|0\rangle$$

where $\mathcal{F}(\phi_z)$ is any functional of ϕ_z will satisfy

$$P^+ \mathcal{F}(\phi)|0\rangle = P^+_{FREE}\mathcal{F}(\phi)|0\rangle = 0$$

In so far as the argument given above is concerned, any such state would be a candidate to be the physical vacuum. Notice that while we have argued that operators from fields initialized on $x^+ = 0$ cannot mix with P^+, operators from fields initialized on $x^- = 0$ could.

2. SCHWINGER MODEL

Here I shall sketch how the zero modes work in the case of the Schwinger model[1][2][3]; details may be found in the references. The field ψ_+ is initialized on $x^+ = 0$, but the field ψ_- cannot be; we shall initialize it on $x^- = 0$. We shall choose antiperiodic boundary conditions on each surface:

$$\psi_+(x^- + 2L) = -\psi_+(x^-)$$

$$\psi_-(x^+ + 2L) = -\psi_-(x^+)$$

We work in the gauge $\partial_- A^+ = 0$. The operator solution is then given by

$$\psi_+ = \frac{1}{\sqrt{2L}}e^{\lambda_+^{(-)}(x)}\sigma_+(x)e^{\lambda_+^{(+)}(x)}$$

$$\psi_- = \frac{1}{\sqrt{2L}} e^{-\lambda_-^*(x^+)} \sigma_-(x) e^{\lambda_-(x^+)}$$

where

$$\lambda_+(x) = -i\sqrt{\frac{\pi}{L}} \sum_{n=1}^{\infty} \frac{1}{\sqrt{p_-(n)}} \left(C(n)e^{-ip(n)x} + C^*(n)e^{ip(n)x} \right)$$

$$\lambda_-(x^+) = \sum_{n=1}^{\infty} \frac{1}{\sqrt{n}} \left(D(n)e^{-ik_+(n)x^+} + D^*(n)e^{ik_+(n)x^+} \right)$$

the σ's are spurions and

$$P_+(n) = \frac{m^2 L}{2n\pi} \quad ; \quad P_-(n) = k_+(n) = \frac{n\pi}{L}$$

The $D(n)$ operators are zero modes. The operator σ_- does pick up an x^- dependence due to the interaction, but it is composed of operators from the ψ_- field and I do not know any way to initialize it on $x^+ = 0$. What role do the operators from the ψ_- field play in the operator solution? For one thing they create states necessary to form the degenerate ground states. The ground states which are eigenvalues of Q_5 are given by

$$(\sigma_+^* \sigma_-)^M |0\rangle = |\Omega(M)\rangle$$

The θ-states are then formed as

$$|\Omega(\theta)\rangle \equiv \sum_{n=-\infty}^{\infty} e^{iM\theta} |\Omega(M)\rangle$$

The operator, σ_- is made up of operators from the ψ_- field and without these operators the θ-states cannot be defined and a fully gauge invariant solution cannot be given. Operators from the ψ_- field also mix with P^+; we have

$$P^+ = \frac{1}{2} \int_{-L}^{L} :2i\left(\psi_+^* \partial_- \psi_+ - \partial_- \psi_+^* \psi_+ \right): dx^- = P_{FREE}^+ - \frac{Lm^2}{4}(A^+)^2$$

The last term results from the need to define a gauge invariant Fermi product by point splitting. The additional term is crucial to the existence of degenerate ground states in that it causes states which are split in free theory to have zero P^+ under the interaction. Recall that we argued above that operators from fields initialized on $x^+ = 0$ cannot mix with P_{FREE}^+; thus A^+ must either be an independent operator (a degree of freedom) or a functional of operators associated with ψ_-. We shall now show that A^+ is a constraint by solving for it — a rather more subtle process than one might guess.

If we take the zero mode of the Maxwell equation

$$\frac{\partial^2 A^-}{\partial x^{-2}} = -\frac{1}{2} J^+$$

we get

$$\frac{1}{2L}Q_+ = \frac{e^2}{2\pi}A^+ \qquad (\textit{this is wrong})$$

The operator Q_+ is formed from ψ_+-modes and so if we use this relation in the expression for P^+ we find immediate conflict with the Heisenberg equations. The source of this problem, and the solution are to be found by considering a problem in classical E&M. A charge density, ρ, which is constant in space cannot create an electric field. On the other hand the Maxwell equation

$$\frac{\partial E}{\partial x} = \rho$$

does not have zero as a possible solution. In E&M, whether classical or quantum, the fields actually couple to the currents

$$J'^0 = J^0 - J^0(0) \quad ; \quad J'^1 = J^1$$

With this correction, we find

$$\frac{1}{2L}Q_- = \frac{e^2}{2\pi}A^+$$

Which is correct. If we had not thought of the argument just given, we might have reached the correct conclusion by observing that the physical states in the Schwinger model must be chargeless and so must satisfy

$$\frac{1}{2}(Q_+ + Q_-)|p\rangle = 0$$

So we have

$$Q_- \approx Q_+$$

From this second point of view we find that we must replace an operator with its weak equivalent in order to avoid conflict with the Heisenberg equations. Such a situation is encountered fairly often in formulating quantum field theories in the light-cone representation and in many cases we do not understand the underlying physics as well as we do in the case of the Schwinger model.

With all this machinery in place one finds a spectrum which agrees with previous calculations, a θ-structure which is isomorphic to the usual formulation, an anomaly of exactly the expected form and a condensate. The condensate has the form

$$\langle\Omega|\overline{\psi}\psi|\Omega\rangle = \frac{1}{L}\cos\theta$$

The condensate goes to zero as L goes to infinity. This effect is due to our boundary conditions; in light-cone gauge in the continuum or with periodicity conditions on a space-like surface one gets the usual condensate either immediately or in the large L limit. For the condensate to survive at large L the spectrum has to be dense in the neighborhood of $p^+ = 0$. That fact is perhaps not surprising if we think of the way the continuum works[4]. In that case the gauge is $A^+ = 0$ and the operator P^+ does not get dressed; rather, the proper definition of the operator products requires the introduction of a ghost field, η. That field combines with the positive norm fusion field from ψ_- to form a zero norm field, B, which can be used to define a kernel of zero norm states according to

$$B \equiv \eta - \phi_2 \quad ; \quad \mathcal{K} \equiv \mathcal{P}(B)|\Omega\rangle$$

The Hilbert space is then formed as the completion of the factor space

$$\mathcal{H} = \frac{\overline{\mathcal{S}}}{\mathcal{K}}$$

All states with nonzero charge or pseudocharge (thus all the θ-states) arise through the completion — they all lie in the bar and thus right at $p^+ = 0$.

I speculate that the effects we have just been discussing will ultimately be found to be related to the resolution of an inconsistency in published solutions to the 't Hooft model[5] (large N QCD in 1+1 dimensions). To formulate his solution 't Hooft excised the small p^+ region from the spectrum. With this regulator he then found the full propagator – a propagator which implies no condensate. With the spectrum and wavefunctions which result from that propagator Zhitnitsky[6] was able to derive an exact, nonzero value for the condensate.

3. QED 3+1

While physics in 1+1 dimensions might be expected to emphasize the role of the zero modes, and the best studied cases are in 1+1 dimensions, they also play a role in four dimensional physics. I shall briefly review what is known of their role in quantizing QED in the light-cone representation and in light-cone gauge[7]. The Maxwell equation for the constrained field A^- is

$$\partial_-^2 A^- = -\partial_- \partial_i A^i - g\frac{1}{2}J^+$$

The most general solution is

$$A^- = -\frac{1}{\partial_-}\partial_i A^i - g\frac{1}{2\partial_-^2}J^+ + (\partial_\perp^2 \phi)x^- + \gamma(x^+, x_\perp)$$

where the form for the coefficient of x^- is chosen for convenience. One might be tempted to set ϕ and γ equal to zero, but that leads to inconsistencies. Indeed we find that

$$[\phi(x^+, x_\perp), \lambda(y^+, y_\perp)] = i\delta(x^+ - y^+)\delta^{(2)}(x_\perp - y_\perp)$$

So neither field is zero. Furthermore, if we define η by

$$\psi \equiv e^{ig\phi}\eta$$

we find that it is the η field that satisfies

$$\{\eta_+(x^-, x_\perp), \eta_+^\dagger(y^-, y_\perp)\} = \Lambda_+ \delta(x^- - y^-)\delta^{(2)}(x_\perp - y_\perp)$$

While it appears that the dependence on the ϕ field could be removed by a gauge transformation, attempting to do so fouls up the Fermi products (one might be able to implement the required gauge transformation by enlarging the representation space but I am not sure about this).

4. ADJOINT SU(2) IN 1+1

A discussion, though not a full operator solution, of SU(2) gauge theory in 1+1 dimensions with matter in the adjoint representation and the theory quantized in the light-cone representation has been given[8]. Here I shall just remark on some of the particular features most relevant to the points I have been discussing. The formulation is greatly simplified by the use of twisted boundary conditions — that is, two color components of the Fermi field (the off diagonal ones) are taken to be antiperiodic while the other is taken to be periodic. The gauge choice is that $A^+ \sim \tau^3$ is a zero mode. If we use the equations of motion to solve for A^+ we again find that it seems that it should be written in terms of the ψ_+ degrees of freedom but again we must make the adjustment of substituting an operator for its weak equivalent so that the whole process looks like

$$A^+ = \frac{\pi Q_+^3}{gL}\tau^3 \implies -\frac{\pi Q_-^3}{gL}\tau^3$$

Similar considerations apply to the zero mode in A^- and we find, after substitution

$$A^-(0) = -\frac{\pi Q_+^3}{gL}\tau^3$$

Again we find that some modes from the ψ_- field mix with the operator P^+ as

$$P^+ = P_{FREE}^+ - \frac{\pi}{2L}(Q_-^3)^2$$

Again that mixing is crucial to the fact that the system has two ground states. A feature which has no obvious analog in the abelian case concerns P^-. Straightforward integration of the appropriate densities over the initial value surfaces gives, for the part of P^- which depends on ψ_+

$$P_R^- = \frac{g^2 L}{4\pi^2}\left[\sum_{N=-\infty}^{\infty}{}' \frac{1}{N^2}C_N^3 C_{-N}^3 + \sum_{n=-\infty}^{\infty}\frac{1}{(n-Q_-^3)^2}\{C_n^+, C_{-n}^-\}\right]$$

This relation again puts the Heisenberg equations in conflict with the initial values and we must make the weak substitution

$$Q_-^3 \implies Q_+^3$$

to get

$$P_R^- = \frac{g^2 L}{4\pi^2} \left[\sum_{N=-\infty}^{\infty}{}' \frac{1}{N^2} C_N^3 C_{-N}^3 + \sum_{n=-\infty}^{\infty} \frac{1}{(n - Q_+^3)^2} \{ C_n^+, C_{-n}^- \} \right]$$

For this substitution we have no physical justification comparable to the one I gave for the case of the Schwinger model.

For the vacuum, the situation is in many ways similar to the Schwinger model. We can define spurion operators out of modes from the 1 and 2 color components of the Fermi field and write

$$(\sigma_+^* \sigma_-)^M |0\rangle = |\Omega(M)\rangle$$

In the nonabelian case, however, it is not enough to write

$$|\theta\rangle = \sum_{N=-\infty}^{\infty} e^{-iN\theta} |V_N\rangle$$

but to have full gauge invariance we must restrict θ to be either 0 or π. Thus the system has two ground states as suggested by topological considerations.

References

[1] McCartor G., *Z. Phys.* C **64**, 349 (1994).
[2] McCartor G.,tex, *Z. Phys.* C **52**, 611 (1991).
[3] McCartor G., *Int. J. Mod. Phys.* **52**, 611 (1991).
[4] Nakawaki Y., Private Communication, and, *Prog. Theor. Phys.* **72**, 134 (1984).
[5] 't Hooft G., *Nucl. Phys.* B **75**, 461 (1974).
[6] Zhitnitsky A., *Phys. Lett.* B **165**, 405 (1985).
[7] McCartor G. and Robertson D., *Z. Phys.* C **68**, 345 (1995).
[8] McCartor G., Robertson D. and Pinsky S., **hep-th/9612083** *in press*

Zero Mode and Symmetry Breaking on the Light Front

Koichi Yamawaki

Department of Physics
Nagoya University
Nagoya 464-01
Japan

1. INTRODUCTION

Much attention has recently been paid to the light-front (LF) quantization [1] as a promising approach to solve the nonperturbative dynamics. The most important aspect of the LF quantization is that the physical LF vacuum is simple, or even trivial [2]. However, such a trivial vacuum, which is vital to the whole LF approach, can be realized only if we can remove the so-called zero mode out of the physical Fock space ("zero mode problem" [3]).

Actually, the Discretized Light-Cone Quantization (DLCQ) was first introduced by Maskawa and Yamawaki [3] in 1976 to resolve the zero mode problem and was re-discovered by Pauli and Brodsky [4] in 1985 in a different context. The zero mode in DLCQ is clearly isolated from other modes and hence can be treated in a well-defined manner without ambiguity, in sharp contrast to the continuum theory where the zero mode is the accumulating point and hard to be controlled in isolation [5]. In DLCQ, Maskawa and Yamawaki [3] in fact discovered a constraint equation for the zero mode (*"zero-mode constraint"*) through which the zero mode becomes dependent on other modes and then they observed that the zero mode can be removed from the physical Fock space by solving the zero-mode constraint, thus *establishing the trivial LF vacuum in DLCQ.*

Such a trivial vacuum, on the other hand, might confront the usual picture

of complicated nonperturbative vacuum structure in the equal-time quantization corresponding to confinement, spontaneous symmetry breaking (SSB),
etc.. Since the vacuum is proved trivial in DLCQ [3], the only possibility to
realize such phenomena would be through the complicated structure of the
operator and only such an operator would be the zero mode. In fact the zero-
mode constraint itself implies that the zero mode carries essential information
on the complicated dynamics. One might thus expect that explicit solution of
the zero-mode constraint in DLCQ would give rise to the SSB while preserv-
ing the trivial LF vacuum. Actually, several authors have recently argued in
(1+1) dimensional models that the solution to the zero-mode constraint might
induce spontaneous breaking of the discrete symmetries [6]. However, the most
outstanding feature of the SSB is the existence of the Nambu-Goldstone (NG)
boson for the continuous symmetry breaking in (3+1) dimensions.

In this talk I shall explain, within the canonical DLCQ [3], how the NG phenomenon is realized through the zero modes in (3+1) dimensions while keeping
the vacuum trivial. The talk is based on recent works done in collaboration
with Yoonbai Kim and Sho Tsujimaru [7, 8]. We encounter a striking feature
of the zero mode of the NG boson: Naive use of the zero-mode constraints does
not lead to the NG phase at all ((false) "no-go theorem") in contrast to the
current expectation mentioned above. It is inevitable to introduce an infrared
regularization by explicit-breaking mass of the NG boson m_π. The NG phase
can only be realized via peculiar behavior of the zero mode of the NG-boson
fields: *The NG-boson zero mode, when integrated over the LF, must have a singular behavior* $\sim 1/m_\pi^2$ *in the symmetric limit* $m_\pi^2 \to 0$. This we demonstrate
both in a general framework of the LSZ reduction formula and in a concrete
field theoretical model, the linear σ model. The NG phase is in fact realized in
such a way that the *vacuum is trivial* while the *LF charge is not conserved* in
the symmetric limit $m_\pi^2 \to 0$.

2. ZERO MODE PROBLEM IN THE CONTINUUM THEORY

Before discussing the zero mode in DLCQ, we here comment that *it is impossible to remove the zero mode in the continuum theory* in a manner consistent
with the trivial vacuum, *as far as we use the canonical commutator* ([1]):

$$[\phi(x), \phi(y)]_{x^+=y^+} = -\frac{i}{4}\epsilon(x^- - y^-)\delta^{(2)}(x^\perp - y^\perp), \qquad (1)$$

where the sign function $\epsilon(x^-)$ is defined by

$$\epsilon(x^-) = \frac{i}{\pi}\mathcal{P}\int_{-\infty}^{+\infty}\frac{dp^+}{p^+}e^{-ip^+x^-} \qquad (2)$$

([1]) We choose the LF "time" as $x^+ = x_- \equiv \frac{1}{\sqrt{2}}(x^0 + x^3)$ and the longitudinal and
the transverse coordinates are denoted by $\vec{x} \equiv (x^-, x^\perp)$, with $x^- \equiv \frac{1}{\sqrt{2}}(x^0 - x^3)$ and
$x^\perp \equiv (x^1, x^2)$, respectively.

through the principal value prescription and hence has no $p^+ \equiv 0$ mode. The point is that the real problem with the zero mode in the continuum theory is *not a single mode* with $p^+ \equiv 0$, which is just measure zero, but actually the *accumulating point* $p^+ \to 0$ [5].

To demonstrate this, let us first illustrate a no-go theorem found by Nakanishi and Yamawaki [5]. The LF canonical commutator (1) gives explicit expression of Wightman two-point function on LF:

$$\langle 0|\phi(x)\phi(0)|0\rangle|_{x^+=0} = \frac{1}{2\pi} \int_0^\infty \frac{dp^+}{2p^+} e^{-ip^+x^-} \cdot \delta^{(2)}(x^\perp), \qquad (3)$$

which is logarithmically divergent at $p^+ = 0$ and local in $x^\perp$ and, more importantly, is independent of the interaction and the mass. It is easy to see that this is wrong, for example, in the free theory where the Lorentz-invariant two-point Wightman function is given at any point x by

$$\Delta^{(+)}(x) = \frac{1}{(2\pi)^3} \int_0^\infty \frac{dp^+}{2p^+} e^{-ip^-x^+ - ip^+x^- + ip^\perp x^\perp} = \frac{m}{4\pi^2\sqrt{-x^2}} K_1(m\sqrt{-x^2}), \qquad (4)$$

where K_1 is the Hankel function. Restricting (4) to the LF, $x^+ = 0$, yields $\frac{m}{4\pi^2\sqrt{x_\perp^2}} K_1(m\sqrt{x_\perp^2})$, which is finite (positive definite), nonlocal in $x^\perp$ and dependent on mass, in obvious contradiction to the above result (3). Actually, the Lorentz-invariant result (4) is a consequence of the *mass-dependent* regularization of $1/p^+$ singularity by the infinitely oscillating (mass-dependent) phase factor $e^{-ip^-x^+} = e^{-i(m^2+p_\perp^2)/2p^+ \cdot x^+}$ *before taking the LF restriction* $x^+ = 0$. Namely, there exists *no free theory on the LF!* This difficulty also applies to the interacting theory satisfying the Wightman axioms [5]. Thus the LF restriction from the beginning loses all the information of dynamics carried by the *zero mode as the accumulating point.*

Next we show another difficulty of the continuum LF theory, that is, *as far as the sign function is used*, the *LF charge does not annihilate the vacuum* in the SSB phase and hence the trivial LF vacuum is never realized [8]. Let me illustrate this in the $O(2)$ σ model where the fields σ, π obey the canonical commutator (1), which yields

$$[Q, \sigma'(x)] = -i\pi(x), \quad [Q, \pi(x)] = i\sigma'(x) + \frac{i}{2}v, \qquad (5)$$

where the LF charge is defined as usual by $Q = \int d^3\vec{x}(\pi\partial_-\sigma' - \sigma'\partial_- pi - v\partial_-\pi)$, and the anti-periodic boundary condition (B.C.) is imposed on π and the shifted field $\sigma' \equiv \sigma - v$ ($v = \langle\sigma\rangle$) in order to eliminate the surface terms in (5), $\phi(x^- = \infty) + \phi(x^- = -\infty) = 0$. The non-zero constant term on the R.H.S. of (5) has its origin in the commutation relation (1), which is consistent with the anti-periodic B.C., $\pi(x^- = \infty) = -\pi(x^- = -\infty) \neq 0$ (^{2}). Then we have

$$\langle 0|[Q, \pi(x)]|0\rangle = i\langle 0|\sigma'(x)|0\rangle + \frac{i}{2}v = \frac{i}{2}v \neq 0. \qquad (6)$$

(^{2}) Without specifying B.C., we would not be able to formulate consistently the LF

Namely, the *LF charge does not annihilate the vacuum*, $Q|0\rangle \neq 0$, due to the zero mode as the accumulating point, even though we have "removed" exact zero mode $p^+ \equiv 0$ by shifting the field σ to σ'. This implies that there in fact *exists a zero-mode state* $|\alpha\rangle \equiv Q|0\rangle$ with zero eigenvalue of P^+ such that $P^+|\alpha\rangle = [P^+, Q]|0\rangle = 0$ (due to P^+ conservation). Our result disagrees with Wilson et al. [10] who claim to have eliminated the zero mode (to be compensated by "unusual counter terms") in the continuum LF theory.

3. NAMBU-GOLDSTONE BOSON ON THE LIGHT FRONT

In contrast to the continuum LF theory mentioned above, we already mentioned in Introduction that the trivial vacuum is always realized in DLCQ [3]. We now discuss how such a trivial vacuum can be reconciled with the SSB phenomena. Here we use DLCQ [3, 4], $x^- \in [-L, L]$, with a periodic boundary condition in the x^- direction, and then take the continuum limit $L \to \infty$ in the end of whole calculation. We should mention here that the above no-go theorem [5] cannot be solved

by simply taking the continuum limit of the DLCQ nor by any other existing method and would be a future problem to be solved in a more profound way.

3.1. "NO-GO THEOREM" (FALSE)

Let us first prove a "no-go theorem" (which will turn out to be false later) that the *naive LF restriction* of the NG-boson field leads to vanishing of both the NG-boson emission vertex and the corresponding current vertex; namely, *the NG phase is not realized in the LF quantization* [7].

Based on the LSZ reduction formula, the NG-boson emission vertex $A \to B + \pi$ may be written as

$$\langle B\pi(q)|A\rangle = i \int d^4x e^{iqx} \langle B|\Box\pi(x)|A\rangle$$

$$= i(2\pi)^4 \delta(p_A^- - p_B^- - q^-)\delta^{(3)}(\vec{p}_A - \vec{p}_B - \vec{q})\langle B|j_\pi(0)|A\rangle, \qquad (7)$$

where $\pi(x)$ and $j_\pi(x) = \Box\pi(x) = (2\partial_+\partial_- - \partial_\perp^2)\pi(x)$ are the interpolating field and the source function of the NG boson, respectively, and $q^\mu = p_A^\mu - p_B^\mu$ are the NG-boson four-momenta and $\vec{q} \equiv (q^+, q^\perp)$. It is customary [11] to take the collinear momentum frame, $\vec{q} = 0$ and $q^- \neq 0$ (not a soft momentum), for the emission vertex of the exactly massless NG boson with $q^2 = 0$.

Then the NG-boson emission vertex should vanish on the LF due to the periodic boundary condition:

$$(2\pi)^3 \delta^{(3)}(\vec{p}_A - \vec{p}_B)\langle B|j_\pi(0)|A\rangle$$

quantization [9, 8]. If we used anti-periodic B.C. for the *full field* σ, we would have no zero mode (no v) and hence no symmetry breaking anyway. On the other hand, vanishing fields at $x^- = \pm\infty$ contradict the commutation relation (1).

$$= \int d^2x^\perp \lim_{L\to\infty} \langle B|\Big(\int_{-L}^{L} dx^- 2\partial_-\partial_+\pi\Big)|A\rangle = 0. \tag{8}$$

Another symptom of this disease is the vanishing of the current vertex (analogue of g_A in the nucleon matrix element). When the continuous symmetry is spontaneously broken, the NG theorem requires that the corresponding current J_μ contains an interpolating field of the NG boson $\pi(x)$, that is, $J_\mu = -f_\pi \partial_\mu \pi + \widehat{J}_\mu$, where f_π is the "decay constant" of the NG boson and $\widehat{J}_\mu$ denotes the non-pole term. Then the current conservation $\partial_\mu J^\mu = 0$ leads to

$$0 = \langle B| \int d^3\vec{x}\, \partial_\mu \widehat{J}^\mu(x)|A\rangle_{x^+=0}$$

$$= -i(2\pi)^3\delta^{(3)}(\vec{q})\frac{m_A^2 - m_B^2}{2p_A^+}\langle B|\widehat{J}^+(0)|A\rangle, \tag{9}$$

where $\int d^3\vec{x} \equiv \lim_{L\to\infty}\int_{-L}^{L} dx^- d^2x^\perp$ and the integral of the NG-boson sector $\Box\pi$ has no contribution on the LF because of the periodic boundary condition as we mentioned before. Thus the current vertex $\langle B|\widehat{J}^+(0)|A\rangle$ should vanish at $q^2 = 0$ as far as $m_A^2 \neq m_B^2$.

This is actually a manifestation of the conservation of a charge $\widehat{Q} \equiv \int d^3\vec{x}\, \widehat{J}^+$ which contains only the non-pole term. Note that $\widehat{Q}$ coincides with the full LF charge $Q \equiv \int d^3\vec{x}\, J^+$, since the pole part always drops out of Q due to the integration on the LF, i.e., $Q = \widehat{Q}$. Therefore the *conservation of $\widehat{Q}$ inevitably follows from the conservation of Q*: $[\widehat{Q}, P^-] = [Q, P^-] = 0$, which in fact implies vanishing current vertex mentioned above. This is in sharp contrast to the charge integrated over usual space $x = (x^1, x^2, x^3)$ in the equal-time quantization: $Q^{et} = \int d^3x J^0$ is conserved while $\widehat{Q}^{et} = \int d^3x \widehat{J}^0$ is not.

Here we should emphasize that the above conclusion is *not* an artifact of DLCQ but is inherent in the very nature of the LF quantization [8], as far as we discuss the exact symmetry limit from the beginning.

3.2. REALIZATION OF NG PHASE

Now, we propose to regularize the theory by introducing explicit-breaking mass of the NG boson m_π and then take the symmetric limit in the end under certain condition. The essence of the NG phase with a small explicit symmetry breaking can well be described by the old notion of the PCAC hypothesis: $\partial_\mu J^\mu(x) = f_\pi m_\pi^2 \pi(x)$, with $\pi(x)$ being the interpolating field of the (pseudo-) NG boson π. From the PCAC relation the current divergence of the non-pole term $\widehat{J}^\mu(x)$ reads $\partial_\mu \widehat{J}^\mu(x) = f_\pi(\Box + m_\pi^2)\pi(x) = f_\pi j_\pi(x)$. Then we obtain

$$\langle B| \int d^3\vec{x}\, \partial_\mu \widehat{J}^\mu(x)|A\rangle = f_\pi m_\pi^2 \langle B| \int d^3\vec{x}\, \pi(x)|A\rangle = \langle B| \int d^3\vec{x}\, f_\pi j_\pi(x)|A\rangle, \tag{10}$$

where the integration of the pole term $\Box\pi(x)$ is dropped out as before. The equality between the first and the third terms is a generalized Goldberger-Treiman relation, if both are non-zero. The second expression of (10) is nothing but the matrix element of the LF integration of the π zero mode (with $P^+ = 0$) $\omega_\pi \equiv \frac{1}{2L} \int_{-L}^{L} dx^-\, \pi(x)$. Suppose that $\int d^3\vec{x}\, \omega_\pi(x) = \int d^3\vec{x}\, \pi(x)$ is regular when $m_\pi^2 \to 0$. Then this leads to the "no-go theorem" again. Thus, in order to have non-zero NG-boson emission vertex (R.H.S. of (10)) as well as non-zero current vertex (L.H.S.) at $q^2 = 0$, the π zero mode $\omega_\pi(x)$ must behave as

$$\int d^3\vec{x}\, \omega_\pi \sim \frac{1}{m_\pi^2} \quad (m_\pi^2 \to 0). \tag{11}$$

This situation may be clarified when the PCAC relation is written in the momentum space:

$$\frac{m_\pi^2 f_\pi j_\pi(q^2)}{m_\pi^2 - q^2} = \partial^\mu J_\mu(q) = \frac{q^2 f_\pi j_\pi(q^2)}{m_\pi^2 - q^2} + \partial^\mu \widehat{J}_\mu(q). \tag{12}$$

What we have done when we reached the "no-go theorem" can be summarized as follows: We first set L.H.S of (12) to zero (or equivalently, assumed implicitly the regular behavior of $\int d^3\vec{x}\, \omega_\pi(x)$) in the symmetric limit in accord with the current conservation $\partial^\mu J_\mu = 0$. Then in the LF formalism with $\vec{q} = 0$ ($q^2 = 0$), the first term (NG-boson pole term) of R.H.S. was also zero due to the periodic boundary condition or the zero-mode constraint. Thus we arrived at $\partial^\mu \widehat{J}_\mu(q) = 0$. However, this procedure is equivalent to playing a nonsense game: $1 = \lim_{m_\pi^2,\, q^2 \to 0} \left(\frac{m_\pi^2 - q^2}{m_\pi^2 - q^2}\right) = 0$ as far as $f_\pi j_\pi \neq 0$ (NG phase). Therefore the "$m_\pi^2 = 0$ theory" with vanishing L.H.S. is ill-defined on the LF, namely, the "no-go theorem" is false. The correct procedure should be to take the symmetric limit $m_\pi^2 \to 0$ after the LF restriction $\vec{q} = 0$ ($q^2 = 0$) although (12) itself yields the same result $f_\pi j_\pi = \partial^\mu \widehat{J}_\mu$, irrespectively of the order of the two limits $q^2 \to 0$ and $m_\pi^2 \to 0$. Then (11) does follow.

This implies that at quantum level the LF charge $Q = \widehat{Q}$ is not conserved, or the current conservation does not hold for a particular Fourier component with $\vec{q} = 0$ even in the symmetric limit:

$$\dot{Q} = \frac{1}{i}[Q, P^-] = \partial^\mu J_\mu|_{\vec{q}=0} = f_\pi \lim_{m_\pi^2 \to 0} m_\pi^2 \int d^3\vec{x}\, \omega_\pi \neq 0. \tag{13}$$

4.   THE SIGMA MODEL

Let us now demonstrate [7] that (11) and (13) indeed take place as the solution of the constrained zero modes in the NG phase of the $O(2)$ linear σ model:

$$\mathcal{L} = \frac{1}{2}(\partial_\mu \sigma)^2 + \frac{1}{2}(\partial_\mu \pi)^2 - \frac{1}{2}\mu^2(\sigma^2 + \pi^2) - \frac{\lambda}{4}(\sigma^2 + \pi^2)^2 + c\sigma, \tag{14}$$

where the last term is the explicit breaking which regularizes the NG-boson zero mode.

In the DLCQ we can clearly separate the zero modes (with $P^+ = 0$), $\pi_0 \equiv \frac{1}{2L} \int_{-L}^{L} dx^- \pi(x)$ (similarly for σ_0), from other oscillating modes (with $P^+ \neq 0$), $\varphi_\pi \equiv \pi - \pi_0$ (similarly for φ_σ). Through the Dirac quantization of the constrained system the canonical commutation relation for the oscillating modes reads [3]

$$[\varphi_i(x), \varphi_j(y)] = -\frac{i}{4}\{\epsilon(x^- - y^-) - \frac{x^- - y^-}{L}\}\delta_{ij}\delta^{(2)}(x^\perp - y^\perp), \quad (15)$$

where each index stands for π or σ. Comparing (15) with (1), we can see that the second term in $\{\}$ corresponds to subtracting the zero mode contribution out of the commutator.

On the other hand, the zero modes are not independent degrees of freedom but are implicitly determined by φ_σ and φ_π through the second class constraints, the zero-mode constraints [3]:

$$\chi_\pi \equiv \frac{1}{2L} \int_{-L}^{L} dx^- \left[(\mu^2 - \partial_\perp^2)\pi + \lambda\pi(\pi^2 + \sigma^2)\right] = 0, \quad (16)$$

and similarly, $\chi_\sigma \equiv \frac{1}{2L} \int_{-L}^{L} dx^- \{[\pi \leftrightarrow \sigma] - c\} = 0$. Thus the zero modes are solved away from the physical Fock space which is constructed upon the trivial vacuum. Note that through the equation of motion these constraints are equivalent to the characteristic of the DLCQ with periodic boundary condition [12]: $\chi_\pi = -\frac{1}{2L} \int_{-L}^{L} dx^- 2\partial_+\partial_-\pi = 0$, (similarly for σ) which we have used to prove the "no-go theorem" for the case of $m_\pi^2 \equiv 0$. Thus the 'no-go" theorem is a consequence of the zero-mode constraint itself in the case of $m_\pi^2 \equiv 0$. Namely, *solving the zero-mode constraint does not give rise to SSB at all in the exact symmetric case* $m_\pi^2 \equiv 0$, in contradiction to the naive expectation [6].

Actually, in the NG phase ($\mu^2 < 0$) the equation of motion of π reads ($\Box + m_\pi^2)\pi(x) = -\lambda(\pi^3 + \pi\sigma'^2 + 2v\pi\sigma') \equiv j_\pi(x)$, with $\sigma' = \sigma - v$ and $m_\pi^2 = \mu^2 + \lambda v^2 = c/v$, where $v \equiv \langle\sigma\rangle$ is the classical vacuum solution determined by $\mu^2 v + \lambda v^3 = c$. Integrating the above equation of motion over $\vec{x}$, we have

$$\int d^3\vec{x}\, j_\pi(x) - m_\pi^2 \int d^3\vec{x}\, \omega_\pi(x) = \int d^3\vec{x}\, \Box\pi(x) = -\int d^3\vec{x}\, \chi_\pi = 0, \quad (17)$$

where $\int d^3\vec{x}\, \omega_\pi(x) = \int d^3\vec{x}\, \pi(x)$. Were it not for the singular behavior (11) for the π zero mode ω_π, we would have concluded $(2\pi)^3\delta^{(3)}(\vec{q}) \langle\pi|j_\pi(0)|\sigma\rangle = -\langle\pi| \int d^3\vec{x}\, \chi_\pi|\sigma\rangle = 0$ in the symmetric limit $m_\pi^2 \to 0$. Namely, the NG-boson vertex at $q^2 = 0$ would have vanished, which is exactly what we called "no-go theorem" now related to the zero-mode constraint χ_π. On the contrary, direct evaluation of the matrix element of $j_\pi = -\lambda(\pi^3 + \pi\sigma'^2 + 2v\pi\sigma')$ in the lowest order perturbation yields non-zero result even in the symmetric limit $m_\pi^2 \to 0$: $\langle\pi|j_\pi(0)|\sigma\rangle = -2\lambda v\langle\pi|\varphi_\sigma\varphi_\pi|\sigma\rangle = -2\lambda v \neq 0$ $(\vec{q} = 0)$, which is in agreement with the usual equal-time formulation. Thus we have seen that naive use of the zero-mode constraints by setting $m_\pi^2 \equiv 0$ leads to the *internal inconsistency* in the NG phase. The "no-go theorem" is again false.

The same conclusion can be obtained more systematically by solving the zero-mode constraints in the perturbation around the classical (tree level) solution to the zero-mode constraints which is nothing but the minimum of the classical potential: $v_\pi = 0$ and $v_\sigma \equiv v$, where we have divided the zero modes π_0 (or σ_0) into classical constant piece v_π (or v_σ) and operator part ω_π (or ω_σ). The operator zero modes are solved perturbatively by substituting the expansion $\omega_i = \sum_{k=1} \lambda^k \omega_i^{(k)}$ into χ_π, χ_σ under the Weyl ordering.

The lowest order solution of the zero-mode constraints χ_π and χ_σ for ω_π takes the form:

$$(-m_\pi^2 + \partial_\perp^2)\,\omega_\pi = \frac{\lambda}{2L} \int_{-L}^{L} dx^- (\varphi_\pi^3 + \varphi_\pi \varphi_\sigma^2 + 2v\varphi_\pi \varphi_\sigma), \qquad (18)$$

which in fact yields (11) as

$$\lim_{m_\pi^2 \to 0} m_\pi^2 \int d^3\vec{x}\, \omega_\pi = -\lambda \int d^3\vec{x}\, (\varphi_\pi^3 + \varphi_\pi \varphi_\sigma^2 + 2v\varphi_\pi \varphi_\sigma) \neq 0. \qquad (19)$$

This actually ensures non-zero $\sigma \to \pi\pi$ vertex through (17): $\langle \pi | j_\pi(0) | \sigma \rangle = -2\lambda v$, which agrees with the previous direct evaluation as it should.

Let us next discuss the LF charge operator corresponding to the current $J_\mu = \partial_\mu \sigma \pi - \partial_\mu \pi \sigma$. The LF charge $Q = \widehat{Q} = \int d^3\vec{x}\,(\partial_- \varphi_\sigma \varphi_\pi - \partial_- \varphi_\pi \varphi_\sigma)$ *contains no zero modes* and hence no π-pole term which was dropped by the integration due to the periodic boundary condition and the ∂_-, so that Q is well defined even in the NG phase and hence annihilates the vacuum simply by the P^+ conservation [3]:

$$Q|0\rangle = 0. \qquad (20)$$

This is also consistent with explicit computation of the commutators: $\langle [Q, \varphi_\sigma] \rangle = -i\langle \varphi_\pi \rangle = 0$ and $\langle [Q, \varphi_\pi] \rangle = i\langle \varphi_\sigma \rangle = 0$ (^{3}), which are contrasted to (6) in the continuum theory. They are also to be compared with those in the usual equal-time case where the SSB charge does not annihilate the vacuum $Q^{\text{et}}|0\rangle \neq 0$: $\langle [Q^{\text{et}}, \sigma] \rangle = -i\langle \pi \rangle = 0, \langle [Q^{\text{et}}, \pi] \rangle = i\langle \sigma \rangle \neq 0$.

Since the PCAC relation is now an operator relation for the canonical field $\pi(x)$ with $f_\pi = v$ in this model, (19) ensures $[Q, P^-] \neq 0$ or a non-zero current vertex $\langle \pi | \widehat{J}^+ | \sigma \rangle \neq 0$ ($q^2 = 0$) in the symmetric limit. Noting that $Q = \widehat{Q}$, we conclude that the regularized zero-mode constraints indeed lead to non-conservation of the LF charge in the symmetric limit $m_\pi^2 \to 0$:

$$\dot{Q} = \frac{1}{i}[Q, P^-] = v \lim_{m_\pi^2 \to 0} m_\pi^2 \int d^3\vec{x}\, \omega_\pi \neq 0. \qquad (21)$$

(^{3}) By explicit calculation with a careful treatment of the zero-modes contribution we can also show that $\langle [Q, \sigma] \rangle = \langle [Q, \pi] \rangle = 0$ [8].

This can also be confirmed by direct computation of $[Q, P^-]$ through the canonical commutator and explicit use of the regularized zero-mode constraints [8].

Here we emphasize that *the NG theorem does not exist on the LF.* Instead we found the singular behavior (11) which in fact *establishes existence of the massless NG boson coupled to the current such that* $Q|0\rangle = 0$ *and* $\dot{Q} \neq 0$, quite analogously to the NG theorem in the equal-time quantization which proves existence of the massless NG boson coupled to the current such that $Q|0\rangle \neq 0$ and $\dot{Q} = 0$ (opposite to the LF case!). Thus the singular behavior of the NG-boson zero mode (11) (or (19)) may be understood as a remnant of the Lagrangian symmetry, an analogue of the NG theorem in the equal-time quantization.

Acknowledgments

I would like to thank Y. Kim and S. Tsujimaru for collaboration. This work was supported in part by a Grant-in-Aid for Scientific Research from the Ministry of Education, Science and Culture (No.08640365).
usual between

References

[1] Dirac P.A.M., Rev. Mod. Phys. **21** (1949) 392.

[2] Leutwyler H., Klauder J.R., and Streit L., Nuovo Cim. **66A**, 536 (1970).

[3] Maskawa T. and Yamawaki K., Prog. Theor. Phys. **56**, 270 (1976).

[4] Pauli H.C. and Brodsky S.J., Phys. Rev. **D32**, 1993 (1985); *ibid* 2001 (1985).

[5] Nakanishi N. and Yamawaki K., Nucl. Phys. **B122**, 15 (1977).

[6] Heinzl T., Krusche S., Simbürger S. and Werner E., Z. Phys. C **56**, 415 (1992); Robertson D.G., Phys. Rev. **D47**, 2549 (1993); Bender C.M., Pinsky S. and Van de Sande B., Phys. Rev. **D48**, 816 (1993).

[7] Kim Y., Tsujimaru S. and Yamawaki K., Phys. Rev. Lett. **74** (1995) 4771.

[8] Tsujimaru S. and Yamawaki K., Heidelberg/Nagoya Preprint, hep-th/9704171.

[9] Steinhardt P. , Ann. Phys. **32**, 425 (1980).

[10] Wilson K.G., Walhout T.S., Harindranath A., Zhang W., Perry R.J. and Glazek S.D., Phys. Rev. **D49**, 6720 (1994).

[11] Weinberg S., Phys. Rev. **150**, 1313 (1966).

[12] McCartor G. and Robetson D.G., Z. Phys. **C53**, 679 (1992).